John Parker

Thermo-dynamics treated with elementary mathematics

Containing applications to animal and vegetable life

John Parker

Thermo-dynamics treated with elementary mathematics
Containing applications to animal and vegetable life

ISBN/EAN: 9783337228736

Printed in Europe, USA, Canada, Australia, Japan

Cover: Foto ©berggeist007 / pixelio.de

More available books at **www.hansebooks.com**

THERMO-DYNAMICS

TREATED WITH

ELEMENTARY MATHEMATICS

AND CONTAINING APPLICATIONS TO

ANIMAL AND VEGETABLE LIFE, TIDAL FRICTION, AND ELECTRICITY.

BY

J. PARKER, M.A.,

AUTHOR OF "ELEMENTARY THERMO-DYNAMICS."

LONDON:
SAMPSON LOW, MARSTON & COMPANY
LIMITED,
St. Dunstan's House,
FETTER LANE, FLEET STREET, E.C.
1894.

PREFACE.

The subject of thermo-dynamics is one in which there is a wide-spread interest. It links together and forms an essential part of many branches of physical science formerly thought to be unconnected. Moreover, it is exceedingly simple. Many of the calculations it involves can be as well performed by elementary methods as by more advanced. Indeed, it often happens that the easiest algebra is sufficient. It is, however, seldom possible to dispense with algebra altogether.

All the mathematical works on thermo-dynamics we possess at present make use of the higher mathematics. There does not appear to have been any attempt made up to now to meet the wants of that large and intelligent class of non-mathematical readers and students who desire to become acquainted with the principal parts of the subject. The present work, it is hoped, will supply the deficiency.

In the following pages, elementary methods of calculation are alone employed. In the two principal chapters and in one or two others, the mathematics are mostly limited to the very easiest algebra and the definitions of the sine and cosine of an angle.

No more space is devoted to descriptions of experiments than is sufficient to render them intelligible. For further information, recourse must be had to the ordinary experimental treatises on heat and electricity.

Care has been taken to make the work a trustworthy

account of the subject. Speaking generally, it contains nothing in addition to well-established propositions but well-considered results which have been more or less tested by publication and otherwise. Furthermore, sufficient notice is always given on the introduction of any argument, however probable, that is not given by our leading physicists.

The work consists of an Introduction, six chapters, and an Appendix. The Introduction contains an account of the metric system, and some useful results in elementary statics, dynamics, and gravitation, which the reader may like to have for reference. Of the six chapters, two are devoted to the foundations of the subject—the principle of Energy and Carnot's principle—and four to applications of those two. The Appendix is a collection of numerical tables.

It will probably be best for the general reader to begin at the beginning of the book and read straight on. The six chapters are, however, largely independent; and those who are already acquainted with thermodynamics may read them in any order that is desired.

The first chapter contains an account of the principle of energy, written with considerable care. Simplicity has been the principal object in preparing the chapter. In fact, the chapter is little more than a translation into the language of elementary algebra of the results of experiment.

The second chapter is devoted to perfect gases, and is made independent of the first by the addition of a brief but sufficient account of the principle of energy. The properties of gases are of great importance in many branches of science. They are also very simple and interesting. Only one proposition is usually obtained by the higher mathematics which cannot be proved by the most elementary. With this one exception, the discussion of perfect gases requires no mathematics higher than the most elementary algebra.

The third chapter contains an account of Carnot's principle. It assumes a previous knowledge of the principle of energy. Those who already possess this knowledge and merely wish to see the nomenclature and notation made use of in the present work, will find all they require at the beginning of Chap. II.

Under the heading "Carnot's principle" are given some general propositions of fundamental importance in thermo-dynamics. These propositions are of great simplicity; and, though originally obtained by means of the higher mathematics, the most elementary algebra is alone employed in the following pages to explain them, and is found amply sufficient in every case. It is therefore believed that in this part of the subject the present work will put the non-mathematical reader on a level with the mathematician.

The principal objects aimed at in the chapter have been simplicity and accuracy. In particular, care has been taken to explain the exact meaning of the symbols in the important general equation $\frac{Q}{\theta} + \frac{Q'}{\theta'} + \frac{Q''}{\theta''} + \ldots \leqq 0$. A little further on, the subject of Entropy is considered; and it has been endeavoured to make the account as simple and complete as possible. We may also mention a deduction from Carnot's principle which includes the Theory of Exchanges and the theory of Tidal Friction. To the majority of readers, Tidal Friction will be a new subject; but treated as a deduction from Carnot's principle, it will be found to be one of great simplicity. In fact, the reader who is sufficiently acquainted with gravitation should have no difficulty, after reading Chap. III., in working out the early parts of the subject himself.

In addition to well-known propositions, two questions of great importance are discussed in Chap. III. One relates to the entropy of the universe, the other to the application of Carnot's principle to animal and vegetable life.

With respect to the entropy of the universe, it is shown that there is no proof that it "tends to a maximum." When the author first ventured to criticize this statement, he was not aware that it was due to the great thermo-dynamicist, Clausius. He has discovered, however, that he is not alone among the followers of Clausius in objecting to the statement; and his criticism appears to be fully justified.

With respect to the second question, it appears to have been hitherto considered doubtful whether Carnot's principle applies to living animals and vegetation or not. The author, however, thinks that the principal difficulties of the question will be found to disappear on taking account of the known facts of "transpiration"—that is, of the exhalation of aqueous vapour from the leaves of plants. In a paper read before the Camb. Phil. Soc. (Proc. Camb. Phil. Soc., Oct., 1892), he endeavoured to show that "transpiration" was sufficient to justify us in concluding that Carnot's principle is true for vegetable life, but he did not consider the more difficult case of animal life. For a reason stated in the text, this paper was received with a "good deal of scepticism." Since then he has examined the case of animal life, and he finds that every difficulty disappears if we take account of "transpiration." In fact, as shown in the text, Carnot's principle shows that transpiration is necessary and experiment shows that transpiration exists to an extent *amply sufficient* to satisfy all the inequalities of thermo-dynamics.

Chap. IV. contains some applications to an unelectrified body at rest. The chapter commences with a brief re-statement of the principal results in Chaps. I. and III. Of the applications, the first relate to evaporation, fusion, etc. Then there are one or two applications to capillarity. Next, the Thermo-dynamic Potential is considered; and it is believed the explanation will be found within the capacity of all who possess a knowledge of elementary algebra. The chapter terminates with an account of Gibbs' geometrical test of stability—which is practically the Thermo-dynamic Potential in a geometrical form;—and it is hoped the account has been made as simple as possible.

Chap. V. is devoted to Tidal Friction. The subject is considered as a deduction from Carnot's principle; but no previous knowledge of Carnot's principle is required. This method of treatment will be found to render some of the earlier propositions in the chapter almost self-evident, and it gives immediately an explanation of the shortening of the period of Encke's comet. It fails, however, to tell us the number of bodies to which the system is ultimately reduced by the tides; but, as shown in the text, this defect is easily supplied to a large extent by other considerations.

Chap. VI. contains some applications to electricity —applications which form the most delightful part of thermo-dynamics. The chapter will not be found difficult by those who possess the requisite previous knowledge of electricity. Those who wish to take it before Chaps. I. and III., should first read the short account of the principle of energy and Carnot's principle at the beginning of Chapter IV.

(Immediately after Chap. VI. there is given in a Note a short account of a curious and novel electrical phenomenon—a spasmodic current in a circuit apparently invariable—which the author has obtained in the course of his experiments.)

After this comes the Appendix containing some useful tables, and then the Index.

J. PARKER.

October, 1894.

A SHORT STATEMENT RESPECTING A NEW WORK

ENTITLED

"THERMO-DYNAMICS,
TREATED WITH ELEMENTARY MATHEMATICS,

AND CONTAINING

APPLICATIONS TO ANIMAL AND VEGETABLE LIFE, TIDAL FRICTION, AND ELECTRICITY,

By J. PARKER, M.A.,

Author of 'Elementary Thermo-dynamics.'"

A WANT has long been felt by non-mathematical students, technical classes, and practical men, for a text-book of Thermo-dynamics which should only use elementary mathematics and yet be fairly exhaustive. Maxwell's "Theory of Heat" is, I believe, the only important elementary work on the subject we possess at present. Maxwell's account of the subject is practically perfect as far as it goes, but it does not go far. It is impossible to do much in the subject without the use of elementary algebra, and Maxwell uses hardly any algebra at all. The present work, it is believed, will supply what is wanted.

As far as I am aware, we possess no complete work on the subject which does not make use of high mathematics. Yet there is nothing in Thermo-dynamics which particularly requires the use of advanced mathematics. The reason why high mathematics are always used in connection with the subject is that the authors of the subject have all been men accustomed to high mathematics and they have naturally adopted the methods and notation familiar to them. Most of their results could, however, have been discovered and explained just as well by elementary methods. This is shown by the fact that in the present work there are four chapters, Chaps. I., II., III., V., in which there are no mathematics beyond the *simplest algebra and the definitions of the sine and cosine of an angle*. In the introduction, use is also made of a "solid angle"; and in the two chapters of *applications*, Chaps. IV. and VI., an algebraic idea is introduced which is not quite so simple as those in the other chapters; but, on the whole, the whole work is elementary.

I have already published a book on the subject entitled "Elementary Thermo-dynamics"; but it makes use of the higher mathematics and was only intended for a limited class of readers. It was produced under very unfavourable circumstances; but it has received some very favourable reviews. Three of these were entirely favourable; the fourth was unfavourable on two chapters, favourable on the other four chapters. I have therefore good reasons for believing the present work will be a success.

The MS. of my "Elementary Thermo-dynamics" was sent in to the Cambridge University Press, on the recommendation of some friends. In due course I was informed by the Chairman of the Syndics that the MS. had been recommended for publication by their reader, Prof. ——. But in the meantime I had made what I considered some important discoveries and had decided to enlarge some parts of the MS., which I did before it was sent to press. The discoveries I had made were in "Tidal Friction"; and without knowing that the subject had been worked at before, I made out most of what is given about it in my "Elementary Thermo-dynamics." Thus in about ten months I made out nearly as much as had been discovered in a generation or two. In enlarging the MS. my principal efforts were expended on "Tidal Friction," and the intensely hard labour of doing this caused several things to be added to the MS.

without much consideration. Also many results were obtained in "Tidal Friction" while I was actually writing the enlarged parts of the MS. for the press: the early articles on "Tidal Friction" therefore do not clearly lead up to them. These facts were unknown to the reviewers, as I made no mention in the book of the way I had worked out the subject of "Tidal Friction." The reviewer in "Nature," who reviewed two chapters unfavourably, seemed to think "Tidal Friction" had no business to appear in a work on Thermo-dynamics. He considered my treatment of the subject to be merely "an elementary exposition of Darwin's calculations in tidal friction." The other things that this reviewer objects to were added during the enlargement of the MS. when my powers were practically exhausted by the hard work on "Tidal Friction." Yet two of the reviewers single out some of these parts for special commendation.

I have discovered an erroneous statement near the beginning of my "Elementary Thermo-dynamics"; and therefore I intend to suppress the work altogether when sufficient copies have been sold, or else I shall practically omit the earlier chapters—the parts which have been reviewed unfavourably—and refer for full information to the present work. Thus there will be *no competition* between my "Elementary Thermo-dynamics" and the present work.

Passing over the unfavourable review of two chapters in "Nature," the following are specimens of the favourable reviews:—

Educational Times.—"The whole subject is handled in a masterly style, and the treatise should become very popular."

Academy, April 23, 1892.—"His book is a most suggestive and valuable one, and the publication is quite worthy of the best traditions of the University Press." . . . "There is a refreshing individuality about Mr. Parker."

Nature, February 4, 1892.—"We are not aware that the general energy methods of Massieu and Helmholtz have ever before been presented in connected form to

English readers. This Mr. Parker has done, and has deservedly earned our tribute of praise. Any one who is familiar only ... will find these two last chapters ... particularly interesting."

Athenæum, February 18, 1893.—"We can cordially recommend Mr. Parker's work.... A long chapter is devoted to an able and exhaustive discussion of Carnot's principle."

The present work has been produced under very favourable circumstances. I had thought out for a year or two the statements it contains *before* beginning to write the book. Many of these statements are well-known propositions and have been carefully compared with *accepted* or the *original* authorities. All that is new has, with NO important exception, been fully *tested* by previous publication or otherwise. A special chapter is devoted to "Tidal Friction," and the subject is treated as a deduction from Carnot's principle in a very simple way that no one can misunderstand. One of the leading chapters in the work is that on Carnot's principle. This is so simple that those who are only acquainted with the simplest algebra will find that in this—one of the most important parts of the subject—they are put on a level with the best mathematicians. On turning to Chaps. IV. and VI., especially Chap. VI., it will, however, be seen that the applications of Carnot's principle are not so simple (algebraically speaking) as the principle itself.

To many readers the most interesting part of the work will be the "Applications to Electricity" (Chap. VI.). My best thermo-dynamical work has always been considered to be done in this part of the subject, and it is now for the first time to be published in a collected form. Two or three of my papers on it have been translated into "Wiedemann's Annalen." They have also been received, etc., in France; and they were the means of gaining me a Fellowship at St. John's College, Cambridge. I may add that some of the work was originally objected to, but that my explanations have produced FULL conviction. The discussion thereby raised has shown me the points that needed most careful explanation.

Not the least important part of the work to many students and practical men will be the *very simple* chapter on perfect gases. The properties of gases have a great fascination for practical men, and they are of great importance in chemistry and physics to the ordinary non-mathematical student. Yet I believe there has hitherto been no attempt made to give a fairly full and elementary account of them. The non-mathematical student has had to be content with such scraps of information as he could pick up here and there in works on chemistry and physics.

I may add that the work is provided with an index. This is probably a matter of some importance, judging from the complaints that have been made of the want of an index to my "Elementary Thermo-dynamics."

J. PARKER, M.A.,
Author of the Work.

CONTENTS.

	PAGES
INTRODUCTION	1–13
CHAPTER I.—THE CONSERVATION OF ENERGY ...	14–44
CHAPTER II.—PERFECT GASES	45–69
CHAPTER III.—CARNOT'S PRINCIPLE	70–138
CHAPTER IV.—APPLICATIONS TO AN UNELECTRIFIED BODY AT REST	139–185
CHAPTER V.—TIDAL FRICTION	186–227
CHAPTER VI.—APPLICATIONS TO ELECTRICITY	228–302
NOTE.—A CURIOUS AND NOVEL ELECTRICAL PHENOMENON	303–304
APPENDIX	305–322
INDEX	323–328

THERMO-DYNAMICS.

INTRODUCTION.

Units of Measurement.

If we wish to express the magnitude of a dynamical or physical quantity, we choose a quantity of the same kind as unit of comparison, and the magnitude of the given quantity is denoted by the number which shows how many times greater or less it is than the unit with which it is compared. The units of different kinds may all be chosen arbitrarily, but a great simplification is effected by a proper choice. It is shown in works on Dynamics that it is best, after choosing the units of length, mass, and time arbitrarily, to drive all the other units from them. Such a system of units is known as "absolute," and there are two such in common use. One is founded on the ordinary English weights and measures, the other on the metric or French. The former system of absolute units, which is chiefly confined to English countries, is known as the "practical" system; the latter may be called the "scientific."

The metric system of weights and measures has a great advantage over the English, because it forms a decimal system, and therefore avoids many of the laborious arithmetical calculations necessary when the English system is used.

The metric system was established in France by law in 1795, and is now adopted in many countries for commercial purposes. Its standards of length and mass are the Metre and Kilogramme respectively. The former is the distance between the ends of a rod of platinum, made by Borda, when the temperature is 0° C. (that of melting ice); the latter is the mass of a piece of platinum, also made by Borda.

The subsidiary measures of the metric system are as follows:—

1 kilometre = 1000 metres.	1 kilogramme = 1000 grammes.
1 hectometre = 100 metres.	1 hectogramme = 100 grammes.
1 decametre = 10 metres.	1 decagramme = 10 grammes.
1 decimetre = $\frac{1}{10}$ metre.	1 decigramme = $\frac{1}{10}$ gramme.
1 centimetre = $\frac{1}{100}$ metre.	1 centigramme = $\frac{1}{100}$ gramme.
1 millimetre = $\frac{1}{1000}$ metre.	1 milligramme = $\frac{1}{1000}$ gramme.

Liquids are measured by the Litre, the volume of which is a cubic decimetre.

When the standards of the metric system were originally chosen, the metre was taken to be the 10-millionth part of the supposed distance from the pole to the equator, and the kilogramme as nearly as possible the mass of a litre of distilled water at 4° C., the temperature of maximum density.

The metric weights and measures may be compared with the English by means of the following data:—

1 metre = 39·370,432 inches.	1 inch = 2·54 centimetres.
1 centimetre = ·393,704 inch.	1 foot = 30·48 centimetres.
1 sq. metre = 1550·03 sq. inches.	1 sq. inch = 6·45 sq. centimetres.

1 litre = 61·025 cubic inches = 1·761,7 pints.
1 pint = ·567,6 litre.

1 kilogramme = 2·204,6 lbs. avoir. = 15,432 grains.
1 lb. avoir. = 453·59 grammes.

In the scientific system of absolute units, the fundamental units of length, mass, and time are the centimetre, gramme, and mean solar second respectively. The system is, therefore, usually spoken of as the centimetre-gramme-second, or C.G.S., system of absolute units. The absolute unit of force in this system, called a Dyne, is defined to be the force which, acting for a second on the mass of a gramme, generates a velocity of one centimetre per second. The C.G.S. absolute unit of pressure is a pressure of one dyne per square centimetre.

Side by side with these "absolute" units, we often use arbitrary units, as the gramme-weight for force and the atmo for pressure.

The weight of a gramme has no definite value unless we specify the place where the weight is to be found, because the weight of a given mass is not quite the same in all parts of the world. To avoid ambiguity when we use the weight of a gramme as a measure of force, we shall agree that the weight is to be found at Paris. The acceleration of a body falling freely at Paris being 980·868 centimetres per second, it follows that the weight of a gramme at Paris is 980·868 dynes.

The Atmo, the name of which is due to Professor J. Thomson, is defined to be the pressure produced (at Paris) by a column of mercury 760 millimetres high, when the temperature is 0° C., and the top of the mercury column acted on by no pressure but that of its own vapour. The value of this arbitrary unit of pressure is found to be 1033·279 gramme-weights, or 1,013,510 dynes, per square centimetre. As the name denotes, the atmo represents the average pressure of the air.

In the English, or practical system of units, the arbitrary units of length, mass, and time, are the foot, the pound avoirdupois, and the mean solar second respectively. The absolute unit of force, to which the name of Poundal has been given by Professor J. Thomson, is the force which in one second generates a velocity of a foot per second in the mass of a pound. As the acceleration produced by gravity at London is 32·1889 ft. per second, the weight of a pound at London will be 32·1889 poundals. For rough purposes, for which the practical units are mostly used, the poundal may be taken to be the weight of half an ounce in any part of the world.

The following comparisons should be noticed:—
Weight of 1 lb. avoir. = 444,900 dynes.
1 lb. per sq. inch = 70·3 gramme-weights per sq. centimetre = 69,000 dynes per sq. centimetre.
1 gramme per sq. centimetre = ·0142 lb. per sq. inch.
1 atmo = 14·7 lbs. per sq. inch.

We shall always work in the C.G.S. system of units, and it is therefore necessary to understand the system thoroughly.

Work and Kinetic Energy.

In applying the laws of motion to a body, it is necessary and convenient to suppose the body to consist of a very large number of very small pieces, or *particles*, each of which is so small in all its dimensions that, in all dynamical considerations, we may treat it as a mere point; the relative motions of the parts being insignificant in comparison with the motion of the particle as a whole.

When a particle under the action of any force receives a displacement which is not at right angles to the force, the force is said to do Work on the particle, or the particle is said to do work against the force, and the amount of work done on the particle is defined to be the product of the displacement into the resolved part of the force in the direction of the displacement. If the displacement be in the direction of the force, the work done on the particle is simply the product of the force and the displacement. If the displacement be in the opposite direction to the force, the work done on the particle is found by multiplying the force by the displacement and prefixing the negative sign. In general, if P be the force, s the displacement, and θ the angle between them, the work done by the force will be $s \cdot P\cos\theta$. As $s \cos \theta$ is the

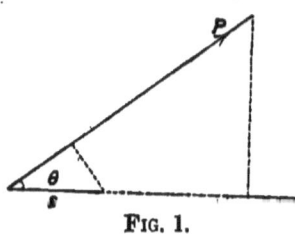

Fig. 1.

necessarily along the line GH), as shown in the accompanying figure.

By resolving each of the two forces at G and H into two parts, one in the line GH and the other at right angles to it, we see that the two forces are equivalent to two pairs of equal and opposite forces, one pair in the line GH, the other at right angles to it, as in the figure.

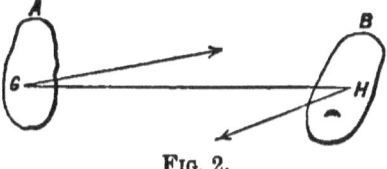

FIG. 2.

The forces at G and H at right angles to the line GH will not usually be zero. The simplest case in which they will be so is that in which both couples which partially replace the mutual attractions are absent. For since the mutual attractions of the two bodies always consist of a number of forces which are equal and opposite in pairs, the sum of their moments about any straight line will be zero. When the mutual attractions are simply equivalent to forces at G and H, it follows that the sum of the moments of these forces about any straight line is zero. This can only happen when the two forces are equal and opposite, that is, when they act along GH, and therefore have no resolved parts at right angles to GH.

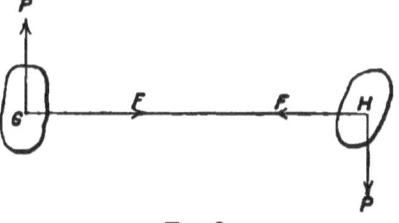

FIG. 3.

If one of the two rigid bodies (A suppose) be a sphere, either homogeneous throughout or formed of homogeneous shells bounded by spherical surfaces concentric with each other, the attraction of B on A will be equivalent to a single force at the centre of mass (or centre) G; in other words, the couple at G will be wanting. If the other body B be a similar sphere, the couple at H will also be wanting. These important propositions are due to Newton, and the proofs of them will appear presently. It will be observed, in the first place, that each sphere may either be solid throughout or contain a spherical cavity concentric with itself. There are therefore two cases to be considered, according as the bodies are external to each other or one inside the other.

The proofs of Newton's propositions, which we shall give, depend on two preliminary propositions or lemmas, the demonstrations of which may be omitted if found too difficult.

(a) Let a single particle of mass m grammes be surrounded by a closed surface of any kind. Take any portion of the surface so small that it may be considered plane, and let σ be the area of the portion in square centimetres. Imagine a particle P whose mass is one gramme placed on the small area σ, and let the force with which the particle m attracts the particle P be resolved along the normal to the surface at P. Suppose the result to be F dynes, and let it be considered positive when it acts inwardly, negative when it acts outwardly. Take the product $F\sigma$, and let similar products be formed for every other small area into which the surface can be divided. Then the sum of these products is independent of the form or size of the surface, and equal to $4\pi\lambda m$. For if θ be the acute angle between the normal at P and the line mP, θ is also the acute angle between the small area σ (or tangent plane at P), and a plane normal to the line mP. Hence $\sigma \cos \theta$ is the projection of the area σ on a plane normal to mP; and this is known to be equal to $r^2 \omega$, where r is the distance mP (in centimetres), and ω the "solid angle" subtended by the area σ at m. Now, since the attraction of m on $P = \dfrac{\lambda m}{r^2}$ dynes, we have,

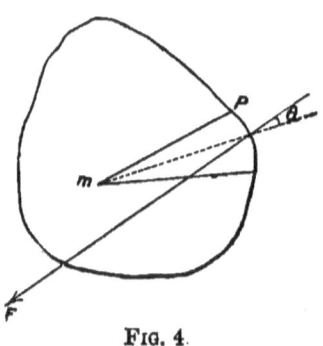

FIG. 4.

when the resolved part of the attraction is to be taken positively, as in the figure

$$F = \frac{\lambda m}{r^2} \cos \theta$$

Hence

$$F\sigma = \lambda \frac{m}{r^2} \cos \theta \times \frac{r^2 \omega}{\cos \theta} = \lambda m \omega$$

But the line mP meets the closed surface on one side of m in an odd number of points. Let these points be P, Q, R, S, T, and let small areas corresponding to σ be cut from the surface at Q, R, S, T, by a cone whose vertex is m and base the small area σ. Then since the force F is to be taken positively at P, negatively at Q, positively at R, and so on alternately, the sum of the products $F\sigma$ for these small areas will be $\lambda m(\omega - \omega + \omega - \omega + \omega)$, or $\lambda m \omega$, and is therefore independent of the number of times the line mP meets the surface on one side of m. Hence we easily see that the value of the sum $F\sigma$ for the whole closed surface is simply λm times the sum of all the "solid angles" about the point m, or $\lambda m \times 4\pi$.

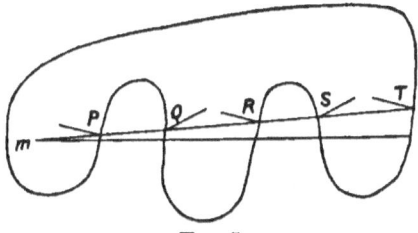

Fig. 5.

If there be two masses m and m' inside the closed surface, the value of the sum of the products $F\sigma$ for the two masses will evidently be $4\pi\lambda m + 4\pi\lambda m'$, or $4\pi\lambda(m + m')$; and in general if a finite mass M be surrounded by a closed surface of any kind, the value of the sum of the quantities $F\sigma$ with respect to this surface is $4\pi\lambda M$.

(b) Let a closed surface of any kind be drawn entirely without the particle m. Then any line through m which meets the surface meets it in an even number of points P, Q, R, S, ... Suppose a cone of small vertical angle whose vertex is at m cuts out small areas at P, Q, R, S, ... , and let ω be the solid vertical angle of the cone (or the solid angle subtended by each of the small areas at m). Then the value of $F\sigma$ for any one of these areas is $+\lambda m\omega$, or $-\lambda m\omega$, and there are as many with one sign as the other. Hence the sum of the products $F\sigma$ for all the small areas on the line mP is zero. From this it follows that the sum of the quantities $F\sigma$ for the whole surface is zero.

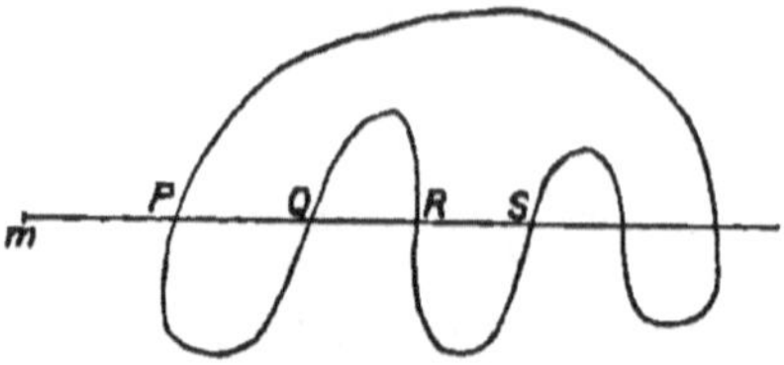

Fig. 6.

Hence also it may be shown that if there be any finite mass without the closed surface, the sum of the products $F\sigma$ with respect to this mass and the closed surface, is zero.

It is now easy to prove Newton's propositions. Let A be a sphere whose mass is M grammes and centre at O, and let it either be homogeneous throughout or formed of homogeneous shells bounded by spherical surfaces whose centre is at O; in the latter case the body A may contain a spherical cavity whose centre is at O. Let a particle whose mass is one gramme be placed at any point P, either outside A or within the cavity. With centre O and radius OP describe a spherical surface through P; and take the sum of the products $F\sigma$ with respect to this surface and the given body A. By symmetry, the attraction of the body A on the particle at P is the same

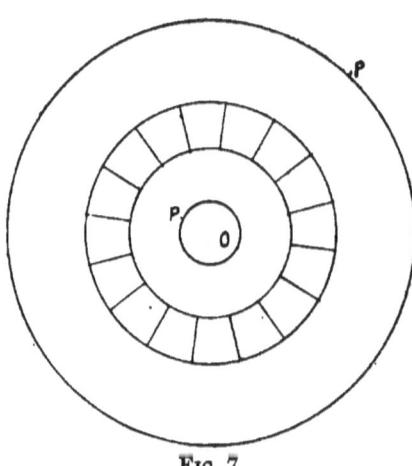

Fig. 7.

wherever P may be situated on the spherical surface, and is moreover everywhere at right angles to the surface. The value of F is therefore the same at every point of the sphere. But if r be the radius OP of the sphere in centimetres, the area of its surface will be $4\pi r^2$ square centimetres. Hence the sum of the quantities $F\sigma$ is $F \cdot 4\pi r^2$.

Here the two cases must be distinguished.

(1) If the point P be outside the sphere A, we have $F \cdot 4\pi r^2 = 4\pi \lambda M$.
Hence

$$F = \lambda \frac{M}{r^2}$$

Thus the attraction of the finite spherical body A (which need not be "rigid") on any external particle P is the same as if the whole body A were concentrated into a single particle at its centre O. If m be the mass of any particle at P, the attraction of the sphere on it acts along PO, and is equal to $\lambda \frac{Mm}{r^2}$ dynes.

If the body A be "rigid," the attraction of the particle P on it will be equivalent to a force at O and a couple. The force at O will be equal and parallel to that at P, and therefore acts in the same straight line OP. Hence since the sum of the moments about every straight line is zero, it follows that the couple must be zero. The attraction of the particle at P on the rigid sphere A therefore reduces to the single force $\lambda \frac{Mm}{r^2}$ dynes at O along OP, just as if the whole sphere were concentrated into a particle at its centre O.

By supposing a number of particles outside the body A to form a finite body of any kind, it is easily seen that the attraction of the external body on the sphere are the same as if the whole sphere was concentrated into a particle at its centre. If the second body be a rigid sphere, such as we have described, it may also be supposed concentrated into a particle at its centre.

Hence the attraction of the first sphere on it reduces to a single force; and if M, M' be the masses of the two spheres in grammes, r the distance between their centres in centimetres, each sphere attracts the other with the same force F dynes along the line joining the centres, where $F = \lambda \dfrac{MM'}{r^2}$

(2) If the point P be inside the cavity in the sphere A, we have $F \cdot 4\pi r^2 = 0$, or $F = 0$. Hence the sphere A exerts no force on a particle in the cavity. If the body A be rigid, the action of any forces on it will be equivalent to a force at the centre O and a couple. In our case, the force is equal to that at P, or zero: hence the couple is also zero. Similar results can easily be seen to hold when there is a finite body of any kind in the cavity of the sphere.

The result that the two rigid external spheres attract each other as if they were concentrated into particles at their centres is very important, because it is approximately the case of the heavenly bodies. It may also be used to obtain a clear idea of the strength of gravitation. Thus let two homogeneous rigid spheres of the same size and density be placed just in contact with one another.

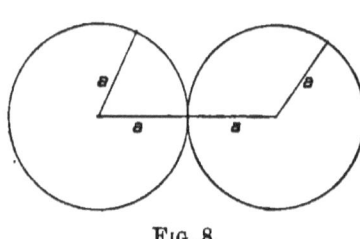

Fig. 8.

If a and ρ be the radius and density of each sphere, we have—

$$M = M' = \tfrac{4}{3}\pi\rho a^3,$$
$$r = 2a.$$

Hence the mutual attraction F, in dynes, will be

$$F = \lambda \dfrac{MM'}{r^2} = \lambda \dfrac{(\tfrac{4}{3}\pi\rho a^3)^2}{(2a)^2} = \lambda \tfrac{4}{9}\pi^2 a^4 \rho^2.$$

If the spheres be of lead, we may take ρ to be $11\tfrac{1}{3}$.

Introduction. 13

The mass of each sphere will then be $46·5 \times a^3$ grammes, and F will be

$$\lambda \tfrac{4}{9}\pi^2 a^4 (3\tfrac{4}{3})^2, \text{ or } 365 \times 10^{-7} \times a^4$$

We have, therefore, the following results for lead spheres in contact:—

Diameter of each sphere.	Mutual attraction.	Mass of each sphere.
One metre	228 dynes, or 0·23 gramme, or 0·000,5 lb. (the 2000th of a lb.).	5,800 kilogrammes, or 12,786 lbs., or 5·7 tons.
One centimetre	228×10^{-8} dynes, or the 400 millionth of a gramme, or the 200,000 millionth of a lb.	5·8 grammes, or 0·013 lb.
One kilometre (1,093·6 yards)	228×10^{12} dynes, or 230 million kilogrammes, or 5×10^8 lbs., or 229,000 tons.	5,800,000 million kilogrammes, or 5,700 million tons.
25·64 centimetres, or 10·1 inches	...	1 dyne.
1·43 metres, or 56·5 inches	...	1 gramme-weight.
2·7 metres, or 9 feet	...	1 poundal.
6·7 metres, or 21·9 feet	...	1 pound-weight.

CHAPTER I.

THE CONSERVATION OF ENERGY.

ART. 1. Heat was formerly thought to be a material substance, to which the name of Caloric was given. The quantity of caloric in a body was supposed to remain constant, except when heat entered or left the body. Experiment has, however, completely disproved the caloric theory of heat, and substituted another, called Mechanical Theory, in its stead.

The caloric theory was able to give an explanation of some of the common phenomena of heat. It was supposed that the conduction of heat was merely the transference of caloric from one body to another. The increase of volume which usually accompanies rise of temperature, was thought to be due to the greater quantity of caloric in the body. The rise of temperature, which is generally produced by sudden compression, was explained by assuming that caloric was squeezed out and rendered sensible to the thermometer and touch. Again, it was supposed that different substances have different capacities for caloric, or require different proportions of caloric to raise their temperature by the same amount. For example, a quantity of water was supposed to require thirty times as much caloric to raise its temperature $1°$ as an equal quantity of mercury. But the most important part of the caloric theory was the doctrine of Latent Heat, first propounded by Black in 1760. When it was found that no change

room for the paddles to pass. After the weight was allowed to run down, the increase in the temperature of the water was observed, and the necessary corrections made. In this way the quantity of work required to produce a known rise of temperature was determined.

Experiments were also made by Joule on other substances; and from the results it is inferred that to produce a given change of temperature by friction in a given body, requires the same quantity of work to be expended whenever and however the experiment is made.

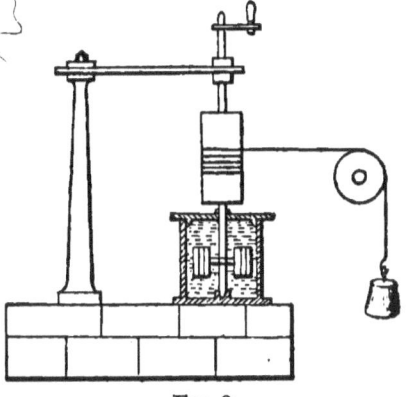

Fig. 9.

Other experiments have been made, in which a given quantity of work was expended on a system in causing its parts to become electrified. When they were allowed to resume their natural unelectrified states, it was found that the rise of temperature was the same as if the same quantity of work had been expended on the system in friction.

Again, it is found that if a quantity of work is expended in friction on a body, A, and then A part with heat to B until its state is the same as before, the change in the temperature of B will be the same as if the work had been expended directly on B instead of A.

These and other more complicated results lead to, and are included in, a very simple principle, known as the Principle of Energy, which will now be explained. This principle, together with Carnot's principle and their applications, constitute the subject of thermodynamics.

Art. 3. Let us take any material system X whatever, which may consist of one body only, or of a number of bodies in contact, or of a number of bodies far apart, and moving about under the influence of their mutual gravitational attractions, like the bodies of the solar system, or of the whole universe. Also let the parts of the system be in any electric and magnetic conditions we please. Then we may give the following definitions of heat absorbed or gained from without in an unelectrified part of the system, and of heat lost or evolved in an unelectrified part of the system.

When the system X changes from any state P to any state P', let an unelectrified part of the system Y be subject to any external thermal influences, due either to conduction or radiation. Imagine the same operation undergone a second time by the system while subject to the same external influences as before, with the sole exception of the external thermal influences which act on the part Y. Let a body A be placed in contact with the part Y, and kept there *without rubbing* during the operation undergone by the given system. Suppose that during this operation the body A is prevented from gaining and losing heat, except by exchange with Y, and that no heat enters or leaves the system X in the part Y, except by conduction with A. Then, if the body A be cooled by the contact, we can bring the body back to the same state as at the beginning of the operation by expending on it a positive quantity W_1 of work in friction. In this case, we say that a quantity of heat W_1 has been evolved or lost by the body A while in contact with Y; and we define the system X to have absorbed or gained in the part Y a quantity of heat W_1 during the change of state from P to P', whether the body A has been made use of during the operation or not. If, however, the body A be heated by the contact, it will evidently be impossible, by expending work on it in friction, to bring it into the same state as at the beginning of the operation[1]. Let therefore W_2 be the

[1] The author finds that in this part of the subject he formerly made an erroneous statement, which he now wishes to withdraw. He asserted that any change of state which could be produced by the joint agency of

quantity of work that must be done on A when in its original state to produce in it the same rise of temperature that is caused by contact with Y. Then we say that a quantity of heat W_2 has been absorbed or gained by A while in contact with Y; and we define the system X to have evolved or lost in the part Y a quantity of heat W_2 during the change of state from P to P′, whether the body A has been made use of during the operation or not.

The foregoing definitions have been simplified by the stipulation that the body A is not to rub the part Y, with which it is put in contact. The case of frictional contact will be considered later on. It is sufficient for the present to state that the conduction of heat is complicated by the simultaneous "production of heat" at the surface across which the "heat" is conducted.

Now let us agree to consider heat evolved to be negative heat absorbed. Also, suppose that during the change of state from P to P′ the system is protected from external electric influences. Let W be the total quantity of work done on the system during the operation, and let (Q_1, Q_2, Q_3, ...) be respectively the algebraic quantities of heat absorbed by it in the same time in the parts (Y_1, Y_2, Y_3, ...). Then it is found that the sum $W + Q_1 + Q_2 + Q_3 + \ldots$ depends only on the states P and P′, and is independent of the way in which the change of state is effected. Writing Q for $Q_1 + Q_2 + Q_3 + \ldots$, we see that $W + Q$ depends only on the states P and P′.

From the above statement it is evident that when the change of state is effected in a given way, the value of Q is the same whatever be the number of parts into which we choose to divide the places where the system is subjected to external thermal influences. We may

---

work and heat could be produced by work alone. It will be seen in Chapter III. that the statement is at variance with the principles of Entropy and of the Degradation of Energy. The statement was, however, only made use of in proving a well-known result, which could easily have been proved without the statement.

therefore call Q the total algebraic quantity of heat absorbed by the system.

It has been supposed in the preceding argument that the places where the system is subjected to external thermal influences are all unelectrified. Now as the system may be electrified and magnetized in any way we please, we may have the case in which the system is subjected to external thermal influences in electrified places (Z). Suppose that the system, still protected from external electric influences, passes from the state P to the state P' in two ways, in each of which the work done on the system is equal to W :—

(1) Let the parts Z be subjected to external thermal influences, and denote the total quantity of heat absorbed in the unelectrified parts of the system by Q'.

(2) Let the parts Z be protected from external thermal influences, and let Q' + Q" be the total quantity of heat absorbed in the unelectrified parts of the system.

We then define the heat absorbed by the system in the electrified parts Z in operation (1) to be Q". If therefore we agree to say that the total quantity of heat Q absorbed by the system is the sum of the quantities absorbed in the electrified and unelectrified parts, it follows that in all cases W + Q depends only on the states P and P'.

The result that W + Q depends only on the states P and P' can be put in a very simple form. Choose any standard state of reference O from which the system can be brought by the joint agency of work and heat into the state P. Let w be the work done on the system, and q the heat absorbed by it when it is brought (in any way) by the joint agency of work and heat from O to P. Then $w + q$ depends only on the states O and P, and is defined to be the Energy of the system in the state P with respect to the standard state O. Let U stand for $w + q$, and denote the energy of the system with respect to the same standard state O when in the state P' by $U + \Delta U$, so that ΔU stands for "increase of U," and not for the product of Δ and U. If therefore we suppose the system brought from O to P′ by being first brought from O to P and then from P to P′, we get

$$U + \Delta U = U + (W + Q)$$

or

$$\Delta U = W + Q \quad \ldots \quad (1).$$

To find how the value of U is affected by the choice of the standard state, take a new standard state O′. Let (U′, U′ + ΔU′) be the quantities corresponding to (U, U + ΔU) for the states P and P′ with reference to the new standard state O′. Let the system be brought from O′ to P′ by the joint agency of work and heat by being first brought from O′ to O, then from O to P, and, lastly, from P to P′. Suppose w' to be the work done on the system, and q' the quantity of heat absorbed by it, during the passage from O′ to O. Then $w' + q'$ depends only on the states O′ and O, and we also have

$$\left.\begin{array}{l}U' = (w' + q') + U \\ U' + \Delta U' = (w' + q') + U + \Delta U\end{array}\right\}.$$

Hence $U' - U = w' + q'$, or a change in the standard state alters the quantity U by an amount which does not depend on the state P and may therefore be called constant. Again, $\Delta U' = \Delta U$, or the change in U in going from one state to another, which is the only thing we are concerned with, is the same whatever the standard state.

The equation (1) is called the "equation of energy." It is also known as the expression of the First Law of Thermo-dynamics,[1] or of the Equivalence of Heat and Work. The reason for the last name is very simple. Suppose the change of state from P to P′ brought about in another way in which the heat absorbed is less than Q by a quantity x. Then the work done on the system must be greater than W by x. Thus a decrease in the heat absorbed is compensated by an equal increase in the work done on the system, and *vice versâ*.

[1] Carnot's principle is often called the Second Law of Thermo-dynamics.

It should be noticed that the energy is zero in the standard state, and that in the C.G.S. absolute system of units heat, like work, is reckoned in ergs.

Art. 4. It is found that any change of state which can be effected by the joint agency of heat and work can be effected in any number of ways, and that the quantities of work required for the different ways are different. Sometimes the change of state and energy may even be produced by heat alone, and by work alone. Hence when the system is merely protected from external electric influences and is at liberty to gain and lose heat, the work done on the system in any change of state will depend on the way in which the change of state is brought about, as well as on the initial and final states themselves. The same will also be true of the heat absorbed during the operation, because the sum of the work done on the system and of the heat absorbed by it is the same however the change of state is effected. These results are so important that a simple practical illustration will be given.

Let a quantity of air at a high temperature and pressure be contained in a cylinder fitted with a smooth air-tight piston, and let it be required to reduce the volume without altering the temperature. This may be done in many different ways, of which we shall only consider two:—

(1) Let the piston be pushed in the required distance so rapidly that there is not time for the cylinder and its contents to cool. Then let the temperature be reduced to its original value by the conduction of heat through the sides of the cylinder.

The pressure being high during the compression, the work done in forcing in the piston will be considerable.

(2) First, let the cylinder be cooled far below the original temperature. As the volume of the air is unaltered, the pressure will have fallen a great deal. If the cylinder be now kept in a very cool place, and the piston pushed in the required distance so slowly that the conduction of heat through the sides of the cylinder

practically keeps the temperature of the air inside constant, the pressure in the cylinder will always be much less than during the compression in the first method. The work done in forcing in the piston will therefore be much less than before. If the change of state be now completed by increasing the temperature of the air by conduction through the sides of the cylinder, there will have been much less work done in bringing about the same change of state in one way than another.

"Heat," in the popular sense of the word, denotes something existing in a body. In any change of state, the increase of this quantity therefore depends only on the initial and final states, and not on the way in which the change of state is effected. The popular meaning of the word "heat" is therefore different from that we have given to it.

Art. 5. Let our system which is protected from external influences be called X. Suppose it to consist of two parts (X_1, X_2) which are electrically independent of one another. Then the definition of energy given in Art. 3 applies to each of the two parts or sub-systems (X_1, X_2). The standard states of these sub-systems may be chosen as we please, but it will obviously be best to choose them so that when the whole system is in its standard state O, the sub-systems are in theirs. With these standard states, let U be the energy of the whole system X in any state P, and let (U_1, U_2) be the energies at the same instant of the sub-systems (X_1, X_2). Then as we cannot assume $U = U_1 + U_2$, we will put $U - (U_1 + U_2) = M$, or $U = U_1 + U_2 + M$, where M is to be found.

Since the quantities (U, U_1, U_2) are all zero when the whole system X is in its standard state O, it is evident that M is also zero in that state.

Let the whole system be brought by the joint agency of work and heat from the standard state O to the state P. Let (Q_1, Q_2, Q) be respectively the quantities of heat absorbed by $X_1, X_2,$ and the whole system during the operation. Here it will not necessarily be true that

$Q = Q_1 + Q_2$. For example, let X_1 and X_2 be two bodies in contact, and suppose they are prevented from gaining and losing heat except by conduction with one another. Then $Q = 0$, because in this case the transference of heat between X_1 and X_2 is merely the passage of heat in the interior of the whole system X. If then $Q = Q_1 + Q_2$, Q_1 and Q_2 must be equal and opposite. This will actually be the case if the two bodies are not rubbed together while in contact.

The equation $Q = Q_1 + Q_2$ will be satisfied and the difficulty of friction avoided if we suppose the sub-systems are never in contact during the operation.

As the two sub-systems are never in contact, the only mutual force between them is the attraction of gravitation. Hence when we consider one of the sub-systems, as X_1, the external forces which act on it may be divided into the two classes following:—(a) those which are external as regards the whole system X_1, and (b) the gravitational attraction of X_2 on X_1. Let W_1 be the work done on X_1 during the operation by the forces (a), and w_1 the work done by the forces (b). Then since the increase of U_1 in going from O to P is simply U_1, we get

$$U_1 = W_1 + w_1 + Q_1.$$

If (W_2, w_2) be similar quantities for X_2

$$U_2 = W_2 + w_2 + Q_2.$$

Now when we consider the whole system X, we do not take account of the work done by the mutual attractions of X_1 and X_2, because these forces are *internal* as regards the whole system. Hence the work done on X is merely $W_1 + W_2$. We therefore have

$$U = W_1 + W_2 + Q.$$

Thus since $U = U_1 + U_2 + M$, we obtain

$$Q = Q_1 + Q_2 + w_1 + w_2 + M.$$

But $Q = Q_1 + Q_2$; hence

$$w_1 + w_2 + M = 0,$$

or
$$w_1 + w_2 = -M.$$

It is therefore evident that the work done by gravitation must be the same however the system passes from O to P. This is well known to be the case, but the proof is not easy enough to be given in this work. We also know that the work done by gravitation as the system returns in any way from P to O, is equal and opposite to that done $(-M)$ in passing from O to P. Thus the value of M in the state P is simply the sum of the quantities of work done on the two sub-systems by their mutual gravitational attractions as the whole system *returns* in any way from P to O.

The part M of the energy of the whole system, which is not possessed by the two parts (X_1, X_2), may be called their Mutual Energy. Its value is evidently greatest when the two sub-systems are furthest apart. Now, if the whole system be protected from external forces and be prevented from receiving and losing heat, $U_1 + U_2 + M$ remains constant. If meanwhile the two parts or sub-systems approach one another, M will decrease. Hence $U_1 + U_2$ will increase at the expense of M. Thus at any instant, M is the part of the energy of the whole system which is not possessed by (or localized in) either of the two sub-systems, but which may be afterwards acquired by them by means of gravitation. It is therefore frequently called the gravitational Potential Energy of the two parts or sub-systems.

The equation $U = U_1 + U_2 + M$ has only been proved to be true when the sub-systems are not in contact with one another. But as the result is true, however near the sub-systems may be without touching, it will evidently be true when they are actually in contact.

Next, suppose that X_2 can also be divided into two parts which are electrically independent. Then if U_b and U_c are the individual energies of these two parts, and N their mutual energy, we have
$$U_2 = U_b + U_c + N$$

If, for symmetry, we put $U_1 = U_a$, we get
$$U = U_1 + U_2 + M$$
$$= U_a + U_b + U_c + M + N$$
$$= U_a + U_b + U_c + M', \text{ say,}$$
where M' is called the "mutual energy" or "gravitational potential energy" of the three sub-systems, and its value in any state P is equal to the sum of the quantities of work done on the three sub-systems by their mutual gravitational attractions as the whole system *returns* in any way from the state P to the standard state O. And so on.

The attraction of gravitation between two moderately-sized bodies is always exceeding small. Hence if our system consist of a moderate number of moderately-sized bodies, the distances between which in the state P do not greatly differ from the distances in the standard state, the value of the mutual energy in the state P will be quite insignificant, and may be neglected.

The method of calculating the value of M is too advanced to be given in this work. The following, however, is a useful case. Let (m_1, m_2) be the masses of two homogeneous rigid spheres in grammes, and r the distance between their centres in centimetres: then the gravitational attraction between them is $\dfrac{\lambda m_1 m_2}{r^2}$ dynes. Thence it can be shown (by a method too advanced to be given here) that if r_0 be the distance between the centres in the standard state, r the distance in the state P, the potential energy of the two spheres in the state P is $\lambda m_1 m_2 \left(\dfrac{1}{r_0} - \dfrac{1}{r}\right)$. This is equal to $-\dfrac{\lambda m_1 m_2}{r} +$ a constant quantity, independent of the state P, which is of no account in the variations of the potential energy (the only thing we usually want to know about the potential energy), and may therefore be generally omitted.

Art. 6. In this article we propose to use the result $U = U_1 + U_2 + M$ in considering the case of contact.

Let (A_1, A_2) be two moderately-sized unelectrified bodies in contact, whose mutual energy may be neglected. Let U_1 and U_2 be their individual energies, and U that of the system of the two bodies in any state P: ($U_1 + \Delta U_1$, $U_2 + \Delta U_2$, $U + \Delta U$) the energies in any state P. Then
$$U = U_1 + U_2,$$
$$U + \Delta U = (U_1 + \Delta U_1) + (U_2 + \Delta U_2);$$
hence
$$\Delta U = \Delta U_1 + \Delta U_2.$$

Now suppose that neither of the two bodies is subjected to external thermal influences except at the surface of contact with the other. Then the heat absorbed by the system of the two bodies is zero. When we take the two bodies separately, the definition of heat absorbed does not generally apply. We will, however, assume an equation to hold of the same form as the equation of energy. The assumption will be justified if it only leads to conclusions in agreement with accepted results.

Let w_1 be the work done on A_1 by the forces arising from the contact[1] of A_2, and let W_1 be the work done by other forces. Then we will write
$$\Delta U_1 = W_1 + w_1 + Q_1.$$
Similarly
$$\Delta U_2 = W_2 + w_2 + Q_2.$$
Also
$$\Delta U = W_1 + W_2.$$
Hence since
$$\Delta U = \Delta U_1 + \Delta U_2,$$
$$w_1 + w_2 + Q_1 + Q_2 = 0.$$

Now w_1 and w_2 will not be equal and opposite, or their sum zero, if the surfaces of contact slip over one another. For example, if A_1 be at rest while A_2 slides,

---

[1] The mutual gravitational attractions of A_1 and A_2 are not considered, being supposed negligible.

the work done by the frictional force on A_1 will be zero, because the place at which this force acts remains (apparently) at rest; and the work done by friction on A_2 is clearly negative. Thus $w_1 + w_2$ is not zero. Hence, when there is slipping, $Q_1 + Q_2$ will not be zero, or Q_1 and Q_2 will not be equal and opposite.

It may be thought that $w_1 + w_2$ must always be zero, because no force acts between any two particles except when they are in contact. This may seem to be proved by the following argument. Let N be the position of two particles in contact, tNT the direction of the equal and opposite action and reaction (F) between them, and let the two particles be displaced to N' without being separated.

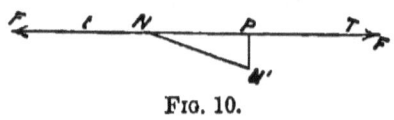

Fig. 10.

Drawing N'P perpendicular to tNT, the work w_1' done on one particle will be F . NP[1], and the work w_2' done on the other — F . NP. Hence $w_1' + w_2'$ is always zero.

It does not, however, follow that $w_1 + w_2 = 0$ because $w_1' + w_2' = 0$. In proving $w_1' + w_2' = 0$, the word "work" has not the same meaning as in the equation of energy. This will be seen by considering the two bodies (A_1, A_2,) of which A_1 is at rest while A_2 slides over A_1. If we use the word "work" with the meaning assigned to it in the equation of energy, the "work" done on A_1 by the force of friction arising from the rubbing of A_2 will be zero. But if we examine the matter a little closer, it will appear that the word "work" admits of a different meaning. The surface of no body is perfectly smooth, but may be supposed covered with a vast number of very small projecting teeth. The resistance experienced when we drag one surface over another is due to the fact that the teeth on one surface have to push the teeth on the other aside. Hence in the case of the two bodies (A_1, A_2,) the "work" done on A_1 by friction is

---

[1] Here the distance NN' is supposed to be so small that the force F remains constant in magnitude and direction during the displacement.

really positive, and not zero, as the equation of energy takes it to be.

It is here desirable to explain modern ideas of the phenomena of heat in the case of a solid body. It is supposed that the visible motions of a solid body are not the only motions which the body possesses; but that in addition to the visible motions, every particle is in a state of brisk agitation about its mean position, the distance of any particle from its mean position being always very small. And it is on these invisible motions that the thermal properties are supposed to depend. The visible motions may be called Mechanical, because they are the only ones taken into account in ordinary mechanics; the invisible may be called Non-mechanical. As the non-mechanical motions of two particles near together may be in any different directions, these motions are often called Irregular.

Again, owing to the irregular motions, the work done on a body will not be the same as it appears to be. Thus if the surface of contact of two bodies (A, B) be at rest, there will appear to be no work done on either body at the surface of contact; but if the surface particles remain in contact for a longer distance when moving from A to B than when moving from B to A, the average force between them being the same in the two cases, it is evident that a positive quantity of work will be done on B, and a negative quantity on A.

The visible work done on a body may be called Mechanical, and the excess of the total work over this, Non-mechanical. The non-mechanical work, done when two bodies are in contact, is supposed to be what is meant when we speak of the conduction of heat.

It will now be clear that in the equation of energy the word "work" means mechanical work; while in proving the proposition $w_1' + w_2' = 0$ as given above, the word "work" means total work. The equation $w_1' + w_2' = 0$ is therefore equivalent to $w_1 + w_2 + Q_1 + Q_2 = 0$.

If there be no rubbing, w_1 and w_2 will both be zero,

and therefore $Q_1 + Q_2 = 0$, or Q_1 and Q_2, equal and opposite.

It only remains to be said that except on a few occasions we shall not indulge in speculations as to the constitution of bodies. Our arguments will generally be based directly on experimental facts of a very simple character. The short forms "work" and "kinetic energy" will therefore be commonly used as equivalent to "mechanical work" and "mechanical kinetic energy," and the full expressions restricted to a few speculative cases.

Art. 7. The equation $\Delta U = W + Q$, which refers to a system protected from external electric influences, shows that when there are no external forces and the system is prevented from receiving and losing heat, $\Delta U = 0$. As ΔU is the difference of the values of the energy U in any two states, the result $\Delta U = 0$ denotes that the energy remains constant. This conclusion is known as the Conservation of Energy. Maxwell gives the principle in the following form:—

"The total energy of any body or system of bodies is a quantity which can neither be increased nor diminished by any mutual action of these bodies, though it may be transformed into any one of the forms of which energy is susceptible."

The equation $\Delta U = W + Q$ is applicable to the solar system. In this case there are practically no external forces and an enormous amount of heat is given out by the sun. Thus $W = 0$ and Q is negative. Hence ΔU is negative, or the energy decreases.

We cannot treat the whole universe in the same way as the solar system, because we do not know sufficiently what the universe is. We must, however, notice a way of considering the question due to Clausius, the great thermo-dynamicist. Assuming that there are no external forces acting on the universe, the equation $\Delta U = W + Q$ shows that the energy can only change on account of the gain or loss of heat. But as the universe includes all material bodies, it is clear that heat can only be gained by it by absorbing radiation, and lost by emitting radiation.

Now when a material body radiates heat into free space, the heat lost by the body may be traced in space in the form of advancing waves, which leave the space passed over in exactly the same condition as they found it. The energy of these waves appears to be exactly equal to the heat radiated by the material body. Similarly, when radiation is absorbed by a material body, the loss of energy in the waves is exactly equal to the heat absorbed by the body.

Hence, if we use the word "universe" to denote not merely all material bodies, but all space as well, it follows that its energy will be constant.

Art. 8. In electrical problems we shall generally suppose the system so chosen that all the electric properties are internal. There will then be no external electric influences, and the equation of energy is $\Delta U = W + Q$. Only one exception will be made to this rule. We shall take the case of a system which an electric current enters and leaves at A and B, respectively, by conductors *of the same metal and in the same condition*, there being no electric actions between the given system and bodies outside it. The limitation that the conductors at A and B are to be of the same metal in the same condition will be found hereafter to be of the utmost importance.

For the future we shall generally leave it for the context to explain what are the external electric influences on our system—whether there are no external electric influences at all, or whether there is an electric current passing through the system and having its circuit completed outside the system, as stated above.

In the present article we propose to consider what form the equation of energy takes in the latter case. It will be found to be

$$\Delta U = W + Q + e \quad . \quad . \quad . \quad (2).$$

The result, however, will be of no service to us except in electrical questions, and may be omitted until required.

We shall make use of some familiar experimental facts. We know that an electric current is produced in a closed metallic circuit of different metals by simply heating the junctions to different temperatures. The current may be varied as we please without making any appreciable change in the non-electric state of the circuit. This may be done by doubling the wire forming any part of the circuit on itself, so that the current in passing from one place A to a neighbouring place A' may have to go a long way round through a large resistance, or, by merely making contact between two adjacent parts of the wire, may be allowed to take a shorter path through a smaller resistance. To stop the current altogether we have merely to interrupt the circuit by slightly separating two pieces of wire formerly in contact. Again, when any part of such a circuit consists of thinnish wire doubled on itself, it is found that the electric condition of this part of the circuit may be supposed the same whatever the current in it may be and whatever the resistance. The state of the part therefore depends only on the non-electric properties, and is the same as when the current is zero. The heat absorbed in this part of the circuit may therefore be found in the ordinary way. It must not, however, be assumed that the heat absorbed when there is a current is the same as when there is none.

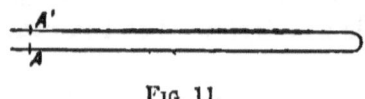

Fig. 11.

Suppose, then, that our system defined above contains a thinnish metallic chain doubled on itself, which is always free from electro-static electrification and magnetism, and is also free from electric currents when the current enters the system from without. Let the chain be used to produce a current *instead* of that which enters from without, by disconnecting external bodies from A and B and coupling up the chain to A and B instead. Then, since the definition of energy (Art. 3) applies to the system when the current is contained entirely within it, and since the state of the system is then exactly the same as when the current enters from without, it follows that the definition of energy applies also to the system in that case.

Now, for simplicity, suppose that the (non-electric) state of the metallic chain is kept the same and that no work is done on it, whether used to produce a current or electrically neutral by being disconnected from A and B. Then when there is no current in the chain, the heat absorbed by it will be zero. Hence when the same change of state is produced in the system by a current in the chain as by a current coming from without, the only difference between the two processes will be that a quantity of heat h will be absorbed by the chain in the former case, and in the latter, none at all. But if ΔU be the increase of energy, W the work done on the system, and Q the heat absorbed by it in addition to that absorbed by the chain, we have in the former case, to which the equation of energy applies in the usual form,

$$\Delta U = W + (Q + h).$$

Hence in the latter case, the principle of energy does not apply in the usual form, but in the form

$$\Delta U = W + Q + e,$$

where e may be called the Electric Energy absorbed.

It only remains to calculate the value of e (or h). We begin by observing that everything in connection with the metallic chain will be the same whether the circuit be completed as we have supposed or in any other way, provided the current is the same in the two cases. We will, therefore, suppose the chain disconnected from A and B and the circuit completed in the simplest way possible, by a homogeneous wire of uniform temperature which is doubled on itself so that Ohm's law holds whether the current be constant or variable. Now in passing from one metal to another, there is generally an *abrupt* change of potential. Also when a current crosses the junction, the temperature of the junction will generally rise or fall unless heat be absorbed or evolved at it. These considerations will be avoided if we suppose the ends of the chain and the homogeneous wire to be of the same metal and in the same condition as A and B.

If I be the current and R the resistance of the homogeneous wire, the rate at which heat is given out by it per second if its temperature remains constant, will (by Joule's law) be $R I^2$. Hence if h be the heat absorbed by the chain in t seconds, during which the current may be considered constant, $h - R I^2 t$ is the heat absorbed by the system of the metallic chain and the connecting wire. There being no work done on this system and no change of state, since the state depends only on the non-electric properties, the total quantity of heat absorbed by the system is zero. Thus $h - R I^2 t = 0$, or $h = R I^2 t$. But if E be the electro-motive force in the homogeneous wire in the direction of the current, Ohm's law gives $E = R I$. Hence $h = R I^2 t = E I t$. Now in travelling completely round a closed circuit, the total change of potential is zero. It is therefore clear that the rise of potential in any one part of the circuit will be exactly equal to the fall in the remaining part. There being no abrupt change of potential in passing from the metallic chain to the homogeneous wire or to A and B, it follows that the rise of potential in the metallic chain is equal to the fall either in the homogeneous wire or in the given system. Hence since electro-motive force is proportional to difference of potential, the electro-motive force from A to B through the given system (the direction of the current) is simply E. Thus e (or h) is obtained in terms of the current and electric states of A and B, without reference to the auxiliary homogeneous wire.

If q be the quantity of electricity that enters the system during the t seconds, $q = It$, and therefore $e = qE$.

The same equation $\Delta U = W + Q + E I t$ holds when the system does not contain a "thinnish metallic chain." This follows from the fact that the metallic chain undergoes no operation whatever and is merely introduced to enable us to calculate e. Thus let there be a second system identical with the first except that it contains no metallic chain, and let it undergo simultaneously the same process as the given system. Then the values of W, Q, and e, are the same for both systems. If therefore we consider a change from the standard state o to any other state P, we see that the two systems have the same energy in any the same state. Hence for a change from a state P to a state P', we have for both systems

$$\Delta U = W + Q + e.$$

If A and B be of different metals, E cannot be found by experimental methods, and moreover e does not take the simple form ItE. For these two reasons it is necessary to restrict A and B to be of the same metal in the same condition. A further reason for the restriction will appear later.

Lastly, if several currents enter a system from without, the equation of energy is easily seen to be

$$\Delta U = W + Q + e_1 + e_2 + e_3 + \ldots,$$

where ($e_1, e_2, e_3, \ldots$) refer to the several currents.

Art. 9. If the operations undergone by a system be such that the final state is the same as the first, the series of operations is called a *cyclical process*. The total change of energy, or ΔU, is zero for such a process, and the usual equation of energy $\Delta U = W + Q$ becomes $W + Q = 0$, or $W = - Q$. As $- Q$ is the heat given out in the process, we see that the work done on a system in a cyclical process is equal to the heat given out. If we call $- W$ the work given out by the system during the cycle, the heat absorbed will be equal to the work given out.

Art. 10. We have seen that in the equation $\Delta U = W + Q$, W and Q generally depend on the way in which the change of state is effected, as well as on the initial and final states themselves. Now there will probably be several methods of effecting the change of state for which W has the same value. So long as we restrict ourselves to these methods, the value of W may be said to depend only on the initial and final states; and from the equation $\Delta U = W + Q$, it follows that the same will also be true of Q.

The value of W can be shown to depend only on the initial and final states, whatever changes of temperature, liquefactions, vaporisations, or other occurrences take place in the system, when the only force which does work on the system is a uniform and constant pressure, as the pressure of the air, acting at right angles to all parts of the surface that are in motion. For simplicity, suppose the system to be a single unelectrified body whose only motion is due to its slow expansion or contraction by heat.

Let the full line be the surface at any instant, the dotted line shortly afterwards. Take a small part XY of the surface—so small that it may be considered plane, and let it come into the position X'Y'. Then if A be the area of XY and p the pressure on it per unit area, the total pressure on the area will be pA. Now it is evident the two areas XY, X'Y' are practically parallel. If, therefore, n be the perpendicular distance between them, the work done on the area A when this part of the surface is pushed out, will be $-p$A.n, but when it shrinks in, the work will be pA.n. But A.n is the volume of the small cylinder XYY'X'.

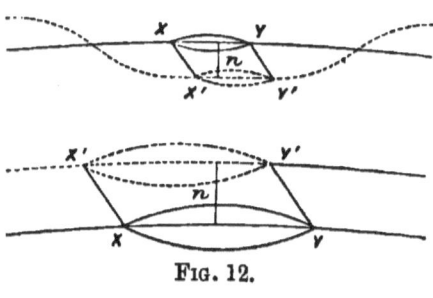

Fig. 12.

Calling this volume v_1, the work done by the pressure on the small area XY is either $-pv_1$ or $+pv_1$. Treating the rest of the surface in the same way, we obtain

$$W = -p(\pm v_1 \pm v_2 \pm v_3 \pm v_4 \pm \ldots).$$

The upper sign must be taken for an area that is pushed out, the lower for one that is pushed in. Thus the quantity in brackets is the excess of the volumes pushed out over the volumes shrunk in—in other words, it is the total increase of volume.[1] Denoting the total increase of volume by Δv, we have $W = -p\Delta v$. As p is a given quantity and Δv depends only on the initial and final states, W does so too. Hence also the heat absorbed in passing from one state to another depends only on those two states.

Now the state of a single unelectrified body at rest will generally depend on various circumstances. For example, the temperature may not be the same throughout the body and the body may contain substances in arbitrary proportions in the three states of solid, liquid, and gaseous. It may often happen, however, that the temperature is the same throughout the body and that the state of the body depends only on the temperature. In this case, the change of state will be known when we are given the initial and final temperatures. For such bodies we are therefore led to give the following definitions :—

(1) At any temperature, the "thermal capacity of a body at constant pressure" is the heat required to raise its temperature one degree C. while the pressure to which it is subjected remains constant.

The pressure which most commonly occurs is that of the air, which may, at any instant, be supposed constant for a short time.

(2) At any temperature, the "thermal capacity of a body at constant volume" is the heat required to raise its temperature one degree C. when the volume is kept constant.

(3) If the body be homogeneous, the thermal capacity of the body per gramme is called its "specific heat."

(4) The specific heat of water at 0° C. under a constant pressure of one atmo is called a Calorie, and is used as an arbitrary unit of heat.

Art. 11. When the only external force acting on a body is a uniform normal pressure, constant or variable, over all the surface, the work done by the body against the force during a change of volume may be represented

---

[1] Decrease of volume being considered negative increase.

graphically by a diagram, known as Watt's Diagram of Energy, or an Indicator Diagram. Two rectangular axes Op, Ov being taken, the pressure and volume at any instant are represented by the position of a point P in the plane of the axes, such that if PM be drawn at right angles to Ov, the volume v is proportional to OM (the abscissa of P), and the pressure p to PM (the ordinate of P). If P' be a consecutive position of P for which the pressure and volume have slightly different values ($p + \Delta p$, $v + \Delta v$), the work done by the body in passing from the state P to the state P' will lie between $p\Delta v$ and $(p + \Delta p)\, \Delta v$, since the pressure always lies between p and $p + \Delta p$. Now if P'M' be the ordinate of P', or the perpendicular from P' on Ov, $p\Delta v$ and $(p + \Delta p)\, \Delta v$ are evidently proportional to the areas of the rectangles PM' and P'M. Thus the work done by the body is represented by an area intermediate between those of these two rectangles.

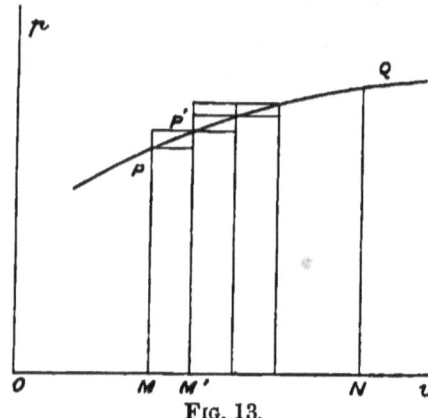

FIG. 13.

It is now clear that if the body pass from one state P to another state Q, by a path represented by the curve PQ, the work done by the body will be intermediate between the sum of the areas of the inscribed and the sum of the areas of the circumscribed rectangles. This is true however numerous the rectangles may be. But when their number is great enough and every one of them very small, the two sums approach to equality with the area of the curved figure PQNM, and no area but that of the curved figure can be intermediate between them. In other words, the area of the curved figure represents the work done by the body.

When the pressure is constant, the curve PQ becomes a straight line parallel to Ov, and the area which represents the work done by the body clearly depends only on the initial and final volumes.

In a cyclical process, the final state is the same as the first. The indicator diagram for such a process is therefore a closed curve. If the point P whose position denotes the pressure and volume, travel round the closed curve in the same direction as the hands of a watch, the upper part of the curve will represent states of increasing volume in which the body does positive work against the normal pressure, and the lower part of the curve will represent states of decreasing volume in which the pressure does positive work on the body, or the body negative work against the pressure. If Am and Bn are the extreme ordinates, the work done by the body in passing along APB will be represented by the area of the figure APBnm. In returning from B to A through Q, the work done by the body will be diminished by an amount proportional to the area BQAmn. Thus the total work done by the body in the cycle is represented by the area of the closed curve.[1]

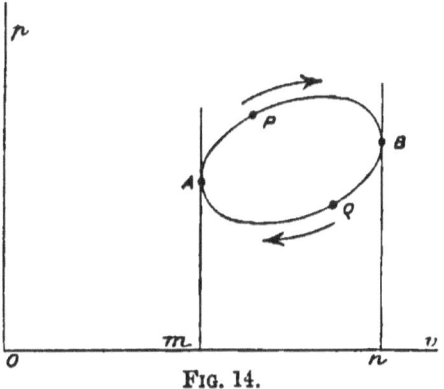

FIG. 14.

Art. 12. In Joule's experiment on the agitation of water, we may suppose the apparatus so arranged that no heat is ever gained or lost by the water during the experiment, either by conduction or radiation. If W_1 be the work done on the water by the paddles, W_2 the work done on it by the constant pressure of the air, and ΔU

---

[1] Those who are only acquainted with elementary mathematics, will probably find this Article rather difficult.

the .increase of energy, we then have $\Delta U = W_1 + W_2$. Now if Q be the heat absorbed by the water when the given change of state is brought about by the action of heat and the same constant pressure of the air as before, we shall have $\Delta U = W_2 + Q$. Hence $W_1 + W_2 = W_2 + Q$, or $Q = W_1$. Thus Joule's experiment gives us the heat in ergs (or the mechanical value of the heat) absorbed by the water under a given pressure of the air when its temperature varies by a known amount.

It is not important to know the exact pressure of the air, because owing to the small expansibility of water by heat, the work done by the pressure of the air under ordinary circumstances will be insignificant. Thus when the pressure of the air is an atmo and a gramme of water is raised from 0° C. to 1° C. by the combined action of the paddles and the pressure of the air, it is calculated that the value of W_1 (in ergs) is over 40 million, while that of W_2 is only about 100. For the present, therefore, we need generally make no mention of the pressure of the air or of the work done by it.

Joule's experiment does not directly give us the mechanical value of a calorie, because it is impossible to arrange the experiment so that the water rises in temperature exactly from 0° C. to 1° C. To deduce the mechanical value of a calorie from Joule's experiments, we must have a comparison of the specific heats of water at different temperatures. This may be obtained by a method known as the "method of mixtures." Thus let a quantity of water at 0° C. be added to an equal quantity at 100° C., and suppose that no work is done except by the pressure of the air. This requires that no work is done in mixing the two quantities of water together. Then if no heat be gained or lost by the mixture, the energy remains constant. Hence the heat gained by one quantity of water is exactly equal to that lost by the other. If therefore the uniform temperature which the water finally assumes be $\theta°$ C., it follows that the heat required to raise the temperature of a quantity of water from 0° C. to $\theta°$ C. is equal to that required to raise the temperature of an equal quantity from $\theta°$ C. to 100° C. In this way the specific heats of water at different temperatures have been compared. The specific heat at 0° C. being a calorie, the specific heats at other temperatures are known in calories, thus:—

at	0° C.	1·000,0,
,,	10° C.	1·000,5,
,,	20° C.	1·001,2,
,,	30° C.	1·002,0.

Suppose, now, that Joule's experiment gives the heat Q in ergs required to raise the temperature of one gramme of water from an observed temperature $t_1°$ C. to an observed temperature $t_2°$ C., when the only force which does work on the water is the pressure of the air. Let J be the number of ergs in a calorie, $(c_1, c_2, c_3, \ldots$ the specific heats of water at $t_1°, (t_1 + 1)°, \ldots$ centigrade. Then evidently $Q = J(c_1 + c_2 + c_3 + \ldots)$. The quantities in the bracket being known, this equation gives J.

It has been found that a calorie is equivalent to about 42,350 gramme-centimetres, or about 42 million ergs,[1] or about 3 foot-pounds. In English measure, the heat required to raise the temperature of one pound of water under a pressure of one atmo from 0° C. to 1° C., is equivalent to about 1390 foot-pounds.

Art. 13. The knowledge of the value of J, the mechanical equivalent of heat, enables us to make some interesting calculations. If a mass of one gramme be moving like a particle (without rotation or vibration) with a speed of v centimetres per second, its kinetic energy will be $\frac{1}{2}v^2$ ergs. If this be equal to a calorie, we have $v^2 = 84,000,000$ nearly, or v greater than 9,100. This velocity is over 91 metres, or about 300 ft., per second: that is, about $5\frac{1}{2}$ kilm., or $3\frac{1}{2}$ miles, per minute. Hence if two equal masses of water at 0° C., moving with this velocity, impinge on one another, and be brought to rest by the collision without the formation of steam, the impact will be sufficient to raise the temperature to 1° C.

---

[1] More strictly, 41,540,000 ergs.

For a rise from 0° C. to 100° C., the velocity would have to be about ten times as great. In the case of iron, the specific heat is only about $\frac{1}{9}$ of a calorie. Hence for a given rise of temperature, the velocity will only be about $\frac{1}{3}$ as large when the two moving masses are iron as when they are water.

Art. 14. We will conclude this chapter by showing how the heat coming from a small body when its temperature sinks to 0° C., under the action of no external force but the pressure of the air, is practically determined by the melting of ice.

The method depends on the simple fact, that at all ordinary pressures of the air, the same quantity of heat is always required to convert one gramme of ice at 0° C. into water at the same temperature. Let this quantity of heat, which is known as the "latent heat" of the liquefaction of ice, be x calories. Suppose that i grammes of ice at 0° C. are added to w grammes of water at θ° C., and that in the final state which the system assumes when prevented from gaining or losing heat, the whole of the ice is melted and the uniform temperature at θ'° C. Then the heat gained by what was originally ice, is approximately $i(x + \theta')$ calories, and that lost by what was water from the first, will be $w(\theta' - \theta)$ calories. These two quantities of heat being equal, we have

$$i(x + \theta') = w(\theta - \theta'),$$

and therefore

$$ix = w\theta - \theta'(w + i),$$

or

$$x = \frac{w\,\theta}{i} - \frac{w + i}{i}\theta'.$$

It is thus found that the latent heat of ice is about 79·25 calories.

In measuring the heat given out by the small body, an instrument called a "calorimeter" is made use of. This is a vessel containing water and ice which is prevented from gaining or losing heat. The temperature of the calorimeter being exactly 0° C., the small body under consideration is dropped into it, care being taken that there is both water and ice in the calorimeter in the final state which it assumes after the small body is dropped into it. The final temperature will then be 0° C. If n be the number of grammes of ice melted in the calorimeter during the cooling of the small body, the heat gained by the calorimeter from the small body, or lost by the small body, will be nx calories, or nxJ ergs.

Bunsen's calorimeter is a small instrument, consisting of three parts a, b, c, formed of glass and sealed together with the blow-pipe. The upper part of the tube c is bent horizontally, calibrated, and graduated. The higher part of b contains distilled water freed from air by boiling, and the lower part of b and the tube c are filled with boiled mercury. To prepare the calorimeter for use, a coating of ice is formed round the tube a by means of alcohol previously cooled below 0° C. in a freezing mixture. The instrument is then placed in a vessel of clean snow, a substance which soon acquires and long maintains the temperature 0° C. when the

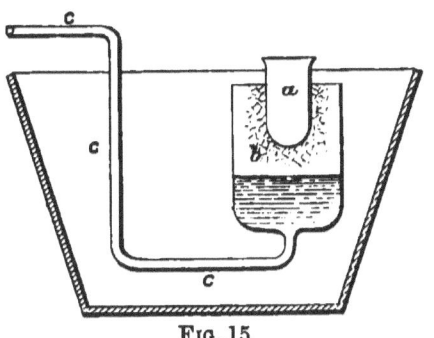

Fig. 15.

temperature of the room is not below 0° C. Lastly, the tube a is partially filled with water or some other fluid which does not dissolve the small body to be experimented on.

When the small body is dropped into the tube a, part of the ice in b will be melted. As the volume of ice is greater than that of the same quantity of water, the melting of ice in b will cause the mercury to rise in b and move along the graduated tube c. The excess of the volume of one gramme of ice at 0° C. over that of one gramme of water at 0° C. being known, the observed movement of the mercury in the graduated tube gives us the number of grammes of ice melted.

It should be noticed that if any air be left unexpelled in the water in b, it will be partially expelled in the form of a small bubble when the ice is formed round the tube a, and partially re-dissolved when the ice is melted. A small error will thus be introduced into the indications of the instrument.

CHAPTER II.

PERFECT GASES.

Art. 15. WE are unable to proceed much further with the principle of energy until we have explained Carnot's principle, except in the case of the more permanent gases, where some simple experiments supply us with all the information we require and make Carnot's principle unnecessary. It appears from these experiments that several gases, commonly called "perfect gases," satisfy two simple results very approximately. The deviations from strict accuracy will be neglected throughout the chapter, and the results taken to be exactly true. The two results are consistent with Carnot's principle. In fact, if one of them be taken for granted, Carnot's principle enables us to deduce the other from it.

We shall assume that the gas never possesses mechanical motions and that its temperature is always uniform throughout the mass. The volume and mass of the gas will be supposed so small that the effect on the gas of its weight may be neglected, and the pressure of the containing vessel taken to be the only external force which acts on it. In consequence, the density of the gas will always be supposed to be uniform; and the pressure of the gas on the containing vessel, or of the vessel on it, to be everywhere the same.

One of the two simple results mentioned above shows that the energy of a given mass of gas is independent of the pressure or volume, and depends only on the temperature. The other is a simple relation between the pressure, volume, and temperature; and is a combination of the two results known in England as the laws of Boyle and Charles, but on the Continent as the laws of Mariotte and Gay Lussac.[1]

Art. 16. The first experiment on perfect gases we shall consider was originally made by Gay Lussac, but repeated in a greatly improved form by Joule in 1844. A vessel of water A contained two strong closed vessels B, C, joined by a pipe in which there was a very perfect stop-cock. Air was compressed in B to a pressure of about 20 atmos, and exhausted from C, and the water was then stirred till its temperature became uniform. Next, the stop-cock was opened, upon which the air rushed from B to C, and the water around B was cooled and that around C heated. When the temperature of the water was again made uniform by stirring and the proper

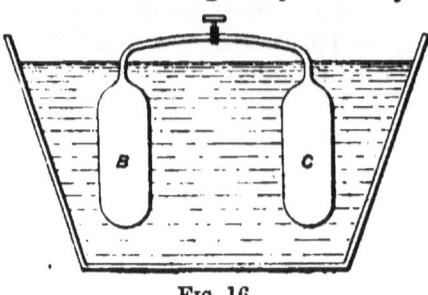

FIG. 16.

[1] In order to make this chapter intelligible to those who do not wish to read Chapter I., it is necessary to explain briefly the principle of energy and the nomenclature and notation made use of in connection with it.

When a body changes from any state P to any state P', let W be the algebraic quantity of work done on it during the operation, and let Q be the algebraic quantity of heat absorbed by it (heat evolved being considered negative heat absorbed). Then it is found that $W+Q$ depends only on the states P and P' and is independent of the way the change of state is effected.

Now let us choose any standard state of reference O. Let U be the sum of the quantities of work and heat required to bring the body in any way from the state O to the state P. Then U is called the Energy of the body in the state P with regard to the standard state O. The energy of the body in the state P' with regard to the same standard state (O) we denote by $U+\Delta U$, so that ΔU stands for "increase of U" and not for the product of Δ and U. It should now be easy for the reader to prove that ΔU depends only on the states P and P', and is the same whatever state be chosen for the standard state; also that $\Delta U = W + Q$.

The equation $\Delta U = W + Q$ expresses the principle of energy, and is called indifferently "the equation of energy" and "the first law of thermo-dynamics."

corrections made, it was found that no perceptible change had been made in the temperature. It was therefore concluded that the quantity of heat yielded to the water by the vessel C was equal to that abstracted from it by B. The total quantity of heat gained by the air during the experiment was therefore zero. Hence as no external work had been done on the air (or by it) during the expansion, it followed from the first law of thermo-dynamics that the energy was unaffected by the process. Thus the energy of a given quantity of air was shown to remain constant so long as the temperature was constant, whatever the pressure and volume might be: in other words, the energy was proved to depend only on the temperature.

This experimental result was carefully tested in a different way a few years later by Joule and Thomson, both for air and some other gases, and found to be very approximately true in every case. It is therefore generally taken to be true for all the more permanent gases.

Art. 17. The result of the last article leads immediately to two important deductions.

First, let a quantity of air or any perfect gas expand in any way from one volume to another, and let the temperature be made the same after the expansion as before. Then since the energy is unaltered, the first law of thermo-dynamics shows that the heat given out by the gas is equal to the work done on it, or the heat absorbed to the work done by it.

Secondly, let the temperature of a quantity of air or any perfect gas be raised 1° C. while the volume is kept constant. Then since no work is done on the gas, the heat absorbed is equal to the increase in the energy, and therefore depends only on the initial and final temperatures (which merely differ by unity). Now the heat absorbed by one gramme of any homogeneous substance when the temperature is raised 1° C. and the volume kept constant, is the specific heat of the substance at constant volume. Thus the specific heat at constant volume of a perfect gas can depend only on the temperature, and may be altogether constant.

Art. 18. Let a quantity of air or any perfect gas be kept in a cylinder fitted with a smooth air-tight piston, so that the volume can be increased or decreased at pleasure. Then it was found by Boyle (and Mariotte) that so long as the temperature is kept constant, the pressure varies inversely as the volume, or the product of the pressure and volume is constant. Thus if p be the pressure at any instant and v the volume, the product pv is constant when the temperature is constant, and therefore its value at different temperatures depends only on the temperature.

Again, a simple relation was found by Charles (and Gay Lussac) to hold between the volume and temperature when the pressure is kept constant. If θ' be the temperature as indicated by a mercury centigrade thermometer, we may write
$$v = V(1 + a\theta'),$$
where V is the volume when $\theta' = 0$, and a has the same value ·003,665, or $\frac{1}{273}$, for all gases and all pressures.

Boyle's law may be written
$$pv = \Theta \quad \ldots \quad (7),$$
where Θ is a quantity which depends only on θ'.

Hence keeping the pressure constant and reducing the temperature to 0° C., we have
$$pV = c,$$
where c is the value of Θ when $\theta' = 0$.

Dividing one result by the other, we get
$$\frac{v}{V} = \frac{\Theta}{c}.$$

Now by the law of Charles, $\frac{v}{V} = 1 + a\theta'$: hence
$$1 + a\theta' = \frac{\Theta}{c},$$

Perfect Gases.

or
$$\Theta = c(1 + a\theta').$$

Thus we obtain from equation (7),
$$pv = c(1 + a\theta')$$
$$= ca\left(\frac{1}{a} + \theta'\right)$$
$$= ca(273 + \theta').$$

If therefore we write θ for $\theta' + 273$, and R for ca, we get
$$pv = R\theta \quad \ldots \quad \ldots \quad (8).$$

Hence if (p_0, v_0, θ_0) be any corresponding values of (p, v, θ),
$$\frac{pv}{\theta} = R = \frac{p_0 v_0}{\theta_0}$$

The equation $pv = R\theta$ may be shown by Carnot's principle to lead to the result that the energy depends only on the temperature.

Art. 19. The following table[1] exhibits some important fundamental experimental results relating to perfect gases, at a pressure of one atmo and at the temperature 0° C.

	Relative Densities.	Relative Specific Volumes.	Mass of a Litre in Grammes.	Volume of a Gramme in Litres.	Volume of a Pound in Cubic Feet.
Air	1	1	1·2932	·7733	12·39
Oxygen (O) ...	1·10563	·90446	1·4298	·6994	11·20
Hydrogen (H)	·06926	14·4383	·08957	11·1645	178·85
Nitrogen (N) ...	·97135	1·02945	1·25615	·7961	12·75
Carbonic oxide (CO)...	·9545	1·0476	1·2344	·8101	12·97
Carbonic acid (CO_2) ...	1·52907	·6540	1·9774	·5057	8·10
Chlorine (Cl) ...	2·4222	·4128	3·1328	·3192	5·11
Cyanogen (NC_2) ...	1·8019	·5550	2·3302	·4291	6·87
Marsh gas (CH_4) ...	·562	1·779	·727	1·375	22·04
Olefiant gas (C_2H_4) ...	·982	1·018	1·270	·787	12·61
Ammonia (NH_3) ...	·5952	1·6801	·7697	1·2992	20·81

[1] Everett's "Units and Physical Constants."

From these data the following table has been calculated, giving the values of $\dfrac{p_0 v_0}{273}$, where v_0 is the volume in cubic centimetres of one gramme of gas at 0° C. under a pressure of one atmo, and (1) p_0 the number of dynes 1,013,510 per square centimetre in a pressure of one atmo; (2) p_0 the number of gramme-weights (at Paris) 1,033·279 in the same pressure.

	(1)	(2)
Air	2,871,000	2,927
Oxygen	2,596,000	2,647
Hydrogen	41,448,000	42,256
Nitrogen	2,955,000	3,013
Carbonic oxide	3,007,000	3,066
Carbonic acid	1,877,000	1,914
Chlorine	1,185,000	1,208
Cyanogen	1,593,000	1,624
Marsh gas	5,105,000	5,205
Olefiant gas	2,922,000	2,980
Ammonia	4,820,000	4,917

Art. 20. The equation $pv = R\theta$ has led to the construction of thermometers in which air is used as the thermometric substance instead of mercury. Let (p_0, v_0, θ_0) refer to 0° C. Then at any temperature, the temperature of the air thermometer is defined to be the value of θ given by the equation

$$\frac{pv}{\theta} = \frac{p_0 v_0}{\theta_0}, \text{ or } \theta = \frac{pv}{p_0 v_0}\theta_0,$$

where θ_0 is taken to be 273.

In air thermometers, either the pressure or the volume is kept constant. In constant pressure thermometers, $p = p_0$ and $\theta = 273\dfrac{v}{v_0}$. In constant volume thermometers,

$$\theta = 273\frac{p}{p_0}.$$

The indications of the air thermometer are nearly but not quite the same as those of the common mercury centigrade thermometer increased by 273. The difference is explained in Kohlrausch's "Physical Measurements."

Since any of the perfect gases may be used as the

thermometric substance instead of air, the scale of the air thermometer is often called "absolute." The name, however, is inappropriate, because there are but a limited number of gases which can be used instead of air. A truly "absolute" scale of temperature, independent of the special properties of any limited class of substances, can only be obtained by means of Carnot's principle. The only "absolute" scale of this kind we shall have to consider is due to Sir W. Thomson[1] and practically coincides with the scale of the air thermometer.

A very simple form of the air thermometer is due to Jolly. The volume of the air is not kept quite constant, but the only change in it is that due to expansion by heat of the glass vessel in which it is contained. The thermometer is made of glass in one piece, and consists of a globe of about 50 cubic centimetres capacity joined by a fine capillary bent tube A to a larger tube B. The globe is filled with dry air, which is confined in it by mercury; the tube B being connected with an open glass tube C by an india-rubber tube D, and the tubes B, D, and the lower part of C filled with mercury.

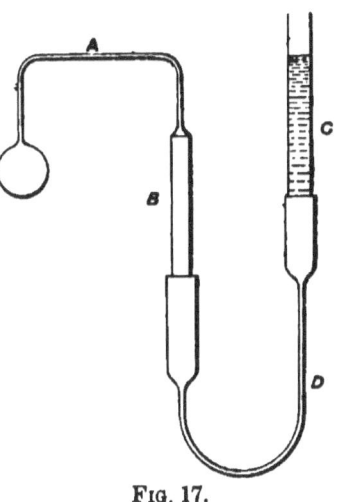

FIG. 17.

In using the instrument, we proceed as follows:—

(1) The temperature of the glass globe is first brought to 0° C. by surrounding it with melting ice. The tube C is then raised or lowered till the surface of the mercury in B comes to a mark near the capillary tube A. Lastly, the pressure p_0 of the air in the glass globe is obtained by observing the height of the barometer and the height of the mercury in C above the mark on B.

[1] Now Lord Kelvin.

Let V_0 be the volume of the glass globe at $0°$ C. Then, since the quantity of air in the capillary tube may be neglected, we have

$$p_0 V = 273 R.$$

(2) The glass globe is brought to any temperature we wish to consider, the mercury adjusted, and the pressure found, as before.

If the temperature be $\theta'°$ C., the capacity of the glass globe, which is practically the same as the volume v of the dry air, is $V(1 + \cdot 000{,}025\theta')$. The value of $\theta - 273$ being approximately the same as that of θ', we may write $\theta - 273$ for θ' in the small term of v: thus

$$v = V\{1 + \cdot 000{,}025(\theta - 273)\}.$$

Hence the equation $pv = R\theta$ becomes

$$pV\{1 + 0\cdot000{,}025(\theta - 273)\} = R\theta.$$

Now $p_0 V = 273 R$: therefore

$$\frac{p}{p_0}\{1 + 0\cdot000{,}025(\theta - 273) = \}\frac{\theta}{273},$$

and

$$\frac{p - p_0}{p_0} + \frac{p}{p_0} \times 0\cdot000{,}025(\theta - 273) = \frac{\theta - 273}{273}.$$

Hence

$$(\theta - 273)\left\{0\cdot003{,}665 - 0\cdot000{,}025\frac{p}{p_0}\right\} = \frac{p - p_0}{p_0},$$

or

$$(\theta - 273)\{0\cdot003{,}665 p_0 - 0\cdot000{,}025 p\} = p - p_0,$$

which gives θ.

Art. 21. A degree of the air thermometer being practically the same as a degree of the mercury centigrade thermometer, the simple relation $pv = R\theta$ for any perfect gas readily enables us to find the difference between the specific heats at constant pressure and constant volume.

Let a gramme of gas be contained in a cylinder fitted with a smooth air-tight piston, and let the temperature be slowly raised 1° C. under two different conditions as to pressure and volume.

(1) Let the volume be kept constant during the rise of temperature. The heat absorbed in this case is the "specific heat of the gas at constant volume," which has already been shown to depend only on the temperature or to be constant. Its value in ergs will be denoted by C_v.

(2) Let the pressure be kept constant. The heat absorbed will be the "specific heat at constant pressure," and its value in ergs will be written C_p.

Since the energy of a given mass of gas depends only on the temperature, the change of energy (ΔU) will be the same in the two cases just given. Hence if W be the work done *on* the gas in the second case, the equation of energy is

$$\Delta U = C_v = C_p + W.$$

But if p be the constant pressure in the second case, v, and v' the initial and final volumes, the work done *by* the gas will be $p(v'-v)$; and therefore the work done *on* the gas, or W, is equal to $-p(v'-v)$. Also if θ be the initial temperature, $pv = R\theta$ and $pv' = R(\theta + 1)$. Hence

$$W = -p(v'-v) = -R.$$

Thus

$$C_v = C_p - R, \text{ or } C_p - C_v = R. \quad . \quad . \quad (9).$$

In the hands of Clausius and Rankine, this simple result and the conclusion that C_v can only depend on the temperature, formed some of the earliest triumphs of the science of thermo-dynamics.

Up to 1850 it was supposed to have been shown that the specific heat of a perfect gas depended on the pressure; but in that year Clausius was led by the theoretical arguments explained above, to assert that it could only depend on the temperature, and he conjectured that it would probably be found to be constant. The fact that what was then a new theory disagreed with the results of accepted experiments, led to an attack upon it by Holtzmann. Three years later, however, the splendid experiments of Regnault on gases were published by which it was clearly shown that the specific heat of a perfect gas was the same, both at all temperatures and all pressures.

A little later in the same year, 1850, Rankine not only obtained the expression for the difference of the specific heats, but actually calculated the specific heats themselves in the case of air. For the latter purpose he made use of the value of $\dfrac{C_p}{C_v}$ which had long been known from a comparison of theory and observation on the velocity of sound in the air.

When a wave of sound travels in the air, the air it passes over is alternately compressed and rarefied, and therefore rises and falls in temperature. Now the space passed over by the wave in a second is so great in comparison with the length of the wave, that the duration of the compression and rarefaction at any place is far less than the hundredth of a second. It is therefore supposed that there is not time for conduction to affect the altered temperature produced at any place by the wave before the air is again in its natural state as to temperature and pressure. Thence a formula is obtained for the velocity of sound in air which contains the ratio of the specific heats, $\dfrac{C_p}{C_v}$—a ratio which is usually denoted by k. On comparing the formula with the velocity of sound in air, according to the determinations of Bravais and Martens, it is found that

$$\frac{C_p}{C_v} \text{ (or } k\text{)} = 1\cdot 410.$$

Hence by equation (9)

$$1 - \frac{1}{k} = \frac{R}{C_p},$$

or

$$k - 1 = \frac{kR}{C_p},$$

or
$$C_p = \frac{k}{k-1}R = \frac{1\cdot 41}{0\cdot 41} \times R$$
$$= \frac{1\cdot 41 \times 2{,}871{,}000}{0\cdot 41}$$
$$= 9{,}873{,}000;$$
and therefore
$$C_v = C_p - R = 7{,}002{,}000.$$

If c_p, c_v be the specific heats in calories, we have $c_p = \frac{C_p}{J}$ and $c_v = \frac{C_v}{J}$. Hence, giving to J the value 41,540,000, we get
$$c_p = \frac{9{,}873{,}000}{41{,}540{,}000} = 0\cdot 2377,$$
and
$$c_v = \frac{7{,}002{,}000}{41{,}540{,}000} = 0\cdot 1686.$$

The value previously accepted for c_p at atmospheric pressure was $0\cdot 2669$. Regnault's experimental result, given a few years later, was $c_p = 0\cdot 2375$.

Art. 22. The specific heats of the perfect gases at constant pressure have been determined by Regnault. Combining his results with the relation $C_p - C_v = R$, we easily get the specific heats at constant volume. The values are generally expressed in calories; but there is another way of considering specific heats, due to Clausius, which is both useful and interesting. We find the ratio of the heat required to raise the temperature of a quantity of gas $1°$ C. at constant pressure or volume to the heat required, under the same conditions as to pressure and volume, to raise the temperature of an equal volume of air by the same amount. These ratios will be denoted by γ_p and γ_v, respectively.

Since we have $C_p - C_v = R$,
we obtain for air—
$$c_p - c_v = \frac{R}{J} = \frac{2{,}871{,}000}{41{,}540{,}000} = 0\cdot 0691.$$

If v' be the volume of one gramme of another gas at the same temperature θ and pressure p as air, and R' the corresponding value of R, we have

$$\left.\begin{array}{l}pv' = R'\theta \\ pv = R\theta\end{array}\right\}.$$

Thus

$$\frac{v'}{v} = \frac{R'}{R},$$

or

$$R' = R\frac{v'}{v} = Rx, \text{ say,}$$

where x is the ratio of the specific volume of the gas to that of air, and is given in the table.

Hence we have, for any gas,

$$C_p' - C_v' = R' = Rx,$$

and

$$c_p' - c_v' = \frac{Rx}{J} = 0{\cdot}0691x \quad . \quad . \quad (10),$$

a formula by which we easily find c_v' from Regnault's experimental determination of c_p'.

Again, the heat absorbed when the temperature of one cubic centimetre of the gas is slowly raised 1° C. at constant pressure, is $\dfrac{C_p'}{v'}$

Hence

$$\gamma_p' = \frac{C_p'}{v'} \div \frac{C_p}{v} = \frac{C_p'}{C_p} \cdot \frac{v}{v'}$$
$$= \frac{c_p'}{c_p} y,$$

where y is the ratio of the specific volume of air to that of the gas, or of the density of the gas to that of the air, and is therefore a known quantity.

Substituting Regnault's experimental value for c_p, we get

$$\gamma_p' = \frac{c_p'}{0{\cdot}2375} y \quad . \quad . \quad . \quad . \quad (11).$$

Similarly

$$\gamma_v' = \frac{C_v'}{v'} \div \frac{C_v}{v} = \frac{C_v'}{C_v} \cdot \frac{v}{v'}$$

$$= \frac{c_v'}{c_v} y$$

$$= \frac{(c_p' - 0{\cdot}0691x)y}{c_v}$$

$$= \frac{c_p'y - 0{\cdot}0691}{c_v}, \text{ since } xy = 1,$$

$$= \frac{c_p'y - 0{\cdot}0691}{c_p} \cdot \frac{c_p}{c_v}$$

$$= \left(\gamma_p' - \frac{0{\cdot}0691}{0{\cdot}2375} \right) \frac{c_p}{c_v}$$

$$= (\gamma_p' - 0{\cdot}2909) \frac{c_p}{c_v} \quad . \quad . \quad (12).$$

The following table is calculated by means of these formulæ from Regnault's experimental determination of the specific heats at constant pressure in calories. The last column gives the value of k, the ratio of the specific heats, a quantity which is often said to be the same for all gases.

	Specific heat at constant pressure.		Specific heat at constant volume.		k.
	In calories.	Compared with an equal volume of air.	In calories.	Compared with an equal volume of air.	
Air	·2375	1	·1684	1	1·410
Oxygen (O)	·21751	1·012	·15501	1·018	1·403
Hydrogen (H) ...	3·40900	·994	2·4114	·992	1·414
Nitrogen (N) ...	·24380	·997	·17266	·996	1·412
Carbonic oxide (CO)	·2450	·985	·1728	·978	1·418
Carbonic acid (CO_2)	·2169	1·396	·1717	1·559	1·263
Chlorine (Cl) ...	·12099	1·234	·0925	1·330	1·308
Marsh gas (CH_4) ...	·5929	1·403	·4700	1·568	1·260
Olefiant gas (C_2H_4)...	·4040	1·670	·3337	1·946	1·211
Ammonia (NH_3) ...	·5084	1·274	·3923	1·386	1·296

Art. 23. We have seen that when a gramme of gas whose volume is kept constant is raised in temperature, the heat absorbed is equal to C_v ergs for each centigrade degree. But when the volume is constant, no work is done on the gas, and the heat absorbed is equal to the increase of energy. If therefore θ_0 and θ be the temperatures of the gas, as indicated by the air thermometer, in the standard state, and in any other state P of the same volume, the energy in the state P will be $C_v(\theta - \theta_0)$. And since the energy depends only on the temperature, this expression will denote the energy at the temperature θ whatever the pressure and volume may be.

The expression $C_v(\theta - \theta_0)$ may be written $C_v\theta + C'$, where C' is a constant. In taking the difference between the values of the energy in two states, the quantity C' does not appear, and we may therefore omit it, or call the energy $C_v\theta$. This is equivalent to supposing that, in the standard state, the temperature of the air thermometer is zero.

If U be the energy in the state P, $U + \Delta U$ in another state P' in which the temperature is $\theta + \Delta\theta$, we have

$$\left.\begin{array}{l} U = C_v\theta \\ U + \Delta U = C_v(\theta + \Delta\theta) \end{array}\right\}.$$

Hence

$$\Delta U = C_v . \Delta\theta.$$

For a perfect gas, the equation of energy therefore is

$$C_v . \Delta\theta = W + Q.$$

The commonest operations we have to consider are the following: (1) the temperature is kept constant, and (2) no heat is lost or gained. In the first, $\Delta\theta = 0$ and the equation of energy is $W + Q = 0$.[1] In the second, $Q = 0$ and the equation of energy $C_v . \Delta\theta = W$.

The temperature of the gas will remain practically

[1] It may be noticed that $W + Q = 0$ is also the equation of energy for any *complete cycle*, whatever the system and changes of temperature may be.

constant when it is contained in an ordinary metallic cylinder situated amongst objects of the same constant temperature, and the compression or rarefaction proceeds so slowly that there is time for the contact of the good-conducting walls of the cylinder to prevent any but the smallest change in the temperature of the gas. The gain or loss of heat cannot be completely prevented in practice. The best way to keep the amount small will be to have the gas in a vessel made of (or lined in the interior with) a feebly-conducting substance and to make the changes of volume so comparatively rapid that there is not time for the feeble conducting power of the containing vessel to produce much effect before the operation is over. If the gas were contained in an ordinary metallic cylinder, the gain or loss of heat could only be kept small by producing the changes of volume so very rapidly that the pressure and density would no longer be the same in all parts of the mass at the same instant, as throughout this chapter we suppose they must.

When the temperature is constant, we have $W + Q = 0$, or $Q = -W$, where W is to be calculated by means of the relation $pv =$ constant. When no heat is gained or lost, $C_v \cdot \Delta\theta = W$ and $pv = R\theta$. The indicator diagram in the first case is known as an Isothermal curve; that in the second case has been called an Adiabatic curve by Rankine, but, by Professor Gibbs, an Isentropic curve, because the entropy, a quantity explained in the next Chapter, remains constant during the operation.

The result obtained for an isothermal process is worthy of being stated in words. When the temperature of a perfect gas is kept constant during any change of volume, the work done *on* the gas is equal to the heat which it *gives out* (work done *by* the gas and heat *absorbed* being considered negative); on the work done *by* the gas is equal to the heat *absorbed* (work done *on* the gas and heat *given out* being now reckoned negative). The equation $W + Q = 0$ which expresses this result, requires that $\Delta U = 0$, a condition which is not generally fulfilled for bodies or systems other than perfect gases, except when the process forms a complete cycle.

The complete consideration of the equations just written requires the use of higher mathematics than we allow ourselves to make use of in the present work. We shall therefore merely state the principal results. Those who wish to see the mathematics by which they are obtained may consult the chapter on Perfect Gases in Parker's "Elementary Thermo-dynamics," or Clausius' "Mechanical Theory of Heat."

When the temperature is constant, let θ' be its value as indicated by the air thermometer. Then when the volume increases from v_1 to v_2, the work done by the gas against the confining pressure, and therefore also the heat *absorbed* by the gas, is $R\theta' \log \frac{v_2}{v_1}$. If the volume be compressed from v_2 to v_1, the work done on the gas, or the heat *given out* by it, is $R\theta' \log \frac{v_2}{v_1}$.

When no heat is gained or lost, the relation between the temperature and volume is $\theta v^{k-1} =$ a constant. Hence, if (p_0, v_0, θ_0) be the values of (p, v, θ) at any instant,

$$\theta v^{k-1} = \text{a constant} = \theta_0 v_0^{k-1},$$

or

$$\frac{\theta}{\theta_0} = \left(\frac{v_0}{v}\right)^{k-1}.$$

For example, let a quantity of air, originally at the freezing point, be compressed adiabatically. Let (θ_0, v_0) refer to the original state, (θ, v) to the compressed state. Then when the volume of the air has been reduced by half, $\frac{v_0}{v} = 2$: hence, since $\theta_0 = 273$, $k = 1\cdot410$,

$$\frac{\theta}{273} = 2^{0\cdot410} = 1\cdot329,$$

or

$$\theta = 273 \times 1\cdot329 = 363.$$

Thus the increase of the temperature caused by the compression is $(363 - 273)°$, or $90°$ centigrade.

When the air has been compressed to a quarter its original volume, $\dfrac{v_0}{v} = 4$, and

$$\frac{\theta}{273} = 4^{0\cdot 410} = 1\cdot 765,$$

or

$$\theta = 482.$$

In this case the rise of temperature is $209°$ C.

To find the relation between the temperature and pressure, or between the pressure and volume, we substitute from $pv = R\theta$ in $\dfrac{\theta}{\theta_0} = \left(\dfrac{v_0}{v}\right)^{k-1}$.

Thus

$$\frac{\theta}{\theta_0} = \left(\frac{\theta_0}{\theta} \cdot \frac{p}{p_0}\right)^{k-1},$$

or

$$\left(\frac{\theta}{\theta_0}\right)^{k} = \left(\frac{p}{p_0}\right)^{k-1}.$$

Also

$$\frac{pv}{p_0 v_0} = \left(\frac{v_0}{v}\right)^{k-1},$$

or

$$\frac{p}{p_0} = \left(\frac{v_0}{v}\right)^{k},$$

or

$$pv^k = p_0 v_0^k.$$

Art. 24. We can easily show that an adiabatic curve is steeper than an isothermal curve for a perfect gas. For, let a point X in the indicator diagram denote the pressure and volume at some particular instant, and let this volume be diminished by compression. If the compression be effected isothermally, a positive quantity of heat will be given out; but if adiabatically, no heat is given out at all. The temperature therefore rises in an adiabatic compression. Now if a given quantity of gas has the same volume in two states, the pressure will be greater in that state in which the temperature is higher. For if V be the volume common to the two

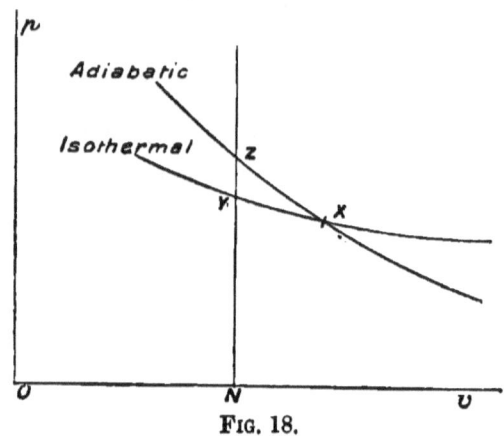

Fig. 18.

states, (p, θ), (p_0, θ_0) the two values of the pressure and temperature, we have $pV = R\theta$ and $p_0V = R\theta_0$, so that $\dfrac{p}{p_0} = \dfrac{\theta}{\theta_0}$. Hence if θ be greater than θ_0, p will also be greater than p_0. Thus if a straight line drawn parallel to Op to the left of X meet the isothermal curve through X in Y, the adiabatic curve through X in Z, and Ov in N, ZN will be greater than YN. The adiabatic curve will therefore be the steeper of the two.

Art. 25. The simple properties of perfect gases will now be used in considering a cycle which will be of service to us in the next chapter. In connection with this cycle, it has been found desirable to add some general remarks not particularly relating to perfect gases.

In a complete cycle, the change of energy in any body or system is zero. We then have $W + Q = 0$, or $W = -Q$, or $Q = -W$. Thus, when a positive quantity of work is done on the system during the cycle, an equal positive quantity of heat is given out, and we may say work is transformed into heat. If, on the other hand, there is, on the whole, a positive quantity of heat absorbed during the cycle, we may say that heat is transformed into work in the cycle.

The equation $W + Q = 0$ holds whenever the change of energy is zero, whether the cycle be complete or not; but we do not speak of the transformation of work into heat, or of heat into work, except when the process considered is a complete cycle. To do so would produce confusion with Carnot's principle, whether the change of energy in the non-cyclical process be zero or not. When the change of energy is not zero, a further and obvious reason may be given. The work done on the system will not be equal to the heat given out. In this case, it would clearly be absurd to say one had been produced from the other.

The question of the production of heat from work has, as we have seen, led to the principle of energy. The converse problem of the production of work from heat was first considered by Carnot in 1824 in a now celebrated essay. In this great work Carnot was led to some erroneous conclusions by following the usual custom of his time in accepting the caloric theory of heat; but his methods have the very great merit that the operations considered constitute a complete cycle, thus rendering the consideration of change of state unnecessary, and making it possible to use the same body (or system) over and over again.

Work may be transformed into heat in several ways. For instance:

1. Work may be expended on the system in friction in rubbing its parts together.

2. When the parts of the system expand and contract, a positive quantity of work may be done on the system during the cycle by expending more work in compression than is restored by expansion. This will be the case in the following cycle. Let a quantity of

gas be compressed at a very high temperature and pressure; then without altering the volume let the temperature and therefore also the pressure, be greatly reduced; next let the gas expand at a low pressure to its former volume; lastly, without further altering the volume, let the temperature be raised to its original value.

3. Work may be expended in producing electric currents within the system which are allowed to waste themselves in heating conductors (of electricity) or electric lamps.

Only one method of transforming into work was known to Carnot, and that was by taking advantage of the changes of volume produced by changes of temperature. A different method is now known. If a closed circuit be formed of different metals and the junctions heated to different temperatures, an electric current is produced which may be used to work an electro-motor or some other kind of electrical machine. But it will be found in the next chapter that Carnot's principle, like the principle of energy, refers to the cycle in the same form whether there be electric actions within the system or not. To illustrate the subject and prepare the way for the next chapter, we will here take the simplest case possible, in which the system consists of a quantity of perfect gas, as explained at the beginning of this chapter, and (external) work is only done on account of change of volume.

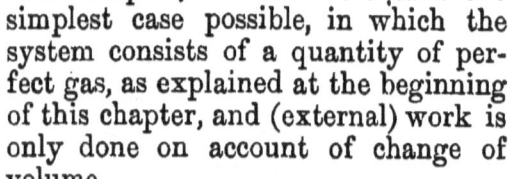

Fig. 19.

Let a gramme of any perfect gas be contained in a cylinder fitted with a smooth air-tight piston, as shown in the sketch. Imagine the piston and every part of the cylinder except the bottom to be perfectly impermeable to heat, and suppose the bottom has so small a capacity for heat that the heat required to raise its temperature may be neglected.

Such a cylinder cannot actually be constructed, and its existence is only supposed possible to shorten the verbiage.

Let there be two bodies A, B, of any kind, each of which is kept at a uniform and constant temperature. Let θ_a, θ_b be the temperatures as indicated by the air thermometer, and suppose A the hotter, so that θ_a is greater than θ_b. Also let a

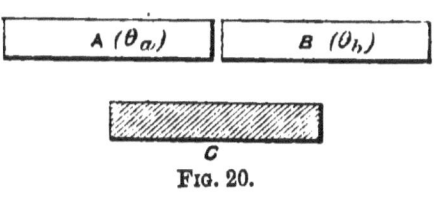

Fig. 20.

third body C be a perfect non-conductor of heat, on which the cylinder can be set any length of time without gaining or losing heat.

Now let the gas inside the cylinder be originally at the temperature of the colder body B, and suppose it to undergo the following complete cycle of four operations:—

(1) Let the cylinder be placed on the non-conducting stand C, and then force the piston slowly down until the temperature has risen from θ_b to θ_a, the indicator diagram for the operation being the adiabatic curve 1 2.

(2) Take the cylinder from C and place it on A, the

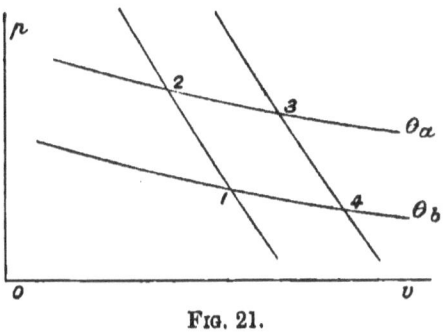

Fig. 21.

temperature of which is equal to that of the gas. Then let the piston be drawn out so slowly that the temperature of the gas is kept constant by the passage of heat from the hot body A through the bottom of the cylinder. Let the operation be represented on the indicator diagram by the isothermal curve 2 3.

(3) Next, transfer the cylinder back to C, and then

let the piston be drawn further out until the temperature again becomes θ_b, as at first; the indicator diagram being the adiabatic curve 3 4.

(4) Lastly, let the cylinder be placed on the cool body B, with which the gas has now the same temperature θ_b. Then let the piston be slowly forced in until the original state of the gas is attained. The indicator diagram will be the isothermal curve 4 1.

In the isothermal operation (2), work is done *by* the gas and an equal positive quantity of heat, Q_a say, will be *absorbed* at the temperature θ_a. In the isothermal operation (4), work is done *on* the gas and an equal positive quantity of heat, Q_b say, will be *given out* at the temperature θ_b.

Denoting the volumes corresponding to the four corners of the curvilinear parallelogram of the indicator diagram by (v_1, v_2, v_3, v_4), we have

$$Q_a = R\theta_a \log \frac{v_3}{v_2},$$

$$Q_b = R\theta_b \log \frac{v_4}{v_1}.$$

Also
$$\theta_b v_1^{k-1} = \theta_a v_2^{k-1}, \; \theta_a v_3^{k-1} = \theta_b v_4^{k-1}.$$

Hence
$$\left(\frac{v_1}{v_2}\right)^{k-1} = \frac{\theta_a}{\theta_b} = \left(\frac{v_4}{v_3}\right)^{k-1},$$

or
$$\frac{v_1}{v_2} = \frac{v_4}{v_3},$$

and therefore
$$\frac{v_3}{v_2} = \frac{v_4}{v_1}.$$

We therefore get
$$\frac{Q_a}{Q_b} = \frac{\theta_a}{\theta_b}.$$

If W be the total algebraical quantity of work done *on* the gas during the cycle, or the excess of the work spent in compressing the gas over the work restored by expansion, the equation of energy is $W + Q_a - Q_b = 0$, or $Q_a - Q_b = -W$. Since Q_a is evidently greater than Q_b, $Q_a - Q_b$ is positive and W therefore negative. Now $-W$ is the total algebraical quantity of work done *by*, or obtained from the gas during the cycle, work done *on* the gas being considered negative work done *by* it. Hence we may say that in the cycle, a quantity of heat is taken from a hot body A, part of it is given out as work or transformed into work, and the remainder rejected at a lower temperature than that of the body A from which it was absorbed.

The hotter body A from which heat is taken, is called the "source"; the cooler body B which receives the waste heat, is called the "refrigerator." The fraction which expresses the part of the heat absorbed that is transformed into work, is called the "efficiency" of the cycle. Its value is $-\dfrac{W}{Q_a}$, or $\dfrac{Q_a - Q_b}{Q_a}$, or $1 - \dfrac{Q_b}{Q_a}$. This is equal to $1 - \dfrac{\theta_b}{\theta_a}$, and depends only on the temperatures of the source and the refrigerator. It is independent of the natures of the two bodies A and B, and it will appear in the next chapter that it is not only the same for all perfect gases but for every body or system of every kind working between the same two temperatures θ_a and θ_b.

The cycle undergone by the gas is evidently reversible—that is, it may be performed in the reverse order. In this case, a positive quantity of heat will be taken from the cooler body B, a larger positive quantity of heat given out to the hotter body A, and a positive quantity of work spent on (not obtained from) the gas. In words, by expending work during the cycle, heat may be taken from a cool body and both this heat and the work done on the gas, given out as heat at a higher temperature.

Art. 26. The equation $\dfrac{Q_a}{\theta_a} = \dfrac{Q_b}{\theta_b}$ may be written $Q_b = \dfrac{\theta_b}{\theta_a} Q_a$. It therefore appears that if we had the refrigerator at the zero of the air thermometer, or $\theta_b = 0$, we should have $Q_b = 0$. The *whole* of the heat absorbed would then be transformed into work. Now it has not yet been found possible to reduce the temperature of any body to the zero of the air thermometer. We may therefore say that θ_b and θ_a never vanish. Thus the *whole* of the heat absorbed in the cycle described cannot be transformed into work, and part of it must always be rejected into the refrigerator.

It will be seen in the next chapter that this result is general. It will be found that in a cycle, a system may give out heat only, or may both absorb and give out heat, but that in no case can a system absorb heat without also giving out heat. If therefore there is, on the whole, a positive quantity of heat absorbed during the cycle, this can only be by the absorption of a greater quantity of heat than is given out. In other words, heat can only be transformed partially into work.

It should be observed that the result $\dfrac{Q_a}{\theta_a} = \dfrac{Q_b}{\theta_b}$ and the expression for the efficiency $1 - \dfrac{\theta_b}{\theta_a}$ (which is always less than unity), have only been obtained for a complete cycle. For example: if the gas expand at constant temperature, the heat absorbed is equal to the work obtained from the gas. If we agreed to speak of the transformation of heat into work in incomplete cycles, we should have to say that in this case the "efficiency" is unity.

Art. 27. As we have already said, Carnot adopted the caloric theory of heat in his essay of 1824. He therefore concluded that the total quantity of heat absorbed in a cycle was zero, or that the heat absorbed in one part of the cycle was exactly equal to the heat given out in the rest of the cycle. Thus in the cycle of Art. 25, he supposed $Q_a = Q_b$. He therefore thought that work was obtained from heat by merely letting it down unchanged in quantity from a higher to a lower temperature. He illustrated this conclusion by the analogy of a water-mill, where work is obtained from water by letting it down from a higher to a lower level.

Carnot did not quite arrange the four operations of the cycle of Art. 25 as we have done. He began with the gas at the temperature of the hot body A, and performed the operations in the order (2), (3), (4), (1). In the operation (4), during which the gas is at the temperature of the cool body B, he directed the compression to be continued until the amount of heat given out to the cool body B was exactly equal to that which had been absorbed from the hot body A. When this was done, he supposed that adiabatic compression to the original volume would raise the temperature to its original value, that of the hot body A. This error was first corrected by Professor J. Thomson in 1849.[1] He pointed out that when the gas was at the temperature of the cool body B, it was to be compressed, not as Carnot directed, but by such an amount that adiabatic compression to the original volume would then raise the temperature to its original value.

[1] Tait's "Sketch of Thermo-dynamics."

CHAPTER III.

CARNOT'S PRINCIPLE.

ART. 28. Carnot in his essay of 1824 adopted the caloric theory of heat. He was not, however, satisfied with the caloric theory; and it is now known that before he died in 1832, he had entirely rejected it. Indeed, he had made a nearer approximation to the mechanical equivalent of heat than was obtained by Mayer in 1842. Reduced to C.G.S. units, Carnot's estimate of the work equivalent to a calorie, that is, the heat required to raise the temperature of one gramme of water from $0°$ C. to $1°$ C., was 37,000 gramme-centimetres: the estimate of Mayer was 36,500 gramme-centimetres. The mechanical value of a calorie adopted in this work from Joule's experiments, is 42,350 gramme-centimetres.

Carnot's work failed to attract notice for some time after its publication. In 1834, Clapeyron recalled attention to it, and exhibited it in a more elegant form by the use of Watt's indicator diagram. On the overthrow of the caloric theory, it was, however, generally thought that there was no alternative but to reject Carnot's theory altogether. This view was not shared by Thomson,[1] Rankine, and Clausius, who showed that by slight modifications, the theory might be brought into accordance with the principle of energy.

In this chapter we shall give Carnot's principle as

[1] Now Lord Kelvin.

accepted and extended at the present time. Carnot only considered the case of a single unelectrified body without mechanical motions and of uniform temperature throughout; but the principle is now applied to the most general system whatever. The bodies forming it may be in contact or at a distance apart like the heavenly bodies; at any instant the temperature may vary in any way from point to point of the system; the mechanical motions may be any whatever; and the system may be electrified and magnetized as we please, provided that all the electric properties are internal. The only case in which external electric influences will be considered, will be that in which an electric current enters and leaves the system by conductors of the same metal and in the same condition, there being no sensible electric force between the system and external bodies. But except when otherwise stated, there will be supposed to be no external electric influences.

Among the reasons for asserting that Carnot's Principle applies to the most general system whatever, we may mention the following:—

(a) Clausius has applied the principle to the whole universe. The form in which he has expressed his result has indeed been objected to, but no one seems to have urged that the system was too general. It therefore appears to be generally agreed that Carnot's principle applies to the case of mechanical motions, to animal and vegetable organisms, and to electric and magnetic actions—in short, that it is universally true.

(b) The principle has been used in electric questions by some of our greatest physicists—Thomson, Gibbs, and Helmholtz—and their results are not only generally accepted, but many of them have been experimentally verified.

In stating Carnot's principle and its extensions, we make use of an axiom based on experience. This was first exactly formulated by Thomson and Clausius, but something very analogous to it had previously been assumed by Carnot. The axiom, however, is not quite sufficient for us except in the case of the body considered by Carnot. To establish the properties of the most general system, we must assume an extension of the axiom. In this work we begin with the simplest case for which the axiom is sufficient—the case of an unelectrified body which has no motion as a whole and whose temperature is uniform when gaining or losing heat—and afterwards extend the axiom to enable us to consider the most general material system.

The principal steps and results given in this chapter referring to the first part of the subject, may be arranged under four heads:—

(1) The fundamental axiom.
(2) Thomson's absolute scale of temperature.
(3) General equation obtained by means of the axiom for a cycle, in which heat is absorbed and given out by the body at two temperatures only.
(4) Conclusion as to the most efficient cycle in (3), and expression for the "efficiency." These results might be deduced from the equation in (3); but owing to their importance, they are demonstrated independently.

The conclusion as to the most efficient cycle is due to Carnot, and the name of "Carnot's principle" is often restricted to it.

An extension of the fundamental axiom is then assumed to enable us to consider a general material system, and three leading propositions are immediately deduced.

(5) The principle of Thomson and Clausius for cyclical processes in which heat is simultaneously absorbed and evolved at any number of temperatures.

(6) The principle of Thomson and Clausius for the case of a non-cyclical process.

(7) Clausius' principle of Entropy.

Two other general results are given in the chapter:—

(8) Thomson's principle of the Degradation of Energy.
(9) The principle of Tidal Friction.

The application of Carnot's principle to radiation is only briefly referred to. In fact, the chapter is chiefly

Carnot's Principle.

occupied with the nine propositions just enumerated and questions arising out of them.

Art. 29. Before we can begin the discussion of Carnot's principle, it is necessary to have exact definitions of equal and different temperatures and of reversible operations.

The present article will be devoted to the consideration of temperature; the next to that of reversibility.

We begin with the simple case of unelectrified bodies at rest.

Let there be any number of bodies in contact so that heat can pass freely between them, and suppose that the system which these bodies form nowhere gains or loses energy, either in the form of work or of heat. Then it is found that the system ultimately attains an invariable state. In this state, the temperature of every body of the system is *defined* to be uniform and the temperatures of all the bodies to be equal; or all the bodies are said to be of the same temperature.

Now let P, A, B be three separate bodies in their ultimate invariable states, like the system just described. Each of the three bodies P, A, B is therefore of a uniform temperature. If we suppose it possible to take a system of bodies in contact and thermal communication containing P and A such that when the whole system is in its invariable state, the bodies P and A are in their *given* invariable states, then the given uniform temperatures of P and A are equal. Similarly, let the temperatures of P and B be equal. Then the body P is in the same invariable state in the invariable system containing P and A as in the invariable system containing P and B; and it appears allowable to suppose that a new invariable system may be formed from these two by merely putting the body P of one system in contact and thermal communication with the body P of the other system, without

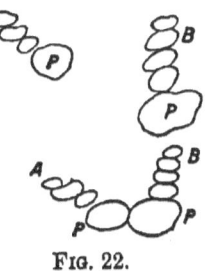

Fig. 22.

making any change in the state of any of the bodies concerned. The given uniform temperatures of A and B are therefore equal. In other words, any two temperatures which are equal to the same third temperature, are equal to one another.

Next, let P and C be two separate bodies in their ultimate invariable states, and therefore of uniform temperatures; and suppose that it is *not* possible to take a system of bodies in contact and thermal communication containing P and C such that when the whole system is in its ultimate invariable state, P and C are in their *given* invariable states. Then the given uniform temperatures of P and C are *defined* to be different.

If D be a third body in an invariable state and if its uniform temperature be equal to that of P, it is easy to show that the temperature of D cannot be equal to that of C. For, if possible, let the temperature of D be equal to that of C. Then it is possible for a system of bodies in its ultimate invariable state to contain D and C in their given states. Combining this system with that containing P and D, we get a new system which in its invariable state contains P, D, and C in their given states. Thus by the definition, the temperatures of P and C are equal. This being contrary to the original supposition about the temperatures of P and C, we conclude that the temperatures of D and C cannot be equal.

In the case of a body (or system of bodies) not in an invariable state, we make use of a *small* testing body C which has the property of *rapidly* assuming an invariable state. If the given body be solid, we bore a small hole into it so that we can place C in the part to be examined; but if the given body be liquid or gaseous, we simply immerse C in it. It is then found that when put in thermal communication with any part of the given body, C rapidly assumes a practically invariable state, like that in the preceding definition of uniform temperature. It is also found that the temperature corresponding to different parts of the given body at any the same instant, varies gradually from place to place. We therefore *define* the temperature of any part of a body to be the uniform temperature assumed by the small testing body which is placed in contact and thermal communication with the part considered.

The reason why the testing body must be small is easy to understand. A large testing body may be supposed made up of several small ones, and we have seen that the temperatures assumed by these small testing bodies are generally different. Thus, in general, a large testing body can give us no information about a body to be tested.

In order to determine whether two given temperatures, P, Q, are equal or different, we may proceed as follows. Take any two bodies, p, q, in ultimate invariable states whose uniform temperatures are respectively equal to P and Q. Then we know that the temperatures P, Q will be equal or different according as the temperatures of p and q are equal or different; and the temperatures of p and q may be compared in the way described in the definitions. It will, however, be convenient to have a body C which gives a visible indication of temperature such that the same indication is always given for the same temperature and different indications for different temperatures. If each of the bodies p, q be such a body C, a glance will be sufficient to tell whether the temperatures P, Q are equal or not.

The condition to be fulfilled by the body C will be sufficiently satisfied if C be a common mercury or air thermometer, protected, when necessary, by being enclosed in a hermetically sealed glass tube. The need for this precaution will be understood if we suppose it required to compare the temperatures at the top and bottom of a deep column of water.

To extend our conceptions of temperature to a body in motion, we assume that the temperature of the small testing body when in an invariable condition in motion is the same as when in the same condition at rest; and it is inferred that the temperature of any part of a body in motion may be found by the same kind of test as when the body is at rest. In the case of an electrified system of bodies, the temperature at any point is taken to be the same as when the system is deprived of its electric properties but otherwise unchanged in state.

The definitions of equal and different temperatures do not enable us to compare different temperatures so as to be able to say whether one body A is hotter or colder than another body B which is at a different temperature. However, it is not necessary for us to be able thus to compare different temperatures at present. A knowledge of equality and difference of temperature, the tests for which have already been given, is sufficient for us until we give Thomson's "absolute scale of temperature," which supplies a true method of comparing temperatures. Accordingly, this is the only knowledge of temperature that will be assumed until we come to the absolute scale. It must, however, be pointed out that it is the usual custom to assume a knowledge of higher and lower temperature before introducing Carnot's principle. This requires an exact definition of higher and lower temperature consistent with experiment which may be given as follows. Let A, B, P, Q be four bodies in ultimate invariable states, as in the definitions of equal and different temperatures, and suppose that the uniform temperature of A is equal to that of P, and the uniform temperature of B equal to that of Q. Then experience shows that if heat begins to flow from A to B when A and B are simply put in contact and thermal communication, heat will also begin to flow from P to Q when these two bodies are put in contact. We may therefore consistently define the temperature of A to be higher than that of B; for it follows that any temperature equal to that of A is higher than any temperature equal to that of B. Another definition may also be given. We have already explained what is meant by heat absorbed from without in any part of an unelectrified body; and by supposing the body divided up into pieces, an exact idea is obtained of the passage of heat in the interior of a body. Suppose now that a line is drawn in any body whatever such that at every point the direction of the line coincides with the direction of the flow of heat at that point. Let P and Q be any two points on this line such that heat flows from P to Q. Then if we take two points A, B on another "line of flow" at which the temperatures are respectively the same as at P and Q, it is found that heat flows from A to B. We may therefore consistently define the temperature at P to be higher than that at Q.

Art. 30. The following definition may be given of a reversible operation. There may either be no external electric influences (that is, the electric properties must all be internal), or there may be an electric current entering and leaving the system, as already explained, by conductors A, B of the same metal and in the same condition.

Suppose that a system passes from a state P to a state P' through intermediate states Q, R, S . . . Then if it can return from P' to P through the same intermediate states in the reverse order in such a way that the energy of any kind (work, heat, or electric energy) absorbed in any part of the system in passing from any state Q to any state R is exactly equal and opposite to the quantity of the same kind of energy absorbed in the same part of the system in returning from R to Q, the passage from P to P' is *defined* to be a reversible operation.

From this definition it is evident that if a quantity of heat q be absorbed in any part of the system in passing from P to P', an equal quantity q will be given out in the reverse process by the same part of the system when at the same temperature as before.

It will be noticed that the definition of reversibility does not require that the system should return from P' to P in the same time as it goes from P to P'.

In the first part of this chapter, we only consider the

case of an unelectrified body which possesses no motion as a whole and whose temperature is uniform when gaining or losing heat. The position of every part of this body must evidently be the same when passing through any state in the return path as when passing through the same state in the direct path. The work done on any part of the body in returning from R to Q being equal and opposite to the work done on the same part of the body in going from Q to R, and the displacements being equal and opposite in the two cases, it follows that the force acting on any part of the body is the same for any state in the reverse path as for the same state in the direct path. Now experience shows that if a body be prevented from gaining and losing heat, and be always acted on by the same forces when in the same state, it cannot undergo an operation and the reverse in the reverse way. If, however, at any instant the body be in an invariable state (or state of equilibrium) Q, a very slight change in the external forces will cause the body to assume a different state of equilibrium R and another very slight change in the forces will bring the body back to Q. Neglecting these slight changes in the forces as too small to be taken into account, we may say that our body which is prevented from gaining and losing heat undergoes an operation and its reverse under the action of the same forces such that at every instant it is practically in a state of equilibrium. For example, let a quantity of gas be contained in a state of equilibrium in a cylinder fitted with a smooth air-tight piston. Then the force exerted on the piston must just balance the pressure of the gas on it,[1] and it is clear that any increase in this force will cause the piston to be pushed in and any decrease in it will allow the piston to be forced out. By taking the increase and decrease small enough, we practically arrive at the case in which the piston is pushed in a finite distance and then allowed to be forced out again in such

---

[1] Here the weight of the piston is left out of account, or the cylinder supposed to be horizontal.

a way that the force on the piston at any instant during the compression is equal to the force when the volume has the same value during expansion. At every instant during both the compression and the expansion, the gas will practically be in a state of equilibrium and the process will be exceedingly slow.

When the body is gaining or losing heat, its temperature is uniform. If the body be then undergoing a reversible operation, it is easy to see that at every instant the state of the body must practically be one of equilibrium and that the operation must be very slow.

In general, it is inferred that if an unelectrified body undergo any reversible operation during which the body need not be at rest, the body must at every instant be practically in a state of equilibrium and the operation be very slow.

The most general system considered in this chapter may contain any number of bodies either separate or in contact, which may be electrified and magnetized in any way we please provided that the electric properties do not extend to external bodies or systems. The system may also possess any mechanical motions whatever, like the solar system or even the whole universe; and heat may be gained and lost simultaneously at different temperatures both by conduction and radiation. If this system undergo a reversible operation, it is inferred that at every instant every body of the system is practically in a state of equilibrium and that the changes in the electric and other conditions of the bodies are very slow. The temperature of any body forming part of the system must then be uniform; and if that heat be allowed to pass freely between any two such bodies, their temperatures must be equal as well as uniform. If, however, two bodies are prevented from exchanging heat with one another, their temperatures, though both uniform, may be different. Also there must be no tides or vibrations in the body (except the irregular vibrations which constitute heat). The existence of tides, etc., may be supposed to be prevented by surrounding the bodies with protecting envelopes, just as the gain or loss of heat is supposed to be prevented by non-conducting envelopes.

Art. 31. We are now prepared to consider Carnot's principle and its extensions.

The fundamental axiom on which the whole theory based is given by Thomson in the form, "It is impossible, by means of inanimate material agency, to derive mechanical effect from any portion of matter by cooling it below the temperature of the coldest of the surrounding objects": by Clausius it is given in the form, "It is impossible for a self-acting machine, unaided by any external agency, to convey heat from one body to another at a higher temperature."[1]

This axiom was tacitly assumed by Carnot, and may be called Carnot's axiom.

The forms in which the axiom is given by Thomson and Clausius presuppose a definition of higher and lower temperature. In the present work, we assume no knowledge of temperature further than to be able to say whether two temperatures are equal or different, until we explain Thomson's absolute scale. We must therefore give the axiom in a slightly different form.

Fundamental Axiom.—Let an unelectrified body undergo any series of operations during which it possesses no motion as a whole and is at a uniform temperature when gaining or losing heat. Then if the operations constitute a cycle and heat be gained and lost at two temperatures only, we assume that, on the whole, no work can be obtained from the cycle if the two temperatures are equal. If work is obtained, the temperatures must therefore be different.

By stipulating in the axiom that the body is to be at a uniform temperature when gaining or losing heat, we secure the simplification that the body does not simultaneously absorb and give out heat at different temperatures.

---

[1] Tait's Historical "Sketch of Thermo-dynamics," Art. 53, 2nd edition.

It may be observed that Thomson and Clausius have shortened their axioms by leaving it for the context to explain that a complete cycle of operations is considered.

Art. 32. From the fundamental axiom we can immediately deduce two important and useful results referring to the cycle described in the axiom.

First, the quantities of heat absorbed by the body at the two temperatures cannot both be positive. For, if possible, let them both be positive, and denote their values by Q, q. Then if W be the total quantity of work done on the body during the cycle, the principle of energy gives $Q + q + W = 0$, or $W = -(Q + q)$. Thus W is negative, or a positive quantity of work $-W$ (or $Q + q$) is obtained from the cycle.

Now let us take two pieces of metal, P, which can give out heat freely at the temperature at which the heat q is absorbed, and in which any amount of "heat" can be produced by friction. Also suppose that when the system P is left to itself with a given energy, there is only one invariable state into which it can ultimately settle down. Then let P and the given body work together in such a way that the given body undergoes exactly the same kind of cycle as before. It (the given body) therefore absorbs the positive quantities of heat Q, q. But since it is immaterial what source heat absorbed comes from, we may suppose the heat q to be obtained by conduction from P. Let this be the case; and imagine P to be prevented from otherwise gaining or losing heat. Suppose the system P to be at first in an invariable state and let the operations undergone by it be as follows:—first, let it be allowed to give out a positive quantity of heat q to the given body; then suppose it protected from further gain or loss of heat and let a quantity of work equal to q be done on it; and lastly, let it be left to itself to settle down into an invariable state. The increase of energy in P is $-q + q$, or 0. Hence the last state of P is the same as the first, and therefore the body compounded of the two pieces of metal and the given body has gone through a cycle of changes. Now the work done on the compound body during the cycle is $W + q$, or $- Q$, and the only heat absorbed by it from sources external to itself is the positive quantity Q absorbed at one temperature. As $-(W + q)$, or $+ Q$, is the quantity of work obtained from the cycle and is positive, we have been able to obtain work from a cycle during which the (compound) body can only gain or lose heat at one temperature, contrary to the fundamental axiom. We therefore conclude that the two quantities of heat absorbed by the body described in the fundamental axiom cannot both be positive.

Next, the quantities of heat absorbed by the given body at the two temperatures cannot both be negative if the cycle be reversible. For if they were, we should only have to reverse the cycle to obtain another cycle in which the quantities of heat absorbed are both positive —a case we have just seen to be impossible. Thus in a reversible cycle, a positive quantity of heat is absorbed at one temperature and a positive quantity evolved at the other.

Art. 33. We can now give an important property of a reversible cycle on which the absolute scale of temperature is based.

Suppose the cycle undergone by the body (X) described in the fundamental axiom, to be reversible. Let Q_a be the positive quantity of heat absorbed at one temperature and Q_b the positive quantity evolved at the other. Then we shall prove that so long as we keep to the same two temperatues, $\dfrac{Q_a}{Q_b}$ has always the same value. It will therefore be independent of the nature of the body which undergoes the reversible cycle, and of the nature of the cycle itself.

Let any other body (X') undergo any reversible cycle between the same two temperatures as X, and suppose that a positive quantity of heat, Q'_a, is absorbed at the same temperature as Q_a, and a positive quantity, Q'_b, given out at the same temperature as Q_b. Then we have to prove that

$$\frac{Q_a}{Q_b} = \frac{Q'_a}{Q'_b}$$

We assume as evident that by merely altering the size of the body X', we obtain a similar body and cycle (X") in which the quantities of heat absorbed and given out are respectively the same as Q'_a and Q'_b increased or decreased in the same ratio. Let this ratio be n, and suppose it so chosen that $nQ'_b = Q_b$. If this cycle be reversed, a positive quantity of heat nQ'_a is given out at the temperature at which Q_a is absorbed by X, and a positive quantity Q_b is absorbed at the temperature at which Q_b is given out by X. Now let the direct cycle of X be combined with the reversed cycle of X". Then the body X gives out a quantity of heat Q_b at a certain temperature and the other body absorbs an equal quantity *at the same temperature*. Since the cycles undergone by X and X" are reversible and therefore very slow, we may suppose the heat Q_b given out by the one body to be absorbed by the other. Hence in the compound body (X, X"), the only heat absorbed during the cycle is the quantity $Q_a - nQ'_a$ absorbed at one temperature only. Now in any complete cycle, the total quantity of heat absorbed is equal to the work given out. Thus the work obtained from the cycle undergone by the compound body (X, X") is $Q_a - nQ'_a$. But according to the fundamental axiom, it is impossible to obtain a positive quantity of work from a cycle in which heat is gained and lost at one temperature only. The quantity $Q_a - nQ'_a$ must therefore be negative or zero. By reversing the reversible cycle of (X, X"), it may in like manner be shown that $nQ'_a - Q_a$ must be negative or zero, or that $Q_a - nQ'_a$ must be positive or zero.

Since the same quantity cannot be positive and negative at the same time, we see that $Q_a - nQ'_b$ must be zero. Thus $Q_a = nQ'_a$. As n has been chosen so as to make $nQ'_b = Q_b$, we find

$$\frac{Q_a}{Q_b} = \frac{Q'_a}{Q'_b}.$$

Art. 34. **Absolute temperature.**—The value of $\frac{Q_a}{Q_b}$ is the same for all bodies and all reversible cycles, and can therefore only depend on the two temperatures. This result will now be used to construct a scale of temperature independent of the special properties of any limited class of bodies.

The different temperatures of bodies may be distinguished by different numbers, which may be chosen in many different ways. For example, the numbers (θ_a, θ_b) belonging to the temperatures of Q_a and Q_b may be chosen so that $\frac{Q_a}{Q_b} = \frac{\theta_a}{\theta_b}$, or so that $\frac{Q_a}{Q_b} = \frac{1+\theta_a}{1+\theta_b}$. But we cannot choose the numbers arbitrarily. Thus we cannot choose θ_a and θ_b so that $\frac{Q_a}{Q_b} = \frac{\theta_a}{1+\theta_b}$; for then we should also have $\frac{Q_b}{Q_a} = \frac{\theta_b}{1+\theta_a}$; and hence $1 = \frac{\theta_a \theta_b}{(1+\theta_a)(1+\theta_b)}$, which can only be true when $(1+\theta_a)(1+\theta_b) = \theta_a \theta_b$, or $1 + \theta_a + \theta_b = 0$

If we assume the absolute accuracy of the laws of perfect gases, as laid down in Chap. II, the value of $\frac{Q_a}{Q_b}$ for a perfect gas is known to be $\frac{\theta'_a}{\theta'_b}$, where ($\theta'_a, \theta'_b$) are the two temperatures as indicated by the air thermometer. Sir W. Thomson (to whom the absolute scale of temperature is entirely due) therefore chooses θ_a and θ_b so that $\frac{Q_a}{Q_b} = \frac{\theta_a}{\theta_b}$. This has the double advantage of simplicity and of enabling us to make the absolute scale of temperature practically agree with that of the ordinary air thermometer. The numbers which denote temperature on the absolute scale are not yet completely fixed, because we have only chosen their ratios to each other. Sir W. Thomson fixed the numbers by requiring that the absolute temperatures of the ordinary boiling and freezing points should differ by 100.

Carnot's Principle. 85

Since Q_a and Q_b are both positive, and $\dfrac{Q_a}{Q_b} = \dfrac{\theta_a}{\theta_b}$, it is clear that all absolute temperatures are of the same sign. Hence since the difference of two of them is chosen to be positive, it follows that all absolute temperatures are positive.

It may perhaps be thought that the scale of the ordinary air thermometer should be sufficient for us, because it leads to the result $\dfrac{Q_a}{Q_b} = \dfrac{\theta'_a}{\theta'_b}$. It must, however, be remembered that the laws of perfect gases, laid down in Chap. II., are not strictly accurate for any gas. For example, the value of pv is not quite constant when the temperature is constant; it is sensibly less at very high pressures than at ordinary pressures. The indications of the air thermometer are therefore not strictly the same as the law $pv = R\theta$ would lead us to suppose. The fact that the laws $pv = R\theta$ and $U = C_v\theta$ give the result $\dfrac{Q_a}{Q_b} = \dfrac{\theta'_a}{\theta'_b}$ merely shows that there has somewhere been a compensation of errors. The explanation of this is that the two laws of perfect gases, though not strictly accurate for any gas, are consistent with one another; and it is easy, from the results of this chapter, to deduce one from the other. The method of doing this requires, however, the use of rather higher mathematics than we make use of in this work.

For further information on this subject, and for the experiments of Joule and Thomson to determine the true absolute temperature of the freezing point, reference may be made to Parker's "Elementary Thermodynamics" or to the article on "Heat" by Sir W. Thomson in the "Encyclopædia Britannica."

Art. 35. In this article we pause to consider the relation $\dfrac{Q_a}{Q_b} = \dfrac{\theta_a}{\theta_b}$.

Since $\dfrac{Q_b}{Q_a} = \dfrac{\theta_b}{\theta_a}$, we have $Q_b = \dfrac{\theta_b}{\theta_a} Q_a$, and therefore when

$\theta_b = 0$, $Q_b = 0$, or the heat absorbed or given out at the temperature θ_b is zero. We are therefore able to obtain work from a reversible cycle during which heat is absorbed at one temperature only, *provided* that during the cycle, the body is reduced to the temperature of absolute zero. This is a contradiction of the fundamental axiom. Now as a matter of fact it has not yet been found possible to reduce any body to the temperature of absolute zero, although temperatures have been obtained so low that the most permanent gases have not merely been liquified but solidified.

The result $Q_b = 0$ combined with the fact that no temperature as measured on the absolute scale can be negative, or below 0, is generally interpreted to mean that at absolute zero the non-mechanical motions of a body which constitute what is popularly called heat, are totally absent. This conclusion enables us to make some interesting calculations.

The specific heat of iron is about $\frac{1}{9}$ of a calorie, and is nearly the same at all temperatures. Hence the energy of a gramme of iron at the freezing point roughly exceeds that at absolute zero by $27\frac{3}{9}$ calories, 1260×10^6 ergs. The principal change produced by rise of temperature in a body which remains solid being a change in the non-mechanical motions, we may take the sum of kinetic energies of these motions and of the potential energies of the displacements for a gramme of iron at $0°$ C to be 1260×10^6 ergs. For simplicity, first suppose the potential energies to be zero. Then if the velocity of each particle be v centimetres per second, we have $\frac{1}{2}v^2 = 1260 \times 10^6$. Hence $v^2 = 2520 \times 10^6 = 25 \times 10^8$ nearly, and therefore $v = 5 \times 10^4 = 50{,}000$.

If the potential energy had not been taken zero, the value of v would have been different. For example, if the potential energy be taken equal to the kinetic energy, v^2 will only be half 25×10^8, and v only $\dfrac{1}{\sqrt{2}}$ (or $\tfrac{5}{7}$) of $50{,}000$. In very rough calculations, such as we propose to make, a factor like $\tfrac{5}{7}$ may be omitted.

Carnot's Principle.

Now it is generally supposed that a body is made up of "atoms," the vibrations of which cause radiation by setting the "ether" in motion, the period of the vibrations of the atoms being equal to that of the corresponding radiation. But if a number of waves of the same constant wave-length λ travelling with the same constant velocity V follow each other at a distance of λ behind each other, the time it takes a wave to pass over any fixed point P, or the period of vibration at P, is $\frac{\lambda}{V}$. This must be equal to the period of vibration of an atom. Now, for simplicity, suppose the centre of mass of the atom to be describing a circle of diameter d with constant velocity u. Then the period of the centre, or the time it takes to describe the circle, is $\frac{\pi d}{u}$.

If this be equal to the period of the wave, $\frac{\pi d}{u} = \frac{\lambda}{V}$. The largest possible value of u is v (5×10^4), and for a given value of λ and V this gives the largest possible value of d $\left(d = \frac{u\lambda}{\pi V}\right)$. Put $u = v$ and give to V its value 3×10^{10} (in C.G.S. units). Then $d = \frac{\lambda}{\pi} \cdot \frac{5 \times 10^4}{3 \times 10^{10}} = \frac{5}{3} \cdot \frac{\lambda}{\pi} \cdot \frac{1}{10^6}$ $= \frac{\lambda}{2} \times \frac{1}{10^6}$ nearly.

Now a solid body emits radiation of very many wave-lengths. The vibrations of the atoms, which synchronise with these waves, must therefore be very complicated. Suppose, however, for the sake of obtaining a result, that we take λ to be the mean of the lengths of the longest and shortest of the commonest waves. In the case of iron at 0° C, every kind of radiation emitted is dark, or every wave-length much greater than those of light. Suppose therefore that λ is 100 times the length of an average wave of light, or $\lambda = \frac{1}{200}$ centimetre. Then $d = \frac{1}{4 \times 10^5}$.

If we suppose the plane of the orbit to change irregularly so that the orbit occupies every possible position on a sphere, and if we take the orbits of different atoms to be equal, it will follow that the distance between any two atoms must be at least equal to d, to allow room for the orbits.

Sir W. Thomson, from electric considerations involving few assumptions, finds the distance between the atoms to be $\dfrac{1}{3 \times 10^8}$ centimetre. Helmholtz takes the distance to be $\dfrac{1}{10^7}$.[1] With Helmholtz' value for d the equation $d = \dfrac{\lambda}{3 \times 10^6}$ gives $\lambda = \dfrac{2 \times 10^6}{10^7} = \tfrac{1}{5}$ (centimetre).

Art. 36. General equation for a cycle in which heat is absorbed and given out at two temperatures only.—The relation $\dfrac{Q_a}{Q_b} = \dfrac{\theta_a}{\theta_b}$, which holds for any reversible cycle, will now be put in a slightly different form, and a corresponding result obtained for any other cycle working between the same two temperatures.

Let heat absorbed be considered positive, heat given out negative. Put Q for Q_a and Q' for $-Q_b$: also, for symmetry, write θ and θ' for θ_a and θ_b. Then the relation $\dfrac{Q_a}{Q_b} = \dfrac{\theta_a}{\theta_b}$ becomes $-\dfrac{Q}{Q'} = \dfrac{\theta}{\theta'}$, or $\dfrac{Q}{\theta} + \dfrac{Q'}{\theta'} = 0$.

We have seen that in no case can Q and Q' both be positive. The value of $\dfrac{Q}{\theta} + \dfrac{Q'}{\theta'}$ can therefore never be positive. For a reversible cycle it has been shown to be zero. It will now be asked, When is $\dfrac{Q}{\theta} + \dfrac{Q'}{\theta'}$ zero and when is it negative for an irreversible cycle? The question cannot be answered directly, and we are obliged to consider in detail the principal kinds of irreversibility to which our body may be subject.

(1) Let the body be a quantity of air contained in a

[1] Mascart's "Electricity" (English Translation): vol. i.

cylinder fitted with a smooth air-tight piston; and suppose that during the cycle it is required to compress it without gain or loss of heat until the temperature rises from θ' to θ. First, let the compression be effected reversibly by slowly forcing the piston in until the volume is diminished from v' to v; and let Q and Q' refer to a reversible cycle of which this reversible compression forms part. Then $\frac{Q}{\theta} + \frac{Q'}{\theta'} = 0$. Next, let the air be compressed with a sensible rapidity from volume v' to volume v without gaining or losing heat. In this operation, the pressure will not remain uniform throughout the cylinder: thus, if the velocity of the piston be great enough, the pressure of the air close to the piston will be much greater than the pressure at the other end of the cylinder. On the whole, we infer that *more work* is expended on the compression than if it had been effected reversibly. The energy of the compressed air will therefore be greater at the end of the irreversible compression than at the end of the reversible compression. Hence in the state of equilibrium assumed at the end of the irreversible compression, the temperature will be higher than θ. To have the temperature equal to θ at the end of the irreversible compression, the compression must be stopped before the volume becomes as small as v. If the volume be now reduced to v and the temperature kept constantly equal to θ, a positive quantity of heat (x) will be given out. Let the rest of the cycle be reversible and the same as in the previous case. Then if (Q_1, Q'_1) refer to the cycle, $Q_1 = Q - x$, $Q'_1 = Q'$: hence $\frac{Q_1}{\theta} + \frac{Q'_1}{\theta'} = \frac{Q-x}{\theta} + \frac{Q'}{\theta'} = -\frac{x}{\theta}$, a negative quantity.

Again, suppose that during the cycle it is required the air should expand without gain or loss of heat until the temperature falls from θ to θ'. When the expansion is reversible, let the volume consequently increase from v to v'. Now it is easy to see that when there is no

gain or loss of heat, *less work* is obtained from the same amount of expansion effected irreversibly than reversibly. Hence when the air expands rapidly without gain or loss of heat from volume v to volume v', the decrease in its energy will be less than if the expansion had been slow, and therefore the temperature assumed at the end of the rapid expansion will be higher than θ'. To make the final temperature equal to θ', let the expansion be continued until the temperature becomes equal to θ'. Then let the volume be made equal to v' without altering the temperature by slowly forcing in the piston and abstracting a positive quantity (y) of heat from the air at the constant temperature θ'. Let the rest of the cycle be reversible and suppose (Q_2, Q'_2) to refer to the cycle: also let (Q, Q') refer to a reversible cycle differing from the preceding only in the expansion of the air as just explained. Then $Q_2 = Q$, $Q'_2 = Q' - y$, and $\dfrac{Q_2}{\theta} + \dfrac{Q'_2}{\theta'}$
$= \dfrac{Q}{\theta} + \dfrac{Q' - y}{\theta'} = - \dfrac{y}{\theta'}$, a negative quantity.

(2) Let work be expended in rubbing two bodies together. Then when the final state is the same as the first, a positive quantity of heat will have been given out. This may be supposed to be given out at one temperature (θ) only. In this irreversible cycle we therefore have Q negative and Q' zero. Hence $\dfrac{Q}{\theta} + \dfrac{Q'}{\theta'}$ is negative.

(3) Take a body P, as a piece of metal, the state of which is invariable and on which no work is therefore done. There being no change of energy, the total quantity of heat absorbed in any time is zero. We will therefore suppose a positive quantity y absorbed in a part A of the body the temperature of which is everywhere (and constantly) equal to θ, and an equal positive quantity y simultaneously given out in a part B at a temperature θ'.

The preceding propositions do not enable us to

consider this case directly, because in the fundamental axiom it has, for simplicity, been supposed that the body considered is at a uniform temperature when absorbing and evolving heat. However, this axiom is true, as will be seen presently, when heat is gained and lost at the two temperatures simultaneously. In the expression $\frac{Q}{\theta} + \frac{Q'}{\theta'}$, θ and θ' now denote the temperature of those parts of the body where the quantities of heat Q and Q' are respectively absorbed. It may also be shown, just as when the temperature of the body is uniform while gaining and losing heat, that Q and Q' cannot both be positive, or that $\frac{Q}{\theta} + \frac{Q'}{\theta'}$ cannot be positive.

In the case of the body P, $Q = y$, $Q' = -y$, and $\frac{Q}{\theta} + \frac{Q'}{\theta'} = y\left(\frac{1}{\theta} - \frac{1}{\theta'}\right) = y\frac{\theta' - \theta}{\theta'\theta}$, which can only be zero when $\theta = \theta'$.

Since $\frac{Q}{\theta} + \frac{Q'}{\theta'}$ cannot be positive, it follows that θ must be equal to or greater than θ'. If we draw a "tube of flow" from A to B, and consider the portion between two cross sections (X, Y), it is easily seen that the temperature is constant in the tube, or decreases in the direction of flow. In other words, as we travel along a "line of flow" in the direction of flow, the temperature is either constant or decreases.

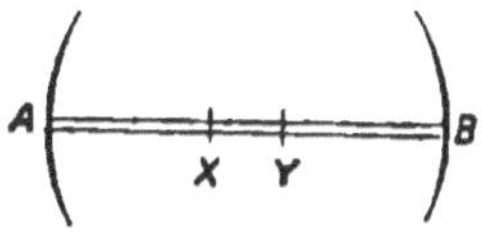

FIG. 23.

We are unable to proceed further without recourse to experiment. We therefore assume as a result of experiment, that if heat be flowing along the tube at a finite rate, the temperature, as indicated on the absolute scale, continually decreases[1] in the direction of flow.[2] When

---

[1] It should be remembered that no knowledge of higher and lower temperature was required in establishing the absolute scale.
[2] Unless the temperature be that of the absolute scale, it is not necessarily true that it decreases continually in the direction of flow.

the flow of heat is reversible, the temperature is evidently constant along the line of flow.

Thus when heat is flowing from A to B at a finite rate, θ is greater than θ', and $\dfrac{Q}{\theta} + \dfrac{Q'}{\theta'}$ or $y\left(\dfrac{1}{\theta} - \dfrac{1}{\theta'}\right) = y\dfrac{\theta' - \theta}{\theta\theta'}$ is negative.

Hence $\dfrac{Q}{\theta} + \dfrac{Q'}{\theta'}$ is negative for each of the three kinds of irreversibility just considered. In general, we infer (but cannot give a general proof) that any departure from equilibrium during a cycle causes $\dfrac{Q}{\theta} + \dfrac{Q'}{\theta'}$ to be negative. This may be expressed by saying that $\dfrac{Q}{\theta} + \dfrac{Q'}{\theta'}$ is zero for a reversible cycle and negative for an irreversible cycle.

It should be noticed that the introduction of the absolute scale of temperature has caused a want of symmetry in the methods of this part of the subject. Thus the value assigned to $\dfrac{Q}{\theta} + \dfrac{Q'}{\theta'}$ for a reversible cycle is partly a definition of absolute temperature, while in other cases the value found for $\dfrac{Q}{\theta} + \dfrac{Q'}{\theta'}$ is entirely a proposition. The want of symmetry does not occur in the methods of Clausius, who simply supposes temperature to be measured on the air thermometer.

Art. 37. Carnot's principle, properly so called.—In this article some results already given are repeated in a slightly different form, on account of their historical importance and intrinsic merit.

Let a body go through a cycle of operations such that the temperature is always uniform when gaining

---

Thus while one thermometer may indicate the temperature of A to be higher than that of B, another thermometer may indicate the temperature of B to be higher than that of A.

Carnot's Principle.

and losing heat. Also let heat be gained and lost at two different temperatures only, and suppose a positive quantity of work (W) obtained from the cycle. Let Q and Q' be the quantities of heat absorbed at the two temperatures, so that the equation of energy is $Q + Q' - W = 0$, or $W = Q + Q'$. Thus since Q and Q' cannot both be positive and their sum is positive, one must be positive and the other negative. In words, a positive quantity of heat (Q_a say) must be absorbed at one temperature and a positive quantity (Q_b say) given out at the other. Since $Q_a - Q_b - W = 0$, we see that Q_a is greater than Q_b.

We now define the "Efficiency of the cycle" to be the ratio of the positive quantity of work obtained from it to the positive quantity of heat absorbed at one of the two temperatures. The efficiency is therefore $\dfrac{W}{Q_a}$, or $\dfrac{Q_a - Q_b}{Q_a}$, or $1 - \dfrac{Q_b}{Q_a}$, which is positive and less than unity, because Q_b and Q_a are both positive and $Q_b < Q_a$.

When a positive quantity of work is obtained from a cycle working between two temperatures, it can easily be shown that the positive quantity of heat absorbed must be obtained at a particular one of the two temperatures: in other words, that a positive quantity of heat cannot be absorbed in one case at the first temperature, and a smaller positive quantity given out at the second temperature, and in another case the positive quantity of heat be absorbed at the second temperature and the smaller positive quantity given out at the first. For let a positive quantity of work (W') be obtained from a second cycle working between the same two temperatures, and suppose, if possible, that in the second cycle a positive quantity of heat (Q'_a) is absorbed at the temperature at which Q_b is given out, and that a positive quantity of heat (Q'_b, less than Q'_a) is given out at the temperature at which Q_a is absorbed. Then by merely altering the dimensions of the body we get a third body and cycle in which the positive quantities of heat nQ'_a and nQ'_b are respectively absorbed and given out. Choose n so that $nQ'_a = Q'_b$, and combine the cycle with the original cycle. We then have a positive quantity of work $W + nW'$ obtained from a cycle in which a quantity of heat $Q_a - nQ'_b$ is absorbed at one temperature and no heat gained or lost, on the whole, at the other temperature. This is contrary to the fundamental axiom: hence the supposition that it is indifferent whether the heat absorbed be obtained at one temperature or the other, must be false.

It will now be proved, as originally shown by Carnot, that in the case of reversible cycles, working between two given temperatures, the efficiency is the same both for all bodies and all cycles.[1] For take two such cycles (W, Q_a, Q_b), (W', Q'_a, Q'_b), and suppose, if possible, that $\dfrac{W'}{Q'_a}$ is less than $\dfrac{W}{Q_a}$. Let the size of the second body be altered until we get a third reversible cycle for which the three quantities are respectively (nW', nQ'_a, nQ'_b), and choose n so that $nQ'_a = Q_a$. Then nW' will be less than W. If therefore the third reversible cycle be reversed and combined with the first reversible cycle, a positive quantity of work $W - nW'$ will be obtained from a cycle and no heat gained or lost, on the whole, at the temperature corresponding to Q_a. This is contrary to the fundamental axiom. Hence we must have

$$\frac{W'}{Q'_a} = \frac{W}{Q_a}.$$

Thus the efficiency $\dfrac{W}{Q_a}$, or $1 - \dfrac{Q_b}{Q_a}$, of a reversible cycle, can only depend on the two temperatures. Now if (θ'_b, θ'_b) be the two temperatures, as indicated on the air thermometer, and if the laws of perfect gases ($pv = R\theta$, $U = C_v \theta$) be taken to be strictly accurate, we know that

---

[1] This is the proposition with which the name of Carnot is most closely associated.

$\dfrac{Q_b}{Q_a} = \dfrac{\theta'_b}{\theta'_a}$. Sir W. Thomson therefore chooses the ratio of the numbers (θ_a, θ_b) which are to represent the two temperatures on his absolute scale, so that $\dfrac{Q_a}{Q_a} = \dfrac{\theta_b}{\theta_a}$, and then makes the absolute temperature of the boiling point exceed that of the freezing point by 100. The absolute scale therefore very nearly coincides with that of the air thermometer, or with that of the mercury centigrade thermometer increased by 273.

Since $Q_b < Q_a$, θ_b is $< \theta_a$. This is expressed by saying that the temperature θ_a is the higher of the two; and we see that work is obtained from a reversible cycle when heat is absorbed at the higher temperature and a smaller quantity of heat given out at the lower temperature. On the absolute scale, the efficiency of a reversible cycle is $1 - \dfrac{\theta_b}{\theta_a}$, or $\dfrac{\theta_a - \theta_b}{\theta_a}$: in words, the ratio of the difference of the two temperatures to the higher temperature.

When work is obtained from a cycle working between two temperatures, the positive quantity of heat absorbed must be obtained at the same temperature whether the cycle be reversible or irreversible. Hence it must always be obtained at the higher temperature.

In the case of an irreversible cycle, the quantities of heat absorbed at the two temperatures may be both negative. When they are both negative, the principle of energy shows that work is not obtained from the cycle: the term "efficiency" has then no meaning. However, a positive quantity of work may be obtained from an irreversible cycle. A positive quantity of heat must then be obtained at the higher temperature and a smaller positive quantity given out at the lower. Here the term "efficiency" is legitimate, and it may be asked, What is the relation between this efficiency and that of a reversible cycle working between the same two temperatures? The question has already been answered in another form, but its importance is sufficient to excuse us treating it *ab initio*.

As no general investigation can be given, we are obliged to consider some of the principal kinds of irreversibility in detail.

(*a*) When air is compressed reversibly in a cylinder, no more force is exerted on the piston than is just sufficient to force it in; and when the air is allowed to expand reversibly, the force exerted on the piston is so large, that the pressure of the air is only just able to force the piston out. Hence it is clear that when the air is prevented from gaining and losing heat, the work required to effect a given amount of compression is greater and the work obtained from a given amount of rarefaction less, when the change of volume is produced irreversibly (or rapidly) than when it is produced reversibly (or very slowly).

Now let the body considered be a quantity of air working between the temperatures (θ_a, θ_b), and let (Q_a, Q_b, W) refer to the reversible cycle it undergoes. Suppose that in this cycle the air is to be compressed without gain or loss of heat until the temperature rises from θ_b to θ_a; and then take another cycle differing from the previous only in the way the air is compressed. In the second cycle let the compression be effected as follows. Compress the air irreversibly without gain or loss of heat, until as much work has been done as is required to effect the given amount of compression reversibly. Then at the end of the operation, the energy will be the same as at the end of the reversible compression; and therefore if allowed to stand a short time, the temperature will become θ_a. But since when there is no gain or loss of heat, a larger quantity of work is required for a given amount of compression when the process is irreversible than when it is reversible, the air will not be compressed sufficiently in the irreversible compression described above. Suppose then that the remaining compression is effected slowly and isothermally, and let y be the positive quantity of work spent in the process and also the positive quantity of heat given out by the air. Then the efficiency of the irreversible cycle is $\frac{W-y}{Q_a-y}$. Now $\frac{W}{Q_a}$ is positive and <1, and we know that such a fraction is diminished if the same positive quantity be subtracted from both numerator and denominator. Thus $\frac{W-y}{Q_a-y}$ is $<\frac{W}{Q_a}$, or irreversibility has diminished the efficiency.

Next, suppose that during the reversible cycle, the air has to be rarefied without gain or loss of heat until the temperature falls from θ_a to θ_b. Take an irreversible cycle differing from the reversible cycle only in the method of rarefaction. Let the air be rarefied without gain or loss of heat by drawing out the piston rapidly until as much work has been obtained as from the reversible rarefaction. Then when the temperature of the air has everywhere become θ_b, let the piston be slowly forced back and the temperature kept constant by abstracting heat until there is no more rarefaction than required. Suppose x to be the positive quantity of work done in forcing the piston back, or the positive quantity of heat given out in consequence by the air at the temperature θ_b. Then the efficiency of the irreversible cycle is $\frac{W-x}{Q_a}$, which is clearly $<\frac{W}{Q_a}$. Thus irreversibility has again diminished the efficiency.

(b) Let (Q_a, Q_b, W) refer to any reversible cycle working between the two temperatures (θ_a, θ_b). Repeat the cycle with the only difference that a positive quantity of work is to be expended in rubbing two parts of the body together and in consequence let additional positive quantities of heat (x, y) be given out at the temperatures (θ_a, θ_b) respectively. Then the quantity of heat absorbed at the higher temperature is now Q_a-x and the work obtained from the cycle is evidently $W-x-y$. Hence the efficiency is $\frac{W-x-y}{Q_a-x}$, which is $<\frac{W-x}{Q_a-x}$ and therefore $< \dfrac{W}{Q_a}$, the efficiency of the reversible cycle.

(c) The fundamental axiom is not suited for the discussion of the conduction of heat at a finite rate, because the axiom supposes that the temperature of the body is uniform when gaining or losing heat. We may, however, speak of the efficiency of a cycle in which heat is simultaneously gained and lost at two temperatures, just as if the temperatures were not simultaneous, provided that we distinguish the quantities of heat according to the two temperatures.

Let heat be flowing at a finite rate through a body whose state is invariable. No work is obtained from this cycle, and we may suppose the parts where heat is absorbed to be all at one temperature and the parts where heat is given out all at another temperature. The efficiency of the cycle is clearly zero, which is less than that of a reversible cycle working between the same two temperatures.

Thus in all these typical cases, irreversibility diminishes the efficiency of the cycle. We may therefore say that all reversible cycles working between the same two temperatures, have the same efficiency, and that the efficiency of any other cycle working between these temperatures is less.

It should be noticed that in the reasoning on efficiency there are two points of importance:—

(1) The cycle must be complete.

(2) Heat is supposed to be absorbed and given out at two temperatures only.

A single body which undergoes a reversible cycle between any two temperatures is called a "Carnot's perfectly reversible heat engine," and we see that the efficiency of a Carnot's perfect heat engine is greater than that of any other heat engine performing a cycle of operations between the same two temperatures.

Art. 38. It has hitherto been supposed that the body which undergoes the cycle possesses no motion as a whole. This restriction will now be removed.

It is proved in works on Rigid Dynamics that if sufficient precautions be taken, the work done on a body originally at rest will be equal to the kinetic energy acquired by the body; and if the body again come to rest, the work expended will be restored. In this case, the change of velocity is a reversible process. If, however, sufficient precautions are not taken, a positive quantity of work will be done (or lost) on the body in a cyclical change of motion.

A little consideration will now show that all the propositions already given for a cycle during which the body has no motion as a whole, are also true for a cycle when the body has any motion whatever, provided that the motion in the final state is the same as at first.

Art. 39. We have supposed the temperature of the body to be uniform when gaining or losing heat. This strictly requires that heat should be gained and lost very slowly; but, practically, it will be possible in some cases for heat to be gained and lost with comparative rapidity. For example, let the body consist of a thin metallic vessel filled with a very volatile liquid and its vapour, so that the least change of temperature produces an almost explosive formation or condensation of vapour. Then the temperature of the body will practically remain uniform at whatever moderate rate heat is being absorbed or evolved.

In considering the case of a single body at rest, it is generally supposed that the body neither gains nor loses heat except by conduction with two external bodies whose temperatures are kept uniform and constant. These two bodies are called the source and refrigerator, because in a reversible cycle, a positive quantity of heat is derived from the one and a positive quantity rejected into the cooler. Their temperatures are evidently the same as those at which heat is gained or lost; and therefore it is usual to speak of the temperatures of the source and refrigerator instead of the temperatures at which heat is gained and lost. Now when heat is absorbed by radiation from a distant body, the temperature of the source of heat may be quite different from the temperature of the body by which the heat is absorbed. Here it may at first perhaps be doubtful whether we should take account of the temperature of the source or of the given body. To solve the difficulty, we observe that if it is necessary to take account of the temperature of the source of heat, it must also be necessary to take account of the temperature of the refrigerator. Now when heat is radiated into vacant space, there is no refrigerator to take account of. We therefore conclude that we need take no notice of the source or refrigerator.

Art. 40. We can now give the results of Thomson and Clausius for the most general system. There may be any number of bodies in the system, separate or in contact, and the temperatures and motions may be any whatever, as in the solar system or even the whole universe. The parts of the system may also possess any electric and magnetic properties, *provided* the electric and magnetic influence does not extend to external systems.

When the system is under consideration care must be taken that it neither gains nor loses *matter*. This precaution will be found to be specially important in discussing animal and vegetable life.

After giving the propositions of Thomson and Clausius, it will be shown that the condition of the electric and magnetic properties should be internal may be removed in a certain case; but we shall suppose throughout the chapter that there are no external electric influences except when it is specially stated otherwise.

The fundamental axiom is not sufficient to enable us to discuss the properties of the general system. We shall therefore have to assume an extension of the axiom in what follows.

Art. 41. **Principle of Thomson and Clausius for a general cyclical process in which heat is absorbed and evolved at any number of temperatures, simultaneously or otherwise.**—Let a system undergo a cyclical process during which the quantities of heat (Q, Q′, Q″, . . .), which may be either positive or negative, are absorbed, either by conduction or radiation, at the various temperatures (θ, θ', θ'', . . .). Consider the quantity Q. It may be supposed to be obtained by conduction from a body X which is placed in contact *without rubbing* with the given system at the place where the heat Q is absorbed. Then the given system obtains a quantity of heat Q from the body X and X loses a quantity Q to the given system. Now suppose that the body X which evolves the heat Q consists of a thin metallic vessel filled with a very volatile liquid and its vapour, so that the least change of temperature produces an almost explosive formation or condensation of vapour. Then however rapidly the heat Q is absorbed by the given system, the evolution of the heat Q from the volatile liquid will be a comparatively slow process and may be supposed reversible. Moreover, the temperature of the volatile liquid will have the same value θ as the part of the given system where the heat Q is absorbed. Lastly, let the body X be made to undergo a reversible cycle during which it is prevented from gaining and losing heat, in addition to the heat Q evolved at the temperature θ, except at a certain fixed temperature θ_0. If Q_0 be the heat absorbed at the temperature θ_0, we have for the reversible cycle, since heat is absorbed at two temperatures only, $\dfrac{Q_0}{\theta_0} - \dfrac{Q}{\theta} = 0$, or $Q_0 = \dfrac{\theta_0}{\theta}Q$.

In like manner, let the heat Q′ be obtained at the temperature θ' from a body X′ undergoing a reversible cycle during which the only other heat it absorbs is a quantity Q'_0 absorbed at the temperature θ_0. Then $-\dfrac{Q'}{\theta'} + \dfrac{Q'_0}{\theta_0} = 0$, or $Q'_0 = \dfrac{\theta_0}{\theta'}Q'$.

Let all the other quantities (Q″, Q‴, . . .) be treated in the same way, and consider the cycle undergone by the system formed of the given system and (X, X′, X″,

...).[1] The quantities of heat (Q, Q', Q'', ...) are absorbed by the given system from sources external to itself; but in the compound system they refer only to the passage of heat in the interior and therefore have not to be taken into account. Hence in a cycle the compound system merely absorbs a quantity of heat $Q_0 + Q'_0 + Q''_0 + \ldots$ at one temperature θ_0. But the work obtained from the compound system during a cycle is equal to the heat absorbed which in our case is $Q_0 + Q'_0 + Q''_0 + \ldots$ *We now assume, as an extension of the fundamental axiom, that when heat is absorbed by the system at one temperature only, the work obtained from it cannot be positive.* Thus $Q_0 + Q'_0 + Q''_0 + \ldots$ is negative or zero.

If the cycle undergone by the given system be reversible the cycle undergone by the compound system will also be reversible. On reversing it, we find that $-(Q_0 + Q'_0 + Q''_0 + \ldots)$ is the heat absorbed. This must be negative or zero. Hence for a reversible cycle, $Q_0 + Q'_0 + Q''_0 + \ldots = 0$, or $\dfrac{Q}{\theta} + \dfrac{Q'}{\theta'} + \dfrac{Q''}{\theta''} + \ldots = 0$.

It has been shown that in no case can $Q_0 + Q'_0 + Q''_0 + \ldots$ be positive. Hence $\theta_0\left(\dfrac{Q}{\theta} + \dfrac{Q'}{\theta'} + \dfrac{Q''}{\theta''} + \ldots\right)$ cannot be positive; and since θ_0 is positive, it follows that $\dfrac{Q}{\theta} + \dfrac{Q'}{\theta'} + \dfrac{Q''}{\theta''} + \ldots$ cannot be positive. For a reversible cycle, it has just been proved zero. The case of an irreversible cycle cannot be discussed in its generality; but on taking some typical cases of irreversibility, as in Art. 37, it is found that $\dfrac{Q}{\theta} + \dfrac{Q'}{\theta'} + \dfrac{Q''}{\theta''} + \ldots$ is negative. We therefore infer that we may put

[1] If the given system exerts a sensible gravitational attraction on the bodies (X, X', X'', ...), the cycle of the compound system will not be complete until these bodies are in the same position relative to the given system as at first. The values of (Q_0, Q'_0, Q''_0, ...) will, however, be as stated in the text, since the bodies are to be moved about reversibly.

$\dfrac{Q}{\theta} + \dfrac{Q'}{\theta'} + \dfrac{Q''}{\theta''} + \ldots \lessgtr 0$, according as the cycle is reversible or irreversible.

A few remarks will be made in connection with this proposition.

(1) There is a defect in our reasoning, since it has been tacitly supposed that all the parts which absorb and evolve heat are unelectrified.

(2) It should be noticed that $(\theta, \theta', \theta'', \ldots)$ are respectively the temperatures of those parts of the material system at which the quantities of heat $(Q, Q', Q'', \ldots)$ are absorbed at the instants of absorption, without regard to where heat absorbed comes from or where heat evolved may go to.

(3) It is clear that the quantities of heat $(Q, Q', Q'', \ldots)$ absorbed in the cycle cannot all be positive: in other words, heat can only be transformed partially into work however many temperatures it may be absorbed at.

Art. 42. Principle of Thomson and Clausius for a non-cyclical process.—Let P and P' be two states of the same system, in each of which every body of the system is in a state of mechanical, thermal, and electric equilibrium. Suppose the system brought from the state P to the state P' by a reversible path, which, it seems, may often be done in any number of ways. Let PAP' be the path actually taken, and let the system be caused to return from P' to P by a reversible path P'XP. Also let the value of $\dfrac{Q}{\theta} + \dfrac{Q'}{\theta'} + \dfrac{Q''}{\theta''} + \ldots$ for any operation (not necessarily cyclical) be denoted by $\Sigma\dfrac{Q}{\theta}$[1] and suppose the operation referred to indicated by a suffix. Then for the reversible cycle PAP'. P'XP we have

$$\left(\Sigma\dfrac{Q}{\theta}\right)_{PAP'} + \left(\Sigma\dfrac{Q}{\theta}\right)_{P'XP} = 0.$$

Now let the system pass from P to P' by a different

---

[1] The letter Σ stands for "summation."

reversible path PBP′ and then return from P′ to P by the same reversible path P′XP as before. Then

$$\left(\Sigma\frac{Q}{\theta}\right)_{PBP'} + \left(\Sigma\frac{Q}{\theta}\right)_{P'XP} = 0.$$

Hence

$$\left(\Sigma\frac{Q}{\theta}\right)_{PAP'} = \left(\Sigma\frac{Q}{\theta}\right)_{PBP'}.$$

Thus the value of $\Sigma\frac{Q}{\theta}$ for a reversible path leading from P to P′ is independent of the particular path, and can therefore depend only on the two states P and P′.

Again, let the system pass from P to P′ by an irreversible path PaP′ and then return from P′ to P by the reversible path P′XP. Then

$$\left(\Sigma\frac{Q}{\theta}\right)_{PaP'} + \left(\Sigma\frac{Q}{\theta}\right)_{PXP} < 0.$$

Hence

$$\left(\Sigma\frac{Q}{\theta}\right)_{PaP'} < \left(\Sigma\frac{Q}{\theta}\right)_{PAP'}.$$

Thus the value of $\Sigma\frac{Q}{\theta}$ for an irreversible path from P to P′ is less than for a reversible path.

Art. 43. **Entropy.**—The results of the last article have led Clausius to the conception of a remarkable quantity called Entropy. Suppose, for shortness, we say that the system is in a state of equilibrium when every body of which it is composed is in a state of equilibrium. Then choosing any standard state of equilibrium 0, Clausius defines the entropy (ϕ) in any state of equilibrium P to be the value of $\Sigma\frac{Q}{\theta}$ for any reversible path leading from 0 to P. It therefore depends only on the states 0 and P. Suppose we take a second standard state of equilibrium 0′ and denote the entropy of the state P with respect to this standard state 0′ by ϕ'.

Then by supposing the system to pass by a reversible path from $0'$ to P through 0, we get

$$\phi' = \left(\Sigma\frac{Q}{\theta}\right)_{o'o} + \phi.$$

Thus ϕ' only differs from ϕ by a quantity independent of the state P and we may say that within a constant ϕ depends only on the state P. If therefore ϕ and $\phi + \Delta\phi$ be the entropies of two states P, P' with reference to the same standard state 0, ϕ' and $\phi' + \Delta\phi'$ the entropies with reference to the same standard state $0'$, we have

$$\left.\begin{array}{l}\phi = \left(\Sigma\dfrac{Q}{\theta}\right)_{o'o} + \phi \\[2mm] \phi' + \Delta\phi' = \left(\Sigma\dfrac{Q}{\theta}\right)_{o'o} + \phi + \Delta\phi\end{array}\right\};$$

hence $\Delta\phi' = \Delta\phi$,
so that $\Delta\phi$ is independent of the choice of the standard state.

Now let the system pass in a reversible way from 0 to P and then in a reversible way from P to P' by the path PAP'. Then

$$\left.\begin{array}{l}\left(\Sigma\dfrac{Q}{\theta}\right)_{\text{OP}} = \phi \\[2mm] \left(\Sigma\dfrac{Q}{\theta}\right)_{\text{OP}} + \left(\Sigma\dfrac{Q}{\theta}\right)_{\text{PAP}'} = \phi + \Delta\phi\end{array}\right\};$$

hence for a reversible path we have the simplest result

$$\left(\Sigma\frac{Q}{\theta}\right)_{\text{PAP}'} = \Delta\phi.$$

Again, let the system pass from P to P' by an irreversible path PaP'. Since the value of $\Sigma\dfrac{Q}{\theta}$ is less for this path than for a reversible path from P to P', we have

$$\left(\Sigma\frac{Q}{\theta}\right)_{\text{P}a\text{P}'} < \Delta\phi.$$

If the temperature of every part of the system possesses the same constant value θ where heat is absorbed or evolved, we have

$$\frac{1}{\theta}(\Sigma Q)_{PP'} \lesseqgtr \Delta\phi,$$

or

$$(\Sigma Q)_{PP'} \lesseqgtr \theta\Delta\phi,$$

according as the operation is reversible or irreversible. In this result $(\Sigma Q)_{PP'}$ is the total quantity of heat absorbed in going from P to P'. Denoting this by the symbol Q, we have, when heat is absorbed and evolved at a constant temperature θ,

$$Q \lesseqgtr \theta\Delta\phi.$$

Lastly, suppose it possible to divide the given system into two parts or sub-systems which are electrically independent. Let the standard states of equilibrium for the sub-systems be so chosen that when the whole system is in its standard state of equilibrium 0, the sub-systems are in theirs. Referred to these standard states, let ϕ be the entropy of the whole system when in any state of equilibrium P, and let (ϕ_1, ϕ_2) be the entropies at the same instant of the sub-systems. Then we shall have $\phi = \phi_1 + \phi_2$; for the heat absorbed by the whole system is equal to the sum of the quantities of heat absorbed by its parts. There is therefore no term in entropy corresponding to mutual energy.

Art. 44. The statement that if P and P' be any two states of equilibrium of the same system, it is often possible to bring the system from P to P' by a reversible path, must be looked upon as a result of experience. In some apparently difficult cases, the reversible process is effected by dissociation; in others, by electrolysis. For example, take a quantity of oxygen and hydrogen in the exact proportion required to form water. These gases, on being burnt, produce nothing but water, and we have to show how water can be resolved into oxygen and hydrogen in a reversible manner.

Take an ordinary voltameter in which the electrodes are both of platinum.[1] Fill with water, and let test-tubes be placed over the platinum electrodes. Then on sending a current through the instrument, which may be done by connecting its terminals with a dynamo and then turning the machine round, the water will be decomposed, bubbles of oxygen appearing at one of the electrodes and of hydrogen at the other. If now the terminals be disconnected from the dynamo and connected with one another, there will be a reverse current in the voltameter and the two gases will begin to combine to form water, as at first.[2]

In these experiments, the only chemical change produced is the decomposition of water and the recombination of the two gases; and if we suppose the system to include the dynamo, all the electric properties are internal.[3] The processes, however, are not quite reversible; for the gases appear to lose their power of recombining if allowed to stand long after their production. Except for this, the two opposite processes could be made as slow as we please, and the decomposition of water into oxygen and hydrogen would be a reversible process.

At any ordinary temperature the energy of water is much less than that of the equivalent quantity of oxygen and hydrogen. For if 1 gramme of hydrogen be burnt in 8 grammes of oxygen, and the temperature made the same after combustion as before, heat is given out to the enormous amount of 34,000 calories, of which the work done by the external pressure only accounts for about $1\frac{1}{4}$ per cent.

Now let us take an unelectrified system of uniform

---

[1] See Dunman's "Electricity;" also Professor S. P. Thompson's "Electricity and Magnetism."

[2] The reverse current may be produced without disconnecting from the dynamo. When sufficient water has been decomposed, let the system be left to itself. Then a reverse current will at once appear and cause the dynamo to run round.

[3] See Art. 28 for the reasons for believing that Carnot's principle applies to an electrified system.

temperature at rest consisting of a voltameter and dynamo. Let ϕ be the entropy of 9 grammes of the water and ϕ' the entropy of the rest of the system. Then by turning the dynamo, let a current be produced which decomposes the 9 grammes of water into gases in a way that may be considered reversible, and suppose the temperature everywhere constant during the operation. When the dynamo stops, the system will again be unelectrified: hence if ϕ_1 be the entropy of 1 gramme of hydrogen and ϕ_2 the entropy of 8 grammes of oxygen, the entropy of the system at the end of the operation will be $\phi_1 + \phi_2 + \phi'$. The increase of entropy is therefore $\phi_1 + \phi_2 - \phi$. But if $\Delta\phi$ be this increase of entropy and Q the heat absorbed during the operation, we have

$$\Delta\phi = \frac{Q}{\theta}.$$

Now in comparison with the change of energy, we may put $Q = 0$. Thus $\Delta\phi$ may be put zero, or $\phi = \phi_1 + \phi_2$.

Thus at any ordinary temperature, the energy of water is considerably less than that of the equivalent quantities of oxygen and hydrogen, while the entropy is, comparatively speaking, the same.

It may here be remarked that we have divided operations into two classes as reversible and irreversible operations. At every instant during a reversible operation, a body is continually in a state of equilibrium. Now it seems from Dr. Guthrie's experiments on saline solutions that a body may sometimes be continually in a state of equilibrium in an irreversible operation. Such an operation appears to fulfil the same equations as a reversible operation. However, as we do not discuss the properties of saline solutions in this work, we shall ignore the existence of irreversible equilibrium operations and suppose that reversible and irreversible processes never satisfy equations of the same form.

Art. 45. In the equation of entropy $\Sigma\frac{Q}{\theta} < \Delta\phi$, it

Carnot's Principle.

is not necessary that the temperature should have the same value at the same instant throughout the system. But when it does happen that the temperature of the system is uniform at every instant, the equation of entropy takes the simple form

$$\frac{1}{\theta}(\Sigma Q) \lesseqgtr \Delta \phi,$$

or

$$\Sigma Q \lesseqgtr \theta \Delta \phi.$$

where the sign of equality or inequality is to be taken according as the operation is reversible or irreversible.

When the temperature is always uniform and the operation reversible, let us take two rectangular axes and denote the temperature and entropy at any instant by the ordinate PM and abscissa OM of a point P in the plane of the axes. Also let AB be the curve traced out by P during the reversible operation. Then in any small part of this operation corresponding to the portion PQ of AB, the heat absorbed will be represented by the area of the figure PQNM; and therefore the heat absorbed in the whole operation will be represented by the area of the whole figure ABnm, which generally depends on the form of the curve AB (or on the nature of the reversible operation) as well as on

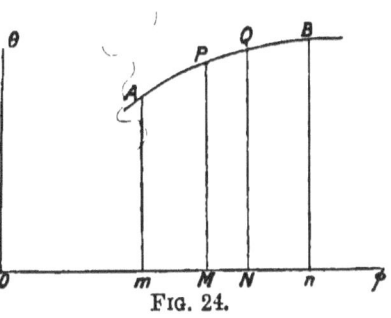

Fig. 24.

the initial and final states A, B. To make the heat absorbed independent of the path leading from A to B, we must introduce some restrictions. The heat absorbed will evidently be independent of the (reversible) path if we restrict the temperature or entropy to be constant during the operation.

Art. 46. Let us suppose, for simplicity, that our system is a single unelectrified body in a state of equilibrium. Then we know, from Rigid Dynamics, that without producing any other change in the body, the motion of the centre of mass may be varied at will by a reversible process without the loss or gain of heat in any part of the body. Hence when two states of equilibrium differ only in the motion of the centre of mass, the entropy has the same value in the two cases, and is therefore independent of the motion of the centre of mass.

If the speed of rotation of the body be altered by a reversible process without gain or loss of heat, the internal state of the body will vary in consequence; but if the body be made of very rigid materials, like a fly-wheel, these variations will be practically insensible. We may therefore say that the entropy of a very rigid body, like a fly-wheel, is independent of the speed of rotation and depends only on the internal state of the body.

Art. 47. **Available Energy, and the Degradation of Energy.**—If a material system be allowed to gain and lose heat, we know that we may obtain a positive quantity of work from it during a cyclical process, the work obtained being exactly equal to the total quantity of heat absorbed by the system during the cycle. By repeating the cycle often enough, we may obtain as much work as we please from the system and the final state be the same as the first. If, however, the system be prevented from gaining and losing heat,[1] work can only be obtained from the system at the expense of its energy, the work obtained being exactly equal to the decrease in the energy of the system. Thus in a cyclical process, the work obtained from the system when no heat is gained or lost, is zero. To obtain work from the system, we must bring it from the given state P to a different state P' in which the energy is less than in the state P.

---

[1] The stipulation that heat is neither to be gained nor lost by the system does not require us to suppose that there is no passage of heat in the interior of the system from one part to another.

Carnot's Principle.

Suppose then that our system is brought from a state P to a different state P' without the gain or loss of heat; and let U be the energy in the state P, U' the energy in the state P'. Then the work obtained from the system during the process is $U - U'$. Again, let us add to our system a number of bodies which may be used as heat engines, and let the passage of heat be permitted between the given system and the heat engines, but suppose there is no gain or loss of heat in any part of the compound system. Let this compound system undergo an operation during which the heat engines perform complete cycles and the state of the given system changes from P to P'. If the given system be large enough to exert a sensible gravitational attraction on the heat engines, let them be brought to such a position at the end of the process that the mutual energy, M, of the original system and the heat engines is the same as at first. Then if u be the energy of the heat engines at the beginning and end of the process, the energy of the compound system is $U + u + M$ at the beginning of the process and $U' + u + M$ at the end. The work obtained in this way from the compound system during the change of state from P to P' is therefore $U - U'$. Hence when the same change of state can be produced with or without the use of heat engines, it is immaterial whether we use the heat engines or not.

In some cases it may be impossible to obtain work from a system without the use of heat engines, but possible with their assistance, as, for example, when the system consists of pieces of metal at different temperatures.

Let us therefore suppose that a system is everywhere prevented from gaining and losing heat, or that it is only allowed to exchange heat with heat engines used as already explained. Also let any restrictions we please be imposed on the system (but not on the heat engines) in addition to the foregoing conditions: for example, in the case of a number of gases, we may have the restriction that the total volume of the gases is to be kept constant. Then we define the Available Energy of the system in any given state P to be the maximum amount of work that can be obtained from the system during a change of state starting from P subject to the conditions and restrictions just stated.

When the system has been brought into a state in which the available energy is zero, it is evident the temperature of the system will be uniform. For if one part (A) of the system was at a higher temperature than another part (B), it would be possible to obtain work from the system by means of a heat engine undergoing cyclical processes during which it absorbs heat from A and gives out heat to B. The question now arises, What is the final uniform temperature of the system? If it were possible for a heat engine to work down to absolute zero, the engine might continually absorb heat from the system and transform it wholly into work until every part of the system was reduced to the temperature of absolute zero. As it is not possible to get down to absolute zero, it is easy to see that the final uniform temperature of the system will depend on various circumstances. For example, let the system be two pieces of metal at different temperatures. Then the temperatures may be reduced to equality by simply putting the two pieces in thermal communication, or we may use a heat engine which absorbs heat from the hotter piece of metal and gives out heat to the colder. Suppose work to be thus obtained from the heat engine. Then, as no work is obtained in the first method of equalizing the temperatures, it is clear that the final uniform temperature will be different in the two cases.

There are some general propositions, of great simplicity, referring to the available energy of a system which fulfils the following conditions:—

(1) Every body forming part of the system is in a state of equilibrium when the system is in the initial state P. The definition of entropy then applies to the system when in the state P.

(2) It is possible to bring the system to a state in which the available energy is zero by a reversible operation during which no heat is gained or lost or is only exchanged with auxiliary heat engines. In such an operation it is easy to see that the entropy of the system remains constant.

Also, for simplicity, suppose that the system possesses no electric or magnetic properties.

(a) Let the system be brought by a reversible path from the initial state P to a final state Y, and by another reversible path from the initial state P to a final state Z. Then since the temperature of the system is uniform in each of the final states (Y, Z), it is clear that these final states are either identical or can be made so by merely expending work in friction so as to raise the temperature of one of them. Hence since the entropy has the same value ϕ in each of the final states as in the initial state P, we easily see that the states Y and Z must be identical.

Thus the final state is the same for all reversible paths. If U_0 be the energy in this final state, U in the initial state, and W the work obtained from the system, we have

$$U - U_0 = W.$$

(b) Next, let the system be brought from the initial state P to a final state Y' by an irreversible path. Then if ϕ' be the entropy of the system in the state Y', we have, by Art. 44,

$$\phi' - \phi > \left(\Sigma \frac{Q}{\theta} \right)_{PY'},$$

so that ϕ' is greater than ϕ, since every one of the quantities Q is zero.

Now it is evident that to make the states Y and Y' identical, it will only be necessary to expend a positive quantity of work in friction so as to raise the temperature of one of them. But a frictional action causes an increase of entropy. Remembering that the entropy is greater in the state Y' than in the state Y, it is easy to see that it is the state Y on which a positive quantity of work must be expended. Denote this work by x, and the energy in the state Y' by U'_0. Then $U'_0 = U_0 + x$. Hence if W' be the work obtained from the irreversible path from P to Y', we have $W' = U - U'_0 = U - U_0 - x$, which is less than W.

Thus to obtain the greatest possible amount of work from the system, the operation it is made to undergo must be reversible.

(c) Let the system be left to itself in the state P so that it can neither gain nor lose energy either in the form of work or of heat; and suppose an action takes place within the system by which the state is changed from P to another "state of equilibrium"[1] P'. Then if (U', ϕ') be the energy and entropy in the state P', we shall have $U' = U$, and since no spontaneous action can be reversible, $\phi' > \phi$. Now let Y'' be the final state into which the system can be brought by a reversible path starting from P', and let U''_0 be the energy of the system in the state Y.'' Then it is easy to prove that $U''_0 = U_0 + y$, where y is a positive quantity. Hence the available energy $U - U''_0$ in the state P' is less than the available energy $U - U_0$ in the state P. Since therefore when a system is left to itself, the entropy has a constant tendency to increase, it follows that there is a constant tendency for the available energy to decrease. This result is the principle of the Degradation of Energy, formerly known as the principle of the Dissipation of Energy, which was first enunciated by Sir W. Thomson.

The consideration and calculation of available energy are generally difficult; but the two following examples, of which the first is taken from Maxwell, are elementary.

(d) Let the system consist of quantities of gas contained in cylinders fitted with smooth air-tight pistons and suppose the sum of the volumes of the gases is to be kept constant. When this system has been brought

---

[1] For the meaning of this expression, see Art. 44.

Carnot's Principle. 115

into a state in which the available energy is zero, it is clear that the different gases will all be at the same temperature and pressure. There are any number of such states into which the system can be brought, and the amounts of work respectively obtained from the system in bringing it into these states from the given state will be different. When the amount of work obtained from the system is a maximum, the operation is reversible, and all reversible operations give the same final state. The system may be brought into this final state without the use of heat engines in the following (reversible) way. First, let the hotter gases be allowed to expand slowly and let the colder gases be slowly compressed without the passage of heat from one gas to another, until the temperature is everywhere the same. Then put all the gases in thermal communication; and let those which are at the higher pressures expand slowly and let the others be slowly compressed, until the pressures are all equal as well as the temperatures.

(e) Let the system consist of a number of bodies A together with a very large body B whose volume is constant and whose temperature is uniform and equal to θ_0. Suppose the maximum amount of work obtained from the system by the following reversible path. Let the bodies A be subjected to isentropic operations until their temperatures are all reduced to θ_0; then let them be put in thermal communication with B and bring the system to its final state by another isentropic operation. In the second isentropic operation, the temperature will remain constantly equal to θ_0, owing to the large size of B. Hence if (U, ϕ) be the original values of the energy and entropy of the bodies A, (U', ϕ') the final values, the heat absorbed by the bodies A while in thermal communication with B will be $\theta_0 (\phi' - \phi)$, and therefore the heat lost is $\theta_0 (\phi - \phi')$. But the decrease in the energy of these bodies during both operations is $U - U'$: the work obtained from them is therefore $U - U' - \theta_0 (\phi - \phi')$. Since no work can be obtained from the large body B whose volume is constant, the expression just found is the available energy of the system.

If we write F for $U - \theta_0 \phi$ and F' for $U' - \theta_0 \phi'$, the available energy can be written in the simple form $F - F'$.

In conclusion, let us consider for a moment the case of a system which cannot be brought by a reversible path from its initial state P to a final state in which the available energy is zero. This appears to be the case with food and fuel. In such a case, the entropy in every final state is necessarily greater than the entropy ϕ in the initial state P; and it is easy to see that the maximum amount of work is obtained from the system by bringing it into a final state in which the entropy exceeds ϕ by the least possible amount.

Art. 48. The definition of entropy only applies when every body of the system is in a state of equilibrium. Thus, for example, we cannot speak of the entropy of the solar system. However, we can generally find a state of equilibrium Q differing very little from a state P' to which the definition of entropy does not apply, and it will usually be permissible to take the entropy of the state Q for the entropy in the state P'. Consider, for instance, a piece of metal in a state P' in which (at the same instant) the temperature varies gradually from point to point. If we suppose the piece of metal broken up into a large number of small pieces, the temperature will have nearly the same value in all parts of any one of the small pieces; and if we replace each of the small pieces by a similar small piece of metal whose temperature is uniform and equal to the average temperature of all parts of the small piece of metal it replaces, we get a state Q to which the definition of entropy is applicable and which differs little from the given state P'. Then without much danger of causing ambiguity, the entropy of the system when in the state Q may be called the entropy of the state P'. And if we use the word "entropy" with this signification, we may speak of the entropy of the sun, or of the solar system, or perhaps even of the whole universe.

If we take the solar system as our material system, we know that the energy continually decreases. The equation referring to the entropy is $\Delta\phi > \Sigma\left(\dfrac{Q}{\theta}\right)$, where the left-hand side is negative. This does not tell us whether $\Delta\phi$ is positive or negative. To find whether the entropy really increases or decreases, we must suppose the system brought from one state Q to a later state Q' by a reversible path in which the quantities of heat absorbed and given out are known. The equation referring to this reversible path is $\Delta\phi = \Sigma\left(\dfrac{Q}{\theta}\right)$, and it appears that the quantities of heat Q are negative. Hence the increase of entropy is negative, or the entropy decreases.

A little consideration will show those who are acquainted with the nebular hypothesis, that the available energy of the solar system decreases. Thus the entropy and available energy decrease simultaneously. This result does not contradict that given in the last article, where an increase of entropy corresponds to a decrease of available energy, because it was there supposed that there was no gain or loss of heat; while in the present case, there is an enormous radiation of heat into space. The result may perhaps be made clearer by remarking that the solar system at present possesses a considerable amount of available energy; and that if the system exist long enough without interference from other systems, the effect of radiation in cooling the sun until the temperature of the system is uniform combined with the effect of tidal friction in diminishing the motions of the planets, will cause a considerable diminution in the available energy.

Clausius has considered the variation of the entropy of the whole universe in the following way. He supposes his system to include "the ether" (or space) as well as all material bodies, and then in the equation $\Delta\phi > \Sigma\left(\dfrac{Q}{\theta}\right)$ he takes every one of the quantities Q to be zero, since no heat can be supposed to enter or leave the "universe." He therefore concludes that the entropy of the universe increases. This argument, however, is inconsistent with his definition of entropy. For in the equation $\Delta\phi > \Sigma\left(\dfrac{Q}{\theta}\right)$, Q denotes a quantity of heat absorbed by a material part of the system and θ the temperature of that material part of the system at which it is absorbed. We cannot suppose $\Sigma\left(\dfrac{Q}{\theta}\right)$ to refer to heat passing in free space, because space does not possess the property of temperature.

If a conjecture may be hazarded about the whole universe, we should say that its energy, entropy, and available energy [1] all decrease.

The assertion that the entropy of the universe increases is not made either by Tait or Maxwell. It seems, however, to be widely accepted. When the author first questioned it, he imagined he was dealing with something universally received as orthodox. In this he finds he was mistaken, as the following extracts from Duhem's "Introduction à la Méchanique Chimique" will show:—

"Imaginons qu'un système matériel soit absolument isolé dans l'espace. Autour de lui, il n'y a ni matière pondérable, ni éther. Rien ne peut lui fournir de chaleur, soit par conductibilité, soit par rayonnement; rien ne peut lui en enlever.

Ce système éprouve une certaine modification qui le fait passer de d'état a à l'état b. Dans chacune des modifications élémentaires en lesquelles se décompose la modification considérée, l'échange de chaleur entre le milieu et l'extérieur, qui est rigoureusement vide, se réduit à 0."

* * * * *

---

[1] The statement that the available energy of the universe decreases, is given in Tait's "Sketch of Thermo-dynamics."

"*Ainsi, si l'on envisage un système absolument isolé dans l'espace, toute modification réalisable de ce système présente les deux caractères suivants :*
(1) *Elle laisse constant son énergie.*
(2) *Elle augmente son entropie.*
Clausius assimile l'univers entier à un semblable système, isolé au milieu d'un espace vide, et il énonce les deux propositions suivantes :
L'énergie de l'univers est constante.
L'entropie de l'univers tend vers un maximum.

Ce dernier énoncé présentait, sous une forme précise, une conséquence que Sir W. Thomson[1] avait déjà déduite du principe de Carnot-Clausius : à savoir que, dans l'univers, toutes les modifications qui se produisent ont, en quelque sorte, une même tendance.

La portée métaphysique des deux lois énoncées par Clausius semblait considérable. Ces lois, il est vrai, prêtaient à des critiques ; le philosophe était en droit de se demander jusqu'à quel point l'univers était assimilable à un système de corps enfermé dans une surface finie et séparé par cette surface d'un espace absolument vide. Mais, si ses critiques étaient de nature à mettre en doute les conséquences qu'une cosmologie hasardée avait énoncées, l'attention du moins ne pouvait manquer d'être vivement attirée vers les propositions solidement établies dont ces conséquences avaient été tirées."

Art. 49. **Tidal Friction and Theory of Exchanges.**—If a material system of any kind be unacted on by external forces, and if the bodies of which it is composed be either all prevented from gaining and losing heat by non-conducting coverings, or if some or all of them be at liberty to radiate heat into infinite space without receiving heat from external systems, it may be assumed as a consequence of Carnot's principle that the whole system will ultimately move like a single rigid body. The parts of the system will then have no motions relative to each other (except the invisible irregular motions

[1] W. Thomson, *On a universal tendency in nature to the dissipation of mechanical energy* (Philosophical Magazine, 1852).

which constitute "heat"). The ultimate temperature of every body will also be uniform; and those bodies which have been at liberty to radiate heat into infinite space will be at a very low temperature—perhaps at absolute zero.

The bodies may contain cavities of any kind. If a thermometer or other object be hung up in one of these closed cavities, it will assume the temperature of the body. This is the foundation of Balfour Stewart's discussion of radiation. The tendency to move as rigid is chiefly due to the action of the tides and is considered at length in Chap. V.

The subject of Tidal Friction will be found to be one of great simplicity when considered as a deduction from Carnot's principle. It is to this method that the author is indebted for his knowledge of the subject. However, the results in Chap. V. have been mostly verified by comparison with accepted authorities. The Theory of Exchanges is also very simple, and is given fully in Balfour Stewart's treatise on Heat. A fair idea of the theory may be obtained from Prof. Tait's "Sketch of Thermo-dynamics."

It should be observed that if the temperatures of different parts of the system are originally unequal, it is not asserted that the colder parts cannot become as hot as the hotter until the final state is attained. For example, let a part A be at a higher temperature than another part B and suppose B to consist of very inflammable materials. Then the heat from A may cause ignition at B, in consequence of which the temperature at B may rise far higher than that at A.

Art. 50. We have hitherto supposed the electric and magnetic properties of our system to be all internal. In the present article, we consider the case in which an electric current enters and leaves the system by conductors of the same metal and in the same condition with the proviso that there is to be no electric force between any part of the given system and external bodies or systems.

Take the simplest form of dynamo or motor consisting of a metallic disc rotating between the two poles of a permanent steel magnet, and touched at its centre and a point of its circumference by fixed contact-pieces of the same metal as the disc and in the same condition. Then if the electric circuit be completed by joining the metallic contact-pieces by a metallic wire of any kind, a current may be produced by forcibly rotating the disc; and conversely, if a current be sent through the machine by joining the contact-pieces to a battery, there will be a force tending to cause the disc to revolve and work may be obtained from it. If the temperature of the machine be uniform and kept constant, the thermal phenomena which take place in it will be of two kinds :—

(1) A positive quantity of heat will be evolved, according to Joule's law. As will be seen hereafter, this positive quantity may be made as small as we please by simply having the size of the disc and contact-pieces large enough.

(2) The rubbing of the contact-pieces will produce a frictional action which may, however, be supposed prevented by substituting rolling for rubbing contact.

When the heat evolved according to Joule's law and owing to the friction has been made small enough, the equation $\Delta U = W + Q + e$ becomes $\Delta U = W + e$, and the operation is reversible. If the cycle be complete, we have $W + e = 0$. Hence when a positive quantity of work is absorbed by (or done on) the machine in a reversible cycle, an equal positive quantity of electric energy is given out; and conversely, if a positive quantity of electric energy be absorbed by the machine, an equal positive quantity of work is given out by (or obtained from) it.

Thus in a complete cycle, electric energy may be transformed wholly into work; and conversely, work may be transformed wholly into electric energy.[1] Hence since heat can only be transformed partially into work,

[1] Machines are actually constructed which are guaranteed to transform 92 per cent. of the work into electric energy.

it follows that heat can only be transformed partially into electric energy.

A machine for transforming electric energy into work or work into electric energy, is called an "electro-magnetic engine," while a machine for transforming heat into work is called a "heat engine." If we define the "efficiency" of an electro-magnetic engine to be the proportion of electric energy it transforms into work, or the proportion of work it transforms into electric energy, during a complete cycle, then the efficiency of a reversible electro-magnetic engine will be equal to unity. This is greater than that of a heat engine, which, we have seen, is always less than unity.

Now take the general case in which an electric current enters and leaves any system whatever by conductors of the same metal and in the same condition, there being no electric force between any part of the system and external bodies or systems. Since it is immaterial how the current which enters the system from without is produced, let it be produced by the rotating disc just described. When the system compounded of the given system and the disc undergoes a complete cycle, we have, since the electric properties are all internal, $\frac{Q}{\theta} + \frac{Q'}{\theta'} + \frac{Q''}{\theta''} + \ldots \lesseqgtr 0$, according as the cycle is reversible or irreversible. But the heat absorbed or evolved in any part of the disc may be supposed zero and the operation undergone by the disc reversible, whatever be the rapidity of the process. Hence if the cycle undergone by the given system be reversible, the cycle undergone by the compound system is also reversible; and if one be irreversible, so will be the other. Thus for the *given system* we have $\frac{Q}{\theta} + \frac{Q'}{\theta'} + \frac{Q''}{\theta''} + \ldots \lesseqgtr 0$, according as the cycle which it undergoes is reversible or irreversible.

It is essential that the conductors by which the current enters and leaves the system should be of the same metal and in the same condition. For example, let the system consist of two pieces of different metal at the same temperature θ, and let the temperature be everywhere kept equal to θ. Then on passing a current through the system in one direction, it is necessary to supply heat to the junction to keep its temperature constant; and if a current pass in the opposite direction, heat must be abstracted from the junction. These quantities of heat are proportional to the quantity of electricity that passes and are the same whether the process be rapid or slow.

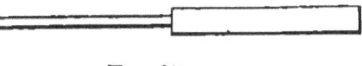

FIG. 25.

Now the heat evolved, according to Joule's law, in the homogeneous parts of the metals, may be made as small as we please by making the current slow enough. Hence the heat absorbed by the system may be altogether positive and we may have $\dfrac{Q}{\theta} > 0$ for a cycle.

Art. 51. We have seen that Carnot's principle has been applied to the whole universe by two of our greatest thermo-dynamicists—Clausius and Tait. Also it seems to be admitted that Carnot's principle applies to every unorganized object in the universe. Several examples of such applications will be given in the next chapter. In Chap. V will be considered the case of the tides, and in Chap. VI it will be shown that Carnot's principle has been used with great success by Sir W. Thomson and others in electrical problems. If, then, Carnot's principle applies to the whole universe, and also to every unorganized part of it, it appears to follow that it must apply to the remaining part of the universe—animals and vegetables.

If Carnot's principle applies to unorganized objects and not to organisms, the laws of Nature cannot be the same in the two cases. Now a living organism may be changed by death and decay into unorganized matter. It would therefore follow that death and decay cause a change in the laws of Nature.

The author was led to study this subject by reading or hearing that the heart works with a greater efficiency than a Carnot's perfectly reversible engine, and by being unable to get the question satisfactorily explained. In the two following articles, he believes he has succeeded in applying Carnot's principle both to animal and vegetable life.[1]

The following extracts will serve to show how small have been the attempts that have hitherto been made to solve the question.

The following passage occurs in Browne's translation of Clausius on the "Mechanical Theory of Heat":—

"Wand considers the process which occurs in nature, when in the growth of plants, under the influence of the sun's rays, carbonic acid and water are absorbed, and oxygen liberated; whilst the organic substances thus formed, if afterwards burnt or serving as nourishment to animals, unite themselves again with oxygen to form carbonic acid and water, and thereby generate heat. This transformation of the sun's heat, he" (Wand) "considers to be in flat contradiction to the second principle" (called the fundamental axiom in the present work, Art. 31). "To the analysis which he gives many objections might be taken; but the author" (Clausius) "considers that a process in which so much is still unknown, as that of the growth of plants under the influence of the sun, is altogether unfit to be used as a proof either for or against the principle in question."

In Tait's Sketch of Thermo-dynamics, the following words occur:—

"Mayer's pamphlet of 1845 adds considerably to the development of the question. He speculates acutely on the merely *directive* agency of the so-called *Vital Force,* and gives some excellently chosen illustrations of his views. Recent researches in chemical synthesis have broken down many of the supports on which the old theory of vital force rested, and the mode of its action

---

[1] As this subject appears to offer a large field for research, he would be glad to see Arts. 51, 52 criticized.

remains in consequence exceedingly obscure. But there can be little doubt that, as Joule suggested (in his paper of 1846), an animal more closely resembles an electro-magnetic, than a heat engine. And it is wonderful that it is a far more economic engine than any which we are yet able to construct. The first idea of this seems to have been entertained by Rumford, for he expressly shows that the amount of work done by a horse is much greater than could be procured by employing its food as fuel in a steam-engine."

From the foregoing passage it is evident that Joule hardly attempted to solve the problem proposed in the present article. The comparison of an animal to an electro-magnetic engine can only be taken to mean that he perceived a difficulty. Now from what has already been explained in this chapter it will be seen that an animal cannot strictly be compared to an electro-magnetic engine, because there are generally no sensible electric manifestations about the animal body and an electro-magnetic engine is a body or system which a sensible electric current enters from without and leaves again in another part of the body or system. An electrified system that has no electric (or magnetic) connection with external systems is as much a heat engine as a system that is electrically (and magnetically) neutral. This will be made clearer in Chap. VI., which is devoted to the application of Carnot's principle to electrical problems.

Again, a little consideration will show us that when we speak of the "efficiency" of an animal, we are not using the word "efficiency" with the meaning it is defined to have in thermo-dynamics. In Carnot's principle, the word "efficiency" only refers to a complete cycle, while the changes which take place in connection with the animal body do not constitute a cycle.

The meaning of the word "efficiency," as commonly used in connection with animal life, seems to be as follows. Let a quantity of food, whose thermal value is Q, be consumed by an animal, and suppose we wait until the animal body is in the same state as before.

Then if W be the work obtained from the animal, $\dfrac{W}{Q}$ is taken to be the efficiency of its body. Here it is evident that the operation undergone by the system of the animal and its food, does not constitute a cycle, for although the animal body undergoes a cyclical process, the food does not. Again, it is clear that Q is the decrease of energy in the system during the operation.

It is easy to find incomplete cyclical operations for which the ratio $\dfrac{W}{Q}$ is greater than the efficiency of a perfectly reversible heat engine undergoing a cycle of operations. For example, let a quantity of gas or any other body expand doing an amount of work W without gain or loss of heat. Then the decrease Q of the energy of the body is equal to W, and therefore the ratio $\dfrac{W}{Q}$ is equal to unity. This is greater than the efficiency of any heat engine.

The fact that it is more economical to serve food to an animal than to burn it as fuel in a steam-engine, must now be considered. In this case Q is the heat absorbed by the engine; and therefore if W be the corresponding amount of work obtained from it, $\dfrac{W}{Q}$ is the "efficiency" of the engine (using the word "efficiency" with the meaning given to it by Carnot).

The temperature of the furnace where the food is burnt will be perhaps 1,500 C. The heat is therefore supposed to fall from this high temperature to the temperature of the boiler (190° C. or lower) without doing any work. If an engine was employed working from 1,500 C. instead of from 190° C., the work obtained from the food by burning it for fuel would be enormously increased. In comparison with such a heat engine the superiority of the animal body as a work-producer would be found to disappear.

But it may perhaps be thought that in making a

comparison between the animal body and a heat engine, the restriction should be imposed that the heat engine is to absorb its heat at the low temperature of the blood (in the human body, this temperature is about 38° C., Huxley's " Elementary Lessons in Physiology "). Now it is difficult to see why the engine is to be always at a low temperature while a high temperature is permissible in the furnace. The vague difficulty will probably be removed by the following reasoning.

If we could consume the food in a reversible operation, it can easily be shown by Carnot's principle that the percentage of work obtained from food would be greater [1] than if the same change of state was effected in an irreversible manner. It would thus be greater than that obtained by the irreversible burning of the food as fuel for a heat engine, even if the engine absorbed heat at the highest temperature possible. It would therefore be enormously greater than the percentage obtained when the heat engine absorbs its heat at the low temperature of the blood.

The percentage of work obtained in the reversible operation would be independent of the temperature to which any part of the system was raised during the operation. If two such reversible operations were possible in one of which the temperature was always low and in the other sometimes high, the percentage would be the same for both.

On the whole, it seems that the reason why so high a percentage of work is obtained from food when it is eaten by animals, is that the processes which take place in the animal body are much more nearly reversible than those which occur in a furnace—and not that Carnot's principle does not apply to the animal body.

Perhaps the following is the chief and last difficulty in connection with the production of work by the animal body. Let our system contain, in addition to the living animal, a sufficient quantity of air, water, vegetation,

[1] Heat being supposed, of course, to be evolved at one given temperature only.

etc., so that while under consideration the system neither gains nor loses matter. Also let the vegetation grow as fast as it is consumed, and suppose that when a quantity of vegetation whose thermal value is Q has been consumed, the whole system is in the same state as at first. Then it may perhaps be thought that Q is the heat absorbed by the system from the sun during the cycle, and that it is legitimate to speak of $\dfrac{W}{Q}$ as the efficiency of the cycle. If such be the case, we have a contradiction of Carnot's principle, because the ratio $\dfrac{W}{Q}$ is found to be greater than the efficiency of a perfectly reversible heat engine working between the same limits of temperature as the cycle just described. It appears, however, from works on Botany, that only a small part of the heat absorbed by a plant is stored up as increased energy in the plant; by far the larger part of the heat absorbed from the sun is required to sustain the process called "transpiration," which is the exhalation of aqueous vapour from the leaves of the plant. Hence when there is an increase Q in the energy of vegetation, the quantity of solar heat absorbed in effecting the process must be written $Q+x$, where x is positive and is found to be equal to or greater than 20 times Q. The efficiency of the cycle described above is therefore $\dfrac{W}{Q+x}$, which is probably less than $\dfrac{W}{21Q}$.

Art. 52. The importance of the process of "transpiration" in vegetable life may be seen by taking a cycle in which animal life does not occur. The system while under consideration must neither gain nor lose matter. It must therefore contain, in addition to the living vegetation, a sufficient quantity of air, carbonic acid, water, soil, etc. Let the only vegetation in the system at first be a twig; then let it grow into a tree; and next let the whole tree except a single twig like the first, be destroyed by fire. Also let the pressure be uniform and constant during the cycle. Then the work done on the system during the cycle is zero, and therefore the total quantity of heat absorbed is also zero. Suppose also that the temperature is uniform and constantly equal to θ during the life of the tree.

Water being one of the products of the burning of wood, the air will contain more aqueous vapour after the products of combustion are reduced to the original temperature θ than it did before the conflagration, unless it was saturated before the conflagration. If the air be saturated with vapour and at the temperature θ before the wood is burnt, the heat given out during and after the conflagration until the temperature is again equal to θ, will be equal to the thermal value of the wood burnt—a quantity denoted in the last article by Q, but which it will now be best to denote by a different symbol q. If the air be not saturated with vapour before the wood is burnt, the heat given out will not be equal to Q. We will therefore suppose the air saturated with aqueous vapour and at the temperature θ before the conflagration. The complete cycle will then consist of the following operations at constant pressure, at the end of each of which the temperature is θ :—

(1) The twig grows into a tree at the temperature θ.
(2) The air is then saturated with aqueous vapour.
(3) The conflagration takes place, and the products of combustion are reduced to the original temperature θ.
(4) The quantity of vapour in the air is made the same as at the beginning of the cycle.

In the third operation, the temperature is always above θ; and since the quantity of heat given out is q, the corresponding value of $\Sigma\left(\dfrac{Q}{\theta}\right)$ may be written $-q$, where $\theta' > \theta$.

Let us now suppose, if possible, that the air neither gains nor loses aqueous vapour during the growth of the tree. Then operations (2) and (4) will be the reverse of one another; and the quantity of heat absorbed and the value of $\Sigma\left(\dfrac{Q}{\theta}\right)$ for the two operations combined, will both be zero. Hence since the total quantity of heat absorbed in the cycle is zero, the heat absorbed in the first operation will be q; and the value of $\Sigma\left(\dfrac{Q}{\theta}\right)$ for the whole cycle is therefore $\dfrac{q}{\theta} - \dfrac{q}{\theta'}$, or $q\left(\dfrac{1}{\theta} - \dfrac{1}{\theta'}\right)$, or $q\dfrac{\theta' - \theta}{\theta\theta'}$, which is positive, contrary to Carnot's principle.

It therefore follows from Carnot's principle that "transpiration" is necessary. In consequence of this process, the air in which the tree grows gradually comes to contain more aqueous vapour. Hence more vapour is deposited in (4) than absorbed in (2); and therefore, in the two operations together, a positive quantity of heat, x say, is given out, and the corresponding value of $\Sigma\left(\dfrac{Q}{\theta}\right)$ is $-\dfrac{x}{\theta_0}$, where $\theta_0 < \theta$. If therefore, as before, q be the quantity of heat given out in (3), $q + x$ will be the quantity absorbed in (1), and the value of $\Sigma\left(\dfrac{Q}{\theta}\right)$ for the complete cycle will be

$$\dfrac{q+x}{\theta} - \dfrac{x}{\theta_0} - \dfrac{q}{\theta'}, \text{ or } q\left(\dfrac{1}{\theta} - \dfrac{1}{\theta'}\right) - x\left(\dfrac{1}{\theta_0} - \dfrac{1}{\theta}\right).$$

It remains to show from experiment that "transpiration" is sufficient to make the value of $\Sigma\left(\dfrac{Q}{\theta}\right)$ for the cycle negative. For this purpose we will consider the case of an oak-tree. Here we may safely take the quantity of aqueous vapour transpired in 24 hours at 20 litres, or about 4½ gallons, of water (Sir J. D. Hooker's "Primer of Botany"). Hence since the latent heat of one gramme of water at ordinary temperatures, is about 600 calories, the value of x for one day's growth will be roughly 12 million calories. If we suppose 50 growing days in a year, the value of x for 20 years will be about 12 thousand million calories. Now if we take the heat given out in the combustion of one gramme of dry wood to be 3,000 calories, it will require 4 million grammes, or 4,000 kilogrammes, or about 4 tons, of dry wood to be burnt to evolve 12 thousand million calories of heat. Hence since we can only suppose the increase of the tree to amount to a small fraction of 4 tons of dry wood in 20 years, q will only be a small fraction of x, perhaps a 20th of x.

If we put $x = 20q$, we find the value of $\Sigma\left(\dfrac{Q}{\theta}\right)$ for the cycle to be

$$q\left(\dfrac{1}{\theta} - \dfrac{1}{\theta'}\right) - x\left(\dfrac{1}{\theta_0} - \dfrac{1}{\theta}\right) = q\left[\left(\dfrac{1}{\theta} - \dfrac{1}{\theta'}\right) - 20\left(\dfrac{1}{\theta_0} - \dfrac{1}{\theta}\right)\right]$$

$$= q\left[\dfrac{\theta' - \theta}{\theta\theta'} - 20\dfrac{\theta - \theta_0}{\theta_0\theta}\right]$$

$$= q\dfrac{\theta - \theta_0}{\theta_0\theta}\left(\dfrac{\theta' - \theta}{\theta - \theta_0}\dfrac{\theta_0}{\theta'} - 20\right).$$

If, for example, we put $\theta_0 = \tfrac{9}{10}\theta$, the above expression becomes

$$q\dfrac{\theta - \theta_0}{\theta_0\theta}\left(9\dfrac{\theta' - \theta}{\theta'} - 20\right),$$

and is therefore negative if $\dfrac{\theta' - \theta}{\theta'} < \tfrac{20}{9}$. As $\dfrac{\theta' - \theta}{\theta'}$ is evidently less than 1, Carnot's principle is satisfied.

It has been urged against the preceding argument that it takes no notice of the fact that a plant cannot grow in the *dark*. The only reply that can be made at present to this objection is, that the argument does not require it to be possible for a plant to grow in the dark. It is merely shown that Carnot's principle is satisfied when the plant grows by absorbing heat without enquiring whether any or what conditions must be satisfied in order that the plant may be able to absorb heat.

It may perhaps be desirable to explain why no account has been taken of the temperature of the sun.

In the equation $\dfrac{Q}{\theta} + \dfrac{Q'}{\theta'} + \dfrac{Q''}{\theta''} + \ldots \leqq 0$, θ denotes the temperature of that part of the given system where the heat Q is absorbed, without any reference to the temperatures of external objects.[1] For example, in the case in which heat is radiated into infinite space, there are no external temperatures to take account of.

It follows from what we have said that "transpiration" is essential to vegetable life. In fact, a plant would be choked if kept in a place constantly saturated with aqueous vapour. On the earth the atmosphere is prevented from being constantly at the point of saturation by the succession of day and night: in other words, "cold and heat, day and night,"[2] are necessary to vegetation.

If a planet be placed so that no heat can reach it from external sources, its temperature will generally become practically constant and its atmosphere saturated with vapour. On such a body vegetable life would be impossible; and we may infer from Carnot's principle that our beds of coal cannot have been produced by the internal heat of the earth alone, but must be due to solar radiation.

In order that the process of transpiration may be kept up, it is not sufficient to place a plant in an atmosphere unsaturated with vapour. It is also necessary to supply the plant with water in the liquid form.

It seems to have been long agreed that the principle of energy is as applicable to organized as to unorganized objects. Carnot's principle appears to be the only great physical law which at present is supposed to hold in the latter case but not in the former. If the foregoing reasoning is correct,[3] then Carnot's principle is true in both cases. It need not, however, be supposed that an

---

[1] This is the meaning assigned to the symbols by Prof. W. Gibbs.
[2] Gen. viii. 22.
[3] The substance of Art. 52 (but not of Art. 51) was given in a paper published in the "Proc. of the Camb. Phil. Soc.," Oct., 1892. When this paper was read, it was "received with a good deal of scepticism," the reason for which seems to have been that in discussing vegetable life, no use was made of the fact that a plant cannot grow in the *dark*.

animal is a mere automaton because the same laws of nature are true for its body that are true for other things. The intelligence or will of an animal has, to some extent, the power of choosing what the history of the animal shall be. The point has been well illustrated by the case of a locomotive engine-driver, who has the power of making a heavy train run backwards or forwards at will by merely turning a handle. The chief defect of the illustration appears to be that the engine-driver must do a *little* work on what he controls, while the intelligence does none.

Art. 53. The application of Carnot's principle to radiation is not discussed in this work. We shall merely give two or three simple results which are of special importance to us. For further information on the subject, reference must be made to Tait's "Sketch of Thermo-dynamics" and Balfour Stewart's "Elementary Treatise on Heat."

An opinion was expressed by Rankine that it would be possible, by concentrating radiant heat into foci by means of reflectors and lenses, to cause the radiation from one body A to raise another body B to a higher temperature than A. This will actually happen in certain cases. For example, let B consist of very inflammable materials and suppose them set on fire by concentrating on them the radiation from A. Then evidently the temperature of B may become higher than that of A. But, in general, Rankine's idea implies a contradiction of Carnot's principle. Thus let a system containing two bodies A, B, and reflecting surfaces and lenses, be at a uniform temperature θ in a state P; and suppose that without any assistance or interference from without, the system comes into another state P' which only differs from the state P in that the temperatures of B and A (θ_b, θ_a) are different from θ. Also let a positive quantity of heat Q have been absorbed by B and an equal positive quantity given out by A,[1] so that

---

[1] The energy is supposed to be the same in the state P' as in the state P.

θ_b is greater than θ_a. Then it is easy to show that Carnot's principle will have been contradicted if it is possible to bring B and A to their original states by simply imparting or abstracting heat. For it would be necessary to abstract a positive quantity of heat Q from B and to impart an equal positive quantity of heat to A. The average temperature of B during this operation is higher than θ and may be written $\theta + x$ where x is positive; and in like manner, the average temperature of A is $\theta - y$ where y is positive. Hence since no heat is gained or lost by A and B during the change of state from P to P' except in exchange with one another, the value of $\Sigma\left(\dfrac{Q}{\theta}\right)$ for the cycle is $-\dfrac{Q}{\theta+x}+\dfrac{Q}{\theta-y}$, or $Q\left(\dfrac{1}{\theta-y}-\dfrac{1}{\theta+x}\right)$, or $Q\dfrac{x+y}{(\theta-y)(\theta+x)}$, which is positive, contrary to Carnot's principle.

Rankine's idea has been carefully considered by Kirchhoff and Clausius, and their conclusion is that Carnot's principle is true whatever concentration of rays there may be: in other words, that when the only effect of heat on a body is such that the cycle can be completed by heat alone,[1] it is impossible, by means of reflectors and lenses but without the help of work, to cause the heat radiated from a hot body of temperature θ to raise the temperature of a cooler body to θ.

A simple illustration may be given. Suppose that rays diverging from a point P on the principle axis of a lens, are refracted to a point Q. Then if small areas be drawn at P and Q perpendicular to the line PQ, rays diverging from a point p of one area will be refracted to a point q of the other, such that pq passes through the

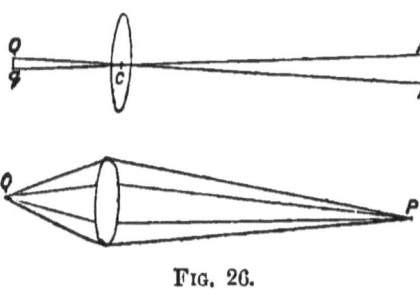

Fig. 26.

[1] As when the only effect of heat is a change of temperature.

centre C of the lens. If CP be much greater than CQ, the area at P will be much greater than the area at Q. It may therefore be thought that as radiation from a large area at P is concentrated on a small one at Q, the temperature at Q should rise above the temperature at P. It must, however, be noticed that the large area at P is very disadvantageously situated for sending radiation to the small area at Q. For a point on the area at P does not send the whole of its radiation to the area at Q, but only that small fraction of its total radiation which falls on the lens. From this it can be shown that the disadvantageousness of the position of the area at P exactly neutralizes its size.

Art. 54. It is easily shown by Carnot's principle that radiation is propagated in the ether (free space) without loss. In the case of sound, we know that a wave propagated in the air is generally destroyed by friction in a few seconds. We therefore infer that the ether is frictionless. But though the ether is frictionless to its own motions, it does not follow that it exerts no resistance to the motion of a material body.

The idea of an ethereal resistance to motion was introduced by Olbers and Encke; and at one time the shortening of the periodic time of Encke's comet seems to have been generally regarded as a proof of the hypothesis. At present there seems to be a strong disposition to reject the hypothesis as insufficient.[1] Again, on turning to the chapter on Tidal Friction, it will be seen that a very simple explanation is discovered of the fact on which the hypothesis of an ethereal resistance chiefly rests—the shortening of the period of the comet. The explanation supplied by tidal friction may, however, only account for part of the observed fact, and the rest may be due to the resistance of the ether. In short, astronomical considerations do not decide the question, but merely enable us to assert that the resistance, if it exists at all, must be very small.

The Theory of Exchanges throws some light on the

[1] Newcomb's "Popular Astronomy."

question. By means of an assumption, Balfour Stewart has shown that when a body emitting radiation is in motion in a vacuum, there is a force tending to stop the body. As it is necessary in this argument that the body should be *emitting* radiation, it may be that the resistance to motion is a consequence of the emission of radiation and may disappear when radiation ceases to be emitted—that is, when the body is at the temperature of absolute zero.

Art. 55. The statements of the preceding article may be illustrated by making a few speculations.

It seems to be often supposed that the vibrating atoms of a hot body produce waves of radiation by exerting force on the ether. As a vibrating tuning-fork produces sound-waves in the air, there is some analogy between the two cases. Let it therefore be assumed that there is no force between a material atom and the ether except when they are in contact, and that when they are in contact, the action and reaction between them are equal and opposite. Then if we use the word "work" to denote "total work," and not that fraction of it usually called work (more strictly, "mechanical work"), it will follow that the quantities of "work" done by the material body and the ether on one another are equal and opposite. Hence the "work" done on the ether is equal to the kinetic energy lost (in consequence) by the material body. If we assume the kinetic energy thus lost by the body to be the heat it radiates into space, we deduce that when a body is radiating heat, the ether exerts forces on the atoms tending to stop their vibrations. In other words, the atoms vibrate in a resisting medium. Similarly, when a body is absorbing radiation, the ether tends to increase the vibrations of the atoms.[1]

If v be the velocity of an atom,[2] the distance it moves

---

[1] The writer has been credited with giving German ideas. He therefore finds it necessary to confess his entire ignorance of the German language. The whole of the present article is merely speculative.

[2] Here for simplicity we are treating an atom as a single particle.

per second is v. Hence if p be the force with which the ether resists its motion, the "work" done on the atom by the ether per second is $-pv$, or the kinetic energy lost by the atom to the ether per second is pv. If there be n such atoms, and if Q be the total quantity of radiation (in ergs) emitted per second, we have $Q = npv$. In this result, Q and v are known: np can therefore be found.

Now let a particle emitting radiation be at rest. Then the resistance of the ether on an atom will be opposite to the irregular motion of the atom and may be supposed to depend only on the irregular velocity. Hence if we choose any fixed direction, the resolved part of the resistance parallel to this direction will be positive in one part of the atom's path or orbit, negative in the other; and the average resistance parallel to the fixed direction may be taken to be zero. A similar result holding for every atom, the force in any direction on the whole body is zero.

If the body be moving with a *small* velocity parallel to the fixed direction, the average velocity of an atom will be rather greater when the atom is moving in this direction than when it is moving in the opposite direction. The resistance may therefore be supposed to be rather greater in the former case than the latter. On the whole, there will be an average resistance on the atom in a direction opposite to that of the body's motion. The whole body will therefore experience a resistance in the opposite direction to motion.

In general, let a body which is emitting radiation be moving with any velocity U parallel to any line Ox; and, for simplicity, suppose the irregular motions of an atom to be always in this direction or the opposite. When the irregular velocity parallel to Ox is v, let the resistance be $kv(U + v)^2$, where k is a constant. This expression vanishes when the total velocity $U + v$ is zero, also when $v = 0$, that is, when the irregular motion is wanting. Moreover, it changes sign with v, and is therefore always opposed to the irregular motion. Consider now two instants when the irregular velocity is $+ u$ and $- u$.

The resisting forces in the direction opposite to Ox are $ku(U + u)^2$ and $- ku(U - u)^2$, and their average is $\frac{1}{2} ku \{(U + u)^2 - (U - u)^2\}$, or $2kUu^2$. From this it is easy to see that the whole body experiences a resistance in the opposite direction to its motion, and that the resistance vanishes both when the velocity U of the body is zero, and also when the irregular motions are absent (or when the absolute temperature is zero).

In conclusion it may be remarked that it is generally supposed the ether explains "actions at a distance"— forces like gravitation and electric and magnetic actions, which cannot be accounted for by the usual pressures, tensions, etc. If the "explanation" be complete, there will be no "actions at a distance" in the ether itself, but merely pressures, etc., between its particles when in contact.

CHAPTER IV.

APPLICATIONS TO AN UNELECTRIFIED BODY AT REST.

ART. 56. In this chapter, Carnot's principle is used to investigate some of the properties of a single unelectrified body at rest whose temperature is uniform when in a state of equilibrium and whose size is such that the mutual gravitational attractions of its parts may be neglected. The condition that the temperature is to be uniform when in a state of equilibrium requires that heat should be free to pass in all parts of the interior of the body. It is therefore not permissible to protect one part of the body from thermal communication with the rest by means of non-conductors of heat.

To make this chapter independent of the rest of the work, some results obtained in Chaps. I. and III. will be briefly restated.

Let the body undergo any operation during which the state changes from any state P to any state P'. Denote by W the work done *on* the body during the operation, and by Q the heat *absorbed*[1] by it. Then W + Q depends only on the states P and P' and is the same however the change of state is effected. Suppose now that we choose a standard state O and let U be the sum of the quantities of work and heat required to bring the body from the state O by any "path" (or process) to the state P. Let $U + \Delta U$[2] be a similar sum for any path from O to P'. Imagine the body brought

[1] Heat evolved is to be reckoned negative heat absorbed.
[2] Here ΔU stands for "increase of U" and does not denote the product of a quantity Δ and the quantity U.

from O to P' by being first brought from O to P and then from P to P'. We get

$$U + \Delta U = U + (W + Q),$$

or

$$\Delta U = W + Q.$$

Thus ΔU depends only on the states P and P', and is the same whatever state O be chosen as the standard state.

The result $\Delta U = W + Q$ is known as the equation of energy.

If W' be the total quantity of work obtained from the body during the operation, work done on the body being reckoned negative work obtained from it, we have $W' = -W$, and $\Delta U = -W' + Q$.

Also if Q' be the total quantity of heat evolved, heat absorbed being considered negative heat evolved, then $Q' = -Q$, and $\Delta U = W - Q'$.

Next, when the body is subjected to a cyclical process, so that the final state is the same as the first, it is assumed as a fundamental axiom, that no work can, on the whole,[1] be obtained from the body during the cycle if the body is always at the same uniform temperature when gaining or losing heat.

Again, when the body undergoes a reversible operation, it must at every instant during the operation be in a state of equilibrium and therefore at a uniform temperature. If the reversible operation constitute a complete cycle, it has been shown by Thomson and Clausius that $\frac{Q_1}{\theta_1} + \frac{Q_2}{\theta_2} + \frac{Q_3}{\theta_3} + \ldots = 0$, where $(Q_1, Q_2, Q_3, \ldots)$ are the quantities of heat absorbed (heat evolved being reckoned negative heat absorbed) by the body when at the corresponding uniform (absolute) temperatures $(\theta_1, \theta_2, \theta_3, \ldots)$. Suppose now that P and P' are states of equilibrium, and let the value of $\frac{Q_1}{\theta_1} + \frac{Q_2}{\theta_2} + \frac{Q_3}{\theta_3} + \ldots$ for any

---

[1] That is, a positive quantity of work cannot be obtained from the cycle.

Applications to an Unelectrified Body at Rest. 141

operation, whether cyclical or not, be denoted by $\Sigma\left(\dfrac{Q}{\theta}\right)$, the path being indicated by a suffix, as in what follows. Then if the body pass from P to P' by a reversible path PAP' and return by a reversible path P'XP, we have for the cycle

$$\left(\Sigma\dfrac{Q}{\theta}\right)_{PAP'} + \left(\Sigma\dfrac{Q}{\theta}\right)_{P'XP} = 0.$$

If the body now pass from P to P' by a different reversible path, PBP', and then return to P by the same reversible path P'XP, as before, we have

$$\left(\Sigma\dfrac{Q}{\theta}\right)_{PBP'} + \left(\Sigma\dfrac{Q}{\theta}\right)_{P'XP} = 0.$$

Hence

$$\left(\Sigma\dfrac{Q}{\theta}\right)_{PAP'} = \left(\Sigma\dfrac{Q}{\theta}\right)_{PBP'},$$

or the value of $\Sigma\left(\dfrac{Q}{\theta}\right)$ is the same for all reversible paths leading from P to P'.

Clausius has put the last result into a simple form by means of a remarkable quantity he calls Entropy. Choosing any standard state of equilibrium O, he defines the Entropy in the state P to be the value of $\Sigma\left(\dfrac{Q}{\theta}\right)$ for any reversible path from O to P. Denote this quantity by the symbol ϕ, and let $\phi + \Delta\phi$ be a similar quantity for any reversible path from O to P'. Then by supposing the body brought in a reversible manner from O to P' by being first brought from O to P and then from P to P' along the path P A P', we get

$$\phi + \Delta\phi = \phi + \left(\Sigma\dfrac{Q}{\theta}\right)_{PAP'},$$

or

$$\left(\Sigma\dfrac{Q}{\theta}\right)_{PAP'} = \Delta\phi.$$

When the body undergoes an irreversible operation, the temperature will not generally be uniform during the process. If the operation constitute a cycle, we knew $\frac{Q_1}{\theta_1} + \frac{Q_2}{\theta_2} + \frac{Q_3}{\theta_3} + \ldots < 0$, where $(\theta_1\ \theta_2\ \theta_3 \ldots)$ are no longer the temperatures of the whole body but of those parts of the body where and when the quantities of heat $(Q_1, Q_2, Q_3, \ldots)$ are absorbed. Let the value of $\frac{Q_1}{\theta_1} + \frac{Q_2}{\theta_2} + \frac{Q_3}{\theta_3} + \ldots$ for any irreversible operation be denoted by $\Sigma\left(\frac{Q}{\theta}\right)$, and suppose the body to pass from the state of equilibrium P to the state of equilibrium P' by an irreversible path PaP', and then let it return to P by the reversible path P' × P. For this cycle we have

$$\left(\Sigma\frac{Q}{\theta}\right)_{PaP'} + \left(\Sigma\frac{Q}{\theta}\right)_{P'\times P} < 0.$$

But

$$\left(\Sigma\frac{Q}{\theta}\right)_{PAP'} + \left(\Sigma\frac{Q}{\theta}\right)_{P'\times P} = 0.$$

Hence for an irreversible process PaP',

$$\left(\Sigma\frac{Q}{\theta}\right)_{PaP'} < \left(\Sigma\frac{Q}{\theta}\right)_{PAP'} < \Delta\phi.$$

If during an operation, heat is absorbed at one temperature (θ) only, the expression $\frac{Q_1}{\theta_1} + \frac{Q_2}{\theta_2} + \frac{Q_3}{\theta_3} + \ldots$ takes the short form $\frac{Q}{\theta}$, where Q is the total quantity of heat absorbed by the body. Thus for a reversible process, we have

$$\frac{Q}{\theta} = \Delta\phi, \text{ or } Q = \theta \times \Delta\phi.$$

For an irreversible process during which the body passes from a state of equilibrium P to a state of equilibrium P', we have, if heat is absorbed at one temperature only,

$$\frac{Q}{\theta} < \Delta\phi, \text{ or } Q < \theta \times \Delta\phi.$$

In the simple case in which no heat is absorbed or evolved by the body, we have for any irreversible process, $\Delta \phi > 0$, so that the increase of entropy is positive. When the entropy has acquired its maximum value, no further irreversible process will be possible. Every action which takes place spontaneously in Nature being irreversible, it follows that the equilibrium of the body is stable when the entropy is a maximum.

When the body experiences a reversible cycle of changes during which heat is absorbed at two temperatures (θ_1, θ_2) only, we have $\dfrac{Q_1}{\theta_1} + \dfrac{Q_2}{\theta_2} = 0$. Hence $\dfrac{Q_1}{Q_2} = -\dfrac{\theta_1}{\theta_2}$, and therefore as θ_1 and θ_2 are both positive, Q_1 and Q_2 are of opposite signs. Suppose Q_1 positive and Q_2 negative; and denote the former by Q_a, the latter by $-Q_b$, so that Q_a and Q_b are both positive. Then $\dfrac{Q_a}{Q_b} = \dfrac{\theta_1}{\theta_2}$. Again, *let a positive quantity of work W be obtained from the cycle.* Then the equation of energy gives $Q_a - Q_b - W = 0$, or $W = Q_a - Q_b$. Thus Q_a is $> Q_b$, and therefore $\theta_1 > \theta_2$. The "efficiency" of the cycle is taken to be the ratio of the quantity of work (W) obtained from the cycle to the heat (Q_a) absorbed at the higher temperature (θ_1). Its value is

$$\frac{W}{Q_a} = \frac{Q_a - Q_b}{Q_a} = 1 - \frac{Q_b}{Q_a} = 1 - \frac{\theta_2}{\theta_1} = \frac{\theta_1 - \theta_2}{\theta_1}$$

$$= \frac{\text{difference of temperatures}}{\text{higher temperature}}.$$

The temperatures referred to above are those belonging to Thomson's absolute scale, which is practically the same as that of the air thermometer or of the mercury centigrade thermometer increased by 273. Strictly, it is proved before using any numbers to denote temperature, that $\dfrac{Q_a}{Q_b}$ is the same for all reversible cycles working between any two given temperatures. The two numbers

(θ_a, θ_b) are chosen for the temperatures such that $\dfrac{Q_a}{Q_b} = \dfrac{\theta_a}{\theta_b}$. This settles the ratio of the numbers: to fix the numbers completely we may either fix one of them or fix their difference. Sir W. Thomson takes the numbers belonging to the freezing and melting points to differ by 100. From this it is easy to see that the number belonging to any temperature is completely fixed. For if θ_c be the number belonging to any temperature C, the ratio $\dfrac{\theta_c}{\theta_a}$ can have but one value. Hence since θ_a is fixed, so is θ_c.

The definition $\dfrac{Q_a}{Q_b} = \dfrac{\theta_a}{\theta_b}$ may be written $\dfrac{Q_a}{\theta_a} = \dfrac{Q_b}{\theta_b}$, or $\dfrac{Q_1}{\theta_a} = -\dfrac{Q_2}{\theta_b}$, or $\dfrac{Q_1}{\theta_a} + \dfrac{Q_2}{\theta_b} = 0$.

The advantage of the absolute scale of temperature is that when temperatures are thus measured, the equation $\Sigma \dfrac{Q}{\theta} = 0$ for a reversible cycle, is strictly correct. If we had been content with the indications of the air thermometer, the equation $\Sigma \dfrac{Q}{\theta} = 0$ would only have been a near approximation. The disadvantages of the absolute scale are (1) that no thermometer is yet in use showing this scale, and (2) that it introduces a want of symmetry into the subject; the equation $\Sigma \dfrac{Q}{\theta} = 0$ being partly a definition when there are only two quantities θ, a proposition when there are more than two; whereas the equation $\Sigma \dfrac{Q}{\theta} = 0$ is a proposition in both cases when the temperatures are simply taken to be those indicated by the air thermometer.

Art. 57. We begin with the case in which the body is in equilibrium in two or more of the three states of aggregation, solid, liquid, and gaseous, at the same

Applications to an Unelectrified Body at Rest. 145

uniform temperature and pressure. For simplicity, each part of the body is supposed to be of uniform density.[1]

When a substance is in equilibrium in two states of aggregation in the same vessel, it appears from experiment that the pressure and temperature are independent of the proportions of the substance in the two states. The pressure and temperature may therefore remain the same while the proportion of the substance in the two states takes every possible value. To fix the ideas, let the substance be water, partly in the liquid and partly in the gaseous, state; and suppose it contained in a cylinder fitted with a smooth air-tight piston. Then if the piston be forced in, the temperature and pressure will tend to rise and will cease to be uniform. When the piston comes to rest, the temperature and pressure will quickly become uniform a second time. If their present values are different from what they were at first, they may clearly be made the same by abstracting a proper quantity of heat from the cylinder and then allowing the temperature and pressure again to become uniform. The forcing in of the piston and the abstraction of heat from the cylinder may obviously take place simultaneously; and if both processes are slow enough, the temperature and pressure may practically be kept uniform and constant. We may therefore suppose the substance changed from one state of aggregation to the other in a reversible manner without any change in the uniform temperature and pressure. The indicator diagram for such an operation is evidently a straight line.

When the liquid and the gaseous states are in equilibrium together, it is clear that there is a relation between the temperature and the pressure. In the case of the solid and liquid states, it is not easy, without the assistance of theory, to see any connection between the

---

[1] This will generally require that each part of the body should be small. For example, if a cylinder 100 metres high be filled with steam, the density of the steam will be greater at the bottom of the cylinder than at the top.

P. D.   L

temperature and pressure. In fact, the temperature of melting ice was supposed to be the same at all pressures until it was pointed out by Prof. J. Thomson, in 1849, that it follows from Carnot's principle than an increase of pressure must lower the melting point.

In considering the relation between the temperature of melting ice and the pressure, it will be shown first of all that there cannot be any number of pressures corresponding to the same temperature, and then that the greater the pressure, the lower the temperature. The same kind of reasoning applies to the melting of all other substances, and even to the evaporation of solids and liquids. In fact, it is applicable whenever there is a change from one of the three states of aggregation to another.

Let us suppose, if possible, that water can co-exist with ice at the same uniform temperature θ whatever the pressure may be. Then let the mixture be made to undergo the following cycle of reversible operations during which the temperature remains uniform and equal to θ.

(1) Compress the mixture slowly until the pressure rises from p_a to p_b, and suppose that water and ice both continue to be present. This is in accordance with the assumption that at a given temperature, the liquid and solid states can exist together at all pressures.

(2) Then by slowly abstracting heat, let a quantity of water be frozen at the constant temperature θ and the constant pressure p_b.

In this operation there will be an increase of volume, and the indicator diagram will be a straight line.

(3) Slowly reduce the pressure from p_b to p_a.

In each of the operations (1) and (3), we may suppose the proportion of water and ice to remain constant; heat being absorbed or evolved, if necessary. Hence there will be more ice at the end of operation (3) than at the beginning of the cycle. The cycle may therefore be completed as follows.

(4) Let heat be applied to the mixture until as much

Applications to an Unelectrified Body at Rest. 147

ice is melted as was formed in (2), and keep the temperature constantly equal to θ and the pressure to p_a during the operation. The mixture will then be in its original state.

As in the second operation, the indicator diagram will be a straight line; but there will be a decrease of volume.

On drawing an indicator diagram, it is seen that there is a gain of work during the cycle. But this is contrary to the fundamental axiom, because the body is at the same temperature throughout the cycle. We therefore conclude that the assumption that any number of pressures can have the same melting point is incorrect. Thus the temperatures corresponding to the two pressures (p_b, p_a) must be different.

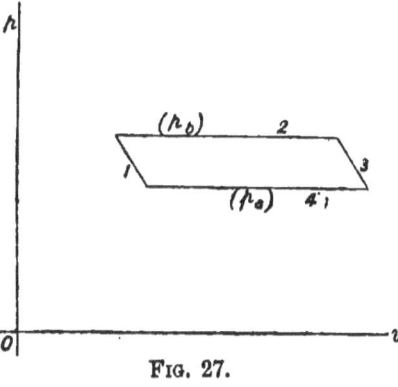

Fig. 27.

It is easy to ascertain which of the two temperatures (θ_b, θ_a) corresponding to the two pressures (p_b, p_a) is the greater. For let a mixture of ice and water undergo a cycle of reversible operations during which the two states of aggregation are always present, and in which heat is absorbed and evolved at two temperatures only. The cycle already described does not satisfy this condition, because we have not imposed the restriction that there is to be no heat gained or lost in operations (1) and (3). The cycle must therefore be the following.

(a) Compress the mixture slowly without allowing it to gain or lose heat until the pressure rises from p_a to p_b, and the temperature changes from θ_a to θ_b.

(β) Then by slowly abstracting heat, let a quantity of the water be frozen at constant temperature θ_b and constant pressure p_b.

In this operation there will be an increase of volume.

(γ) Slowly reduce the pressure from p_b to p_a without gain or loss of heat. The temperature will consequently change from θ_b to θ_a.

The proportions of water and ice are not to be supposed constant in operations (α) and (γ); but we may evidently cause so much water to be frozen in operation (β) that there is more ice in the mixture at the end of operation (γ) than at the beginning of the cycle. The cycle may then be completed as follows.

(δ) Let heat be imparted to the mixture so as to melt some of the ice at the constant temperature θ_a and constant pressure p_a.

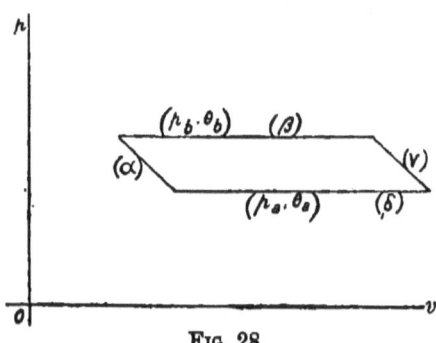
Fig. 28.

In the last operation there is a decrease of volume. Hence in the cycle a positive quantity of work is obtained. Heat being absorbed and evolved at two temperatures only, we know from Carnot's principle that a positive quantity of heat is absorbed at the higher temperature and a positive quantity evolved at the lower. Hence the temperature θ_a corresponding to the lower pressure is higher than the temperature θ_b corresponding to the greater pressure. In other words, an increase of pressure lowers the melting point of ice.

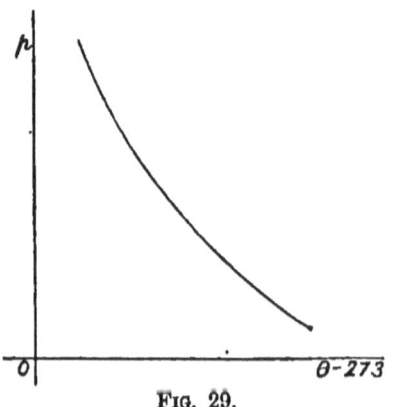
Fig. 29.

If we take two rectangular axes as axes of pressure and temperature, the relation between the pressure and temperature in the case of melting ice may evidently be represented by a curve whose inclination is shown in an exaggerated way in the accompanying diagram.

The conclusion that the melting point of ice is lowered by an increase of pressure was verified by W. Thomson in 1850—the year after the deduction was made.

In the case of substances which expand during fusion, like wax, sulphur, and most kinds of stone, the melting point can be shown to depend on the pressure by reasoning similar to that employed in the case of ice. If, however, heat be abstracted in the 2nd operation and imparted in the 4th, a positive quantity of work will be lost during the cycle. To contradict the fundamental axiom by obtaining a positive quantity of work from a cycle performed at constant temperature, it is necessary that heat should be imparted in the 2nd operation and abstracted in the 4th.

Again, it can be shown that in these substances, wax, sulphur, etc., an increase of pressure raises the melting point. If therefore we take two rectangular axes of pressure and temperature, the relation between the pressure and temperature of the melting point may be represented by a curve inclined as in the diagram.

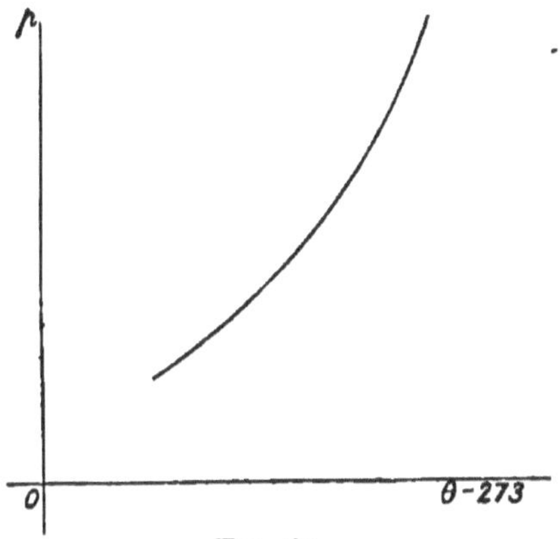

The theoretical result that the melting point is raised by pressure has been experimentally verified by Bunsen and Hopkins for spermacetti, paraffin, wax, sulphur, and stearine.

Fig. 30.

The conclusions respecting the effect of pressure on the melting point have some important applications.[1]

---

[1] These applications are taken from Maxwell's "Theory of Heat" and Tait's "Sketch of Thermo-dynamics."

(1) The temperature of the earth increases as we descend, and at a depth which is moderate compared with the distance to the centre of the earth, the temperature must be so high that all substances with which we are acquainted would be melted if raised to this temperature and subjected to no greater pressure than that of the atmosphere. It does not, however, follow that the interior of the earth is in the liquid state, because at depths at which the temperature is high, the enormous pressure may be more than sufficient to prevent liquefaction.

(2) When two pieces of ice are squeezed together, the pressure causes melting to take place at the portions of the surfaces in contact. The small quantity of water so formed is at a temperature below $0°$ C., the ordinary freezing point. Hence when the pressure is removed, the two pieces of ice are frozen together. This phenomenon is known as Regelation.

(3) The conclusion that the melting point of ice is lowered by pressure "has been successfully applied by the Thomsons and Helmholtz to explain the extraordinary plasticity of glacier ice." It was noticed by Forbes in 1842 that a glacier descends along its bed with a motion similar to that of a very viscous fluid, such as tar. His observations were summed up in the statement: "A glacier is an imperfect fluid, or a viscous body, which is urged down slopes of a certain inclination by the mutual pressure of its parts."

The explanation is as follows. A glacier is a very porous mass of ice, and the pressure therefore varies greatly from place to place. The temperature being always about $0°$ C., the ice will be continually melting at the points where the pressure is most intense. The form of the glacier will consequently change without the necessity of rupture.

Passing on to other changes of aggregation, the same kind of reasoning that has shown the melting point of a solid to depend on the pressure, also shows us that the temperature at which a solid or liquid

Applications to an Unelectrified Body at Rest. 151

substance is evaporated, depends on the pressure. The evaporation of a solid or liquid resembles the melting of wax or sulphur in being attended by an increase of volume. Hence it is easy to see that an increase of pressure raises the temperature of evaporation.

Taking two rectangular axes as axes of pressure and temperature, the relation between p and θ corresponding to a change of aggregation, will be represented by a curved line. That which refers to the change from the solid to the liquid state is called by Prof. J. Thomson, the "ice line"; that which refers to the change from the liquid to the gaseous state, the "steam line"; and that which refers to the change from the solid to the gaseous state, the "hoar-frost line."

An important observation respecting the three thermal lines has been made by Prof. J. Thomson. Suppose a quantity of any substance contained in a state of vapour in a closed vessel. Then on reducing the temperature sufficiently, part of the vapour will be condensed to the liquid state. On still further reducing the temperature, the liquid will begin to freeze. The three states of aggregation are then in equilibrium together in the same vessel. The three thermal lines therefore meet in a point which Prof. J. Thomson calls the Triple Point.

The three thermal lines and the triple point, for water, are shown in the accompanying diagram.

The ice line meets the axis Op in a point P corresponding to a pressure of one atmo and the temperature 0° C. A large change of pressure being required to produce a small change in the temperature at which ice

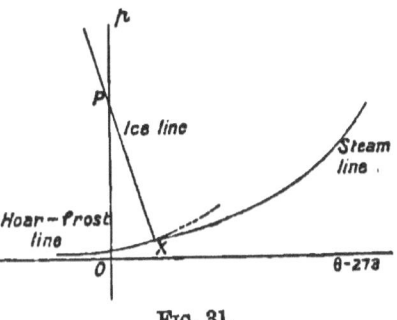

Fig. 31.

melts, it follows that when we travel along the ice line to the right of P until the pressure is very little above zero, the corresponding temperature will have risen but slightly above 0° C. Hence at the triple point X, at which ice, water, and steam co-exist, the pressure will be nearly zero, and the temperature only just above 0° C. If a tangent (shown in the figure by the dotted line) be drawn at X to the hoar-frost line, it will be shown further on that this tangent lies between the ice and steam lines.

Art. 58. The transformation from the liquid to the gaseous state has been most studied experimentally. The phenomena it presents appear to be the same for all substances, merely differing in degree. As an example, we will describe the isothermals for steam and water.

Suppose a gramme of water contained in a cylinder fitted with an air-tight piston. When the interior of the cylinder is large enough, the whole of the water is found to be in the form of steam satisfying approximately the relation $pv = R\theta$, where R is equal to 4,752,000. On forcing in the piston but keeping the temperature constantly equal to θ, the product pv at first is constantly equal to $R\theta$, but at length begins to diminish and continues to diminish, until the volume is so far diminished that the steam would be in equilibrium with water at the same temperature. The smallest decrease in the volume of the steam is then sufficient to cause some of it to condense into water, and by carefully decreasing the volume, all the steam may be converted into water without any perceptible variation in either the pressure or temperature.

When the steam is in equilibrium with water, it is said to be "saturated." Before it arrives at that state, it is said to be "superheated."[1]

Let v be the volume of one gramme of superheated steam far enough removed from the saturated state, and let v' be the volume of one gramme of air at the same temperature and pressure. Then we have

$$\left. \begin{array}{l} pv = \theta \times 4{,}752{,}000 \\ pv' = \theta \times 2{,}871{,}000 \end{array} \right\}.$$

[1] That is, "superheated" steam is steam at the same temperature as "saturated" steam, but at a larger volume (or smaller density).

Hence
$$\frac{v}{v'} = \frac{4{,}752}{2{,}871} = \tfrac{8}{5}, \text{ nearly.}$$

The density of highly superheated steam is therefore about $\tfrac{5}{8}$ of that of air at the same temperature and pressure.

On the indicator diagram, the isothermal curve for steam is a curved line until the volume is so far diminished that the steam is saturated. At this point the isothermal suddenly changes into a horizontal line and continues such until the whole of the steam is condensed into water.

The accompanying figure exhibits several isothermals for different temperatures. The dotted line B B' B" . . . represents the pressures and volumes of one gramme of saturated steam at different temperatures, and is therefore called the "steam line." The term "steam line" has been already used in a slightly different sense as the name of the curve on the diagram of pressure and temperature which denotes the relation between the pressure and temperature of saturated steam.

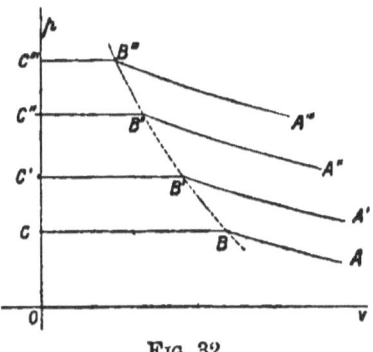

Fig. 32.

Rankine finds that the following empirical formulæ are approximately satisfied by saturated steam:—

$$\left. \begin{array}{l} pv^{\tfrac{17}{16}} = \text{constant} \\ p = (\theta - 233)^5 \times \text{constant} \end{array} \right\}.$$

In mathematical language, these are the "equations" of the steam line on the two diagrams—the indicator diagram and the diagram of pressure and temperature.

The curved part of an isothermal on the indicator diagram represents the state of the substance when it is wholly gaseous: the horizontal part when it is partly gaseous and partly liquid. When the substance is wholly liquid, its state is practically represented by the axis Op, the volume of a given quantity of water at any ordinary temperature being negligible in comparison with the volume of the same quantity of saturated steam. The axis of p is therefore the "water line."

The volume of a given quantity of saturated steam diminishes as the temperature increases. Hence on the indicator diagram, the steam and water lines approach one another. The question whether they meet has been investigated experimentally by Cagniard de la Tour and Dr. Andrews,[1] who find that there is a certain temperature, known as the "critical point" of the substance, above which the distinctions between the gaseous and liquid states disappear.

Art. 59. When a liquid is converted into vapour by an operation during which the temperature and pressure remain constant, there is a positive quantity of heat absorbed during the process. If the mass of liquid thus evaporated be a gramme, the heat absorbed is known as the Latent Heat of evaporation of the substance; and hence it is obvious that when a gramme of vapour is condensed at constant temperature and pressure, the heat evolved is equal to the latent heat of evaporation.

Similarly, when a gramme of a solid substance is melted at constant temperature and pressure, a positive quantity of heat called the Latent Heat of fusion is absorbed, and in the converse operation, an equal quantity of heat is evolved.

In what follows, the value of the latent heat in ergs is denoted by the symbol L.

In establishing the mathematical formulæ relating to latent heat, it is best to begin with the case of the evaporation of a liquid; but the very same methods and formulæ will be seen to be applicable when a solid is melted or evaporated.

The quantity L can only depend on the pressure

[1] See Maxwell's "Theory of Heat."

and temperature. Now we know that there is a relation between the pressure and temperature of saturated vapour. We may therefore take L to depend on the pressure alone, or on the temperature alone.

Let (p, θ) be simultaneous values of the pressure and temperature; and suppose that the pressure is p_1 when the temperature is $\theta + \tau$, where τ is a positive quantity. Then if the rate of increase of the pressure with respect to the temperature be the same at all temperatures, $p_1 - p$ will simply be τ times this rate of increase. If, however, the rate of increase is not the same at all temperatures, $p_1 - p$ will lie between τ times the greatest rate of increase and τ times the least rate of increase that the pressure has for temperatures between θ and $\theta + \tau$. If τ becomes smaller and smaller, the greatest and least rates of increase of the pressure while the temperature varies from θ to $\theta + \tau$, will approximate to equality with one another, and therefore also with the rate of increase when the temperature is θ. Denoting the rate of increase when the temperature is θ by p_θ, we see that when τ is small enough, $p_1 - p$ may, without sensible error, be written $\tau \times p_\theta$.

The result $p_1 - p = \tau p_\theta$ shows how we can find p_θ, the rate at which the pressure is increasing with respect to the temperature when the temperature is θ. We have merely to take any *small* increase of temperature and divide the corresponding increase of pressure by it.

Similarly, if p_0 be the pressure when the temperature is $\theta - \tau$, $p - p_0$ lies between τ times the greatest rate of increase and τ times the least rate of increase that the pressure has when the temperature lies between θ and $\theta - \tau$. But if τ be small enough, the greatest and least rates of increase will approximate to equality with one another, and therefore with p_θ. Hence when τ is sufficiently small, $p - p_0 = \tau p_\theta$, or $p_0 - p = - \tau p_\theta$.

Thus
$$\left. \begin{array}{l} p_1 = p + \tau p_\theta \\ p_0 = p - \tau p_\theta \end{array} \right\}.$$

These two results may be written as one. Let τ' be any

small increase, positive or negative, of θ; and let p' be the pressure corresponding to the temperature $\theta + \tau'$: then

$$p' = p + \tau' p_\theta,$$

or

$$p' - p = \tau' p_\theta.$$

Now let us suppose a mixture of water and steam at the pressure p and temperature θ contained in a cylinder fitted with a smooth air-tight piston; and let it be made to undergo the following cycle of reversible operations during which water and steam are always present and in which heat is only gained and lost at two temperatures.

(1) By slowly drawing out the piston and imparting heat, let a gramme of water be evaporated at the constant temperature θ and constant pressure p.

The amount of heat, measured in ergs, absorbed during the operation will be L; and the indicator diagram for the operation will be a straight line.

(2) Then let the piston be slowly drawn further out without allowing the contents of the cylinder to gain or lose heat until the pressure falls from p to p_0 and the temperature from θ to $\theta - \tau$, where τ is any small positive quantity.

The corresponding indicator diagram will be the short curve BC.

In this operation, the steam will evidently continue saturated;[1] but it cannot be assumed that it remains constant in quantity. However, it is not important for us to know at present whether it does so or not.

(3) Next, let the piston be slowly forced in and keep the temperature constantly equal to $\theta - \tau$, so that some of the vapour is condensed at constant pressure and temperature into water. This process (represented on the diagram by the horizontal line CD) is to be continued so far that an adiabatic compression (DA) then brings the mixture into its original state.

[1] Because water and steam are always present.

During this cycle a positive quantity of work (W, say) is obtained, represented by the area of the figure ABCD. But since τ is very small, the short lines AD, BC may be considered straight, and CD may be supposed equal to AB. The figure ABCD is therefore practically a parallelogram; and if the dotted lines Aa, Bb be drawn from A and B at right angles to CD, its area will be equal to that of the rectangle Aa bB, or to Aa × AB. Now Aa is proportional to $p - p_0$, or to τp_θ. Again, if s be the volume of one gramme of saturated steam at the temperature θ and σ the volume of one gramme of water at the same temperature and pressure, AB is proportional to $s - \sigma$. Hence $W = (s - \sigma)\tau p$.

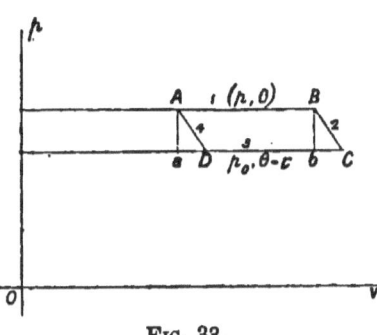

FIG. 33.

Now when a positive quantity of work is obtained from a reversible cycle during which heat is absorbed and evolved at two temperatures only, the ratio of the work obtained to the heat absorbed at the higher temperature is the "efficiency," which is the ratio of the difference of the two temperatures to the higher temperature. In our case therefore

$$\frac{W}{L} = \frac{\tau}{\theta}.$$

Hence

$$\frac{(s - \sigma)\tau p_\theta}{L} = \frac{\tau}{\theta},$$

or

$$L = (s - \sigma)\theta p_\theta. \quad . \quad . \quad . \quad (13).$$

It is usual to put u for $s - \sigma$, so that the important equation (13) takes the form

$$L = u\theta p_\theta \quad . \quad . \quad . \quad . \quad . \quad . \quad . \quad (13)'.$$

Art. 60. Regnault has determined experimentally the latent heat and the pressure of saturated steam at a great number of temperatures as measured on the ordinary scale. If we suppose the absolute temperature to be the same as the centigrade temperature increased by 273, the value of θ will be known and p_θ may be found from Regnault's experiments. Substituting these known values of L, θ, and p_θ in (13)', Clausius and Rankine have calculated the values of u.

The following tables give some results calculated in this way for saturated steam.

In the first table, the C.G.S. absolute units are used exclusively. The values of p are obtained from Clausius; and the values of $\dfrac{ps}{\theta}$ are given to enable us to see the deviation of saturated steam from the laws of a perfect gas for which $\dfrac{ps}{\theta}$ is constant. The value of s is equal to that of u increased by unity: it is the ratio of the volume of a quantity of saturated steam to that of an equal quantity of water.

In the second table the units employed are various. It is sufficient to state here that the column headed Latent Heat gives the latent heat of a gramme of steam in calories or the latent heat of a pound of steam in terms of the heat required to raise 1 lb. of water from 0° C. to 1° C.

Art. 61. The theory will next be used to determine whether a saturated vapour is condensed into the liquid form when compressed or expanded adiabatically.

For this purpose the equation $L = u\theta p_\theta$ is not sufficient. An additional result must be obtained by means of the principle of energy. Now the principle of energy cannot be applied to the cycle in Art. 60. One of the adiabatic operations BC, DA is not the reverse of the other. During the operation DA, a considerably larger proportion of the mixture is in the liquid form than during the operation BC. We cannot therefore

TABLE I.

$\theta-273$	p	θp_θ	Latent Heat in ergs.	u	$\dfrac{ps}{\theta}$
0	6,134·40	120,021	25193,860000	209,910	4,717,000
5	8,713·52	166,696	25048,660000	150,270	4,710,000
10	12,222·1	229,373	24905,070000	108,580	4,689,000
15	16,934·9	312,054	24760,580000	79,347	4,667,000
20	23,192·0	420,073	24616,030000	58,599	4,638,000
25	31,405·5	557,431	24471,400000	43,900	4,627,000
30	42,071·3	732,128	24326,680000	33,227	4,614,000
35	55,779·1	952,166	24181,860000	25,397	4,600,000
40	73,220·8	1,225,550	24032,780000	19,610	4,588,000
45	95,203·3	1,561,610	23891,890000	15,300	4,581,000
50	122,661	1,973,680	23746,720000	12,032	4,570,000
55	156,661	2,471,100	23601,420000	9,551·4	4,562,000
60	198,416	3,068,540	23455,970000	7,644·0	4,555,000
65	249,294	3,781,990	23310,370000	6,163·5	4,547,000
70	310,830	4,626,140	23164,610000	5,007·1	4,538,000
75	384,734	5,618,310	23018,670000	4,097·1	4,533,000
80	472,904	6,777,180	22872,560000	3,374·9	4,521,000
85	577,437	8,121,400	22726,250000	2,798·3	4,515,000
90	700,645	9,684 360	22579,740000	2,331·6	4,502,000
95	845,070	11,479,300	22433,020000	1,954·2	4,486,000
100	1,013,510	13,530,400	22286,080000	1,647·1	4,478,000
105	1,208,760	15,853,400	22138,910000	1,396·5	4,469,000
110	1,434,080	18,499,200	21991,510000	1,188·8	4,455,000
115	1,692,840	21,473,100	21843,860000	1,017·3	4,443,000
120	1,988,720	24,816,300	21695,950000	874·3	4,430,000
125	2,325,580	28,551,700	21547,780000	754·7	4,416,000
130	2,707,510	32,716,400	21399,330000	654·1	4,401,000
135	3,138,850	37,335,900	21250,590000	569·1	4,387,000
140	3,624,140	42,444,700	21101,570000	497·1	4,372,000
145	4,168,130	48,072,400	20952,200000	435·8	4,358,000
150	4,775,810	54,249,500	20802,590000	383·5	4,343,000
155	5,452,370	61,020,000	20652,630000	338·5	4,330,000
160	6,203,240	68,402,600	20502,330000	299·7	4,310,000
165	7,033,950	76,436,000	20351,690000	266·3	4,300,000
170	7,950,270	85,156,200	20200,410000	237·2	4,280,000
175	8,958,140	94,595,200	20049,360000	211·9	4,260,000
180	10,063,600	104,766,000	19897,650030	189·9	4,250,000
185	11,272,900	115,725,000	19745,550000	170·6	4,230,000
190	12,592,500	127,496,000	19593,070000	153·7	4,220,000
195	14,028,600	140,094,000	19440,190000	138·8	4,200,000
200	15,588,000	153,565,000	19286,910000	125·6	4,180,000

TABLE II.

Temp. (Centigr.)	Pressure in millimetres of mercury.	Pressure in grammes per square centimetre.	Pressure in atmos.	Pressure in pounds (at London) per square inch.	Latent Heat of a gramme in calories.	Volume of a pound of saturated steam in cubic feet.
0°	4·600	6·254	·006	·089	606·5	3363
5°	6·534	8·883	·0086	·126	603·0	2407
10°	9·165	12·460	·0120	·177	599·5	1739
15°	12·699	17·265	·0168	·246	596·0	1271
20°	17·391	23·644	·0229	·336	592·6	939
25°	23·550	32·018	·0309	·455	589·1	703
30°	31·548	42·892	·0415	·610	585·6	532
35°	41·827	56·867	·0550	·809	582·1	407
40°	54·906	74·649	·0722	1·062	578·6	314
45°	71·390	97·060	·0939	1·380	575·1	245
50°	91·980	125·054	·1210	1·779	571·6	193
55°	117·475	159·716	·1545	2·27	568·1	153
60°	148·786	202·262	·1957	2·88	564·6	122
65°	186·938	254·247	·246	3·61	561·1	98·7
70°	233·082	316·893	·306	4·51	557·6	80·2
75°	288·500	392·238	·379	5·58	554·1	65·7
80°	354·616	482·128	·467	6·86	550·6	54·1
85°	433·002	588·700	·570	8·37	547·1	44·8
90°	525·392	714·311	·704	10·16	543·6	37·4
95°	633·692	861·553	·833	12·25	540·1	31·3
100°	760	1,033·279	1	14·697	536·5	26·4
105°	906·41	1,232·33	1·19	17·53	533	22·4
110°	1,075·37	1,462·05	1·41	20·80	529	19·1
115°	1,269·41	1,725·86	1·67	24·55	526	16·3
120°	1,491·28	2,027·51	1·96	28·84	522	14·2
125°	1,743·88	2,370·94	2·29	33·72	519	12·1
130°	2,030·28	2,760·32	2·67	39·26	515	10·5
135°	2,353·73	3,200·08	3·09	45·5	512	9·1
140°	2,717·63	3,694·83	3·57	52·6	508	8·0
145°	3,125·55	4,249·43	4·11	60·4	504	7·0
150°	3,581·23	4,868·97	4·71	69·3	501	6·2
155°	4,088·56	5,558·71	5·38	79·1	498	5·4
160°	4,651·62	6,324·24	6·12	90·0	494	4·8
165°	5,275·54	7,171·15	6·94	102·0	490	4·3
170°	5,961·66	8,105·34	7·84	115·3	486	3·8
175°	6,717·43	9,132·87	8·84	129·9	483	3·4
180°	7,546·39	10,259·9	9·93	145·9	479	3·1
185°	8,453·23	11,492·8	11·12	163·5	475	2·7
190°	9,442·70	12,837·1	12·42	182·6	472	2·5
195°	10,519·63	14,302·2	13·84	203·4	468	2·2
200°	11,688·96	15,892·0	15·38	226·0	464	2·0

Applications to an Unelectrified Body at Rest.

assume as much vapour to be formed in one of the two adiabatic operations as is condensed in the other. Thus we cannot find how much heat is absorbed during the cycle; for although we know that the heat absorbed during the operation AB is L, and that heat is neither absorbed nor evolved during BC and DA, we have no means of determining the amount of heat evolved (or of vapour condensed) during the operation CD.

It was permissible in finding the area of the figure ABCD, to take the length of the line CD equal to the length of the line AB; but we cannot assume the heat evolved in CD equal to the heat (L) absorbed in AB. If we did make this assumption, we should find the total quantity of heat absorbed during the cycle to be zero; and therefore, by the principle of energy, the work obtained from the cycle would also be zero, which is plainly absurd.

The difficulty is avoided by causing the mixture of liquid and saturated vapour contained in the cylinder to undergo the following reversible cycle in which there is no adiabatic compression or expansion, but in which both liquid and vapour are always present.

(1) Let the piston be slowly drawn out until a gramme of liquid is evaporated at the constant temperature θ and constant pressure p.

The heat absorbed in the process will be L.

(2) Let the piston be drawn further out until the temperature falls from θ to $\theta - \tau$, where τ is a small positive quantity, *just enough heat being imparted to the contents of the cylinder or abstracted from them to prevent the liquid from evaporating and the vapour from condensing.*

To express the heat absorbed in this operation, C' is used to denote the quantity of heat (positive or negative) that must be imparted to one gramme of saturated vapour to keep it from condensing or becoming superheated when it is slowly compressed without further gain or loss of heat until its temperature rises 1°. If C' is positive, it is necessary for the vapour to absorb heat to continue saturated during compression; and therefore during an adiabatic compression, the vapour begins to condense. If C' is negative, the vapour must evolve heat to continue saturated during compression, and therefore an adiabatic compression superheats it. Again, when the vapour expands slowly without ceasing to be saturated, C' is the heat evolved while the temperature falls 1°.

The quantity C' is called the "specific heat of the saturated vapour." The specific heat of the liquid under the same conditions of temperature and pressure is denoted by C. Since the properties of a liquid are nearly independent of the pressure (at least when the pressure is not very great), it follows that C is practically equal to C_p, the specific heat of the liquid at constant pressure.

Lastly, let us denote by H the thermal capacity under the same conditions of the contents of the cylinder when the proportions of liquid and vapour are the same as at the beginning of the cycle. Then the thermal capacity of the contents of the cylinder during the second operation is $H - C + C'$. Remembering that there is a fall of temperature τ during the operation, we see that the heat absorbed (reckoned algebraically is)

$$- (H - C + C') \tau.$$

(3) Next, let the piston be slowly pushed in until a gramme of vapour is condensed at the constant temperature $\theta - \tau$ and constant corresponding pressure.

The latent heat of the saturated vapour being L when the temperature is θ, the latent heat at the temperature $\theta - \tau$ will be $L - \tau L_\theta$. The heat absorbed during the present operation is therefore

$$- (L - \tau L_\theta).$$

(4) The proportion of liquid and vapour in the cylinder being now the same as at first, let the temperature and pressure be also made the same as at first.

The heat absorbed will be $H\tau$.

The total quantity of heat absorbed in the whole

Applications to an Unelectrified Body at Rest. 163

cycle is the sum of the quantities absorbed in the four operations of which it consists. It is therefore

$$\tau (L_\theta + C - C').$$

If now W be the work obtained from the cycle, the principle of energy gives

$$\tau (L_\theta + C - C') - W = 0.$$

To find W, draw the indicator diagram *abcd* representing the cycle of operations. Then W is proportional to the area of *abcd*. This figure is practically a parallelogram, and its area is therefore $ab \times PQ$, which is proportional to $(s - \sigma) \times \tau p_\theta$, or $u\tau p_\theta$. Hence $W = u\tau p_\theta$. Remembering that $L = u\theta p_\theta$, we obtain $W = \dfrac{\tau}{\theta} L$.

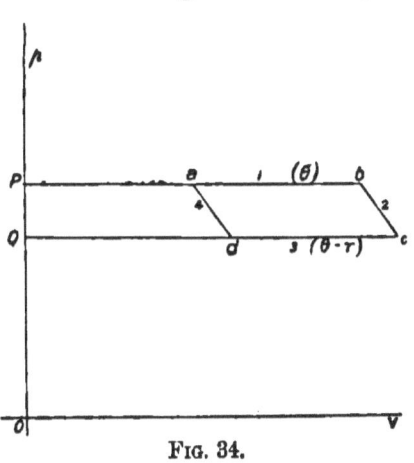

Fig. 34.

The equation of energy for the cycle is therefore

$$\tau\left(L_\theta + C - C' - \frac{L}{\theta}\right) = 0,$$

or

$$L_\theta + C - C' = \frac{L}{\theta} \quad \ldots \quad (14).$$

Art. 62. Before the principle of energy was established, it was supposed that the heat absorbed in any complete cycle was zero. In the cycle just described, the expression $L_\theta + C - C'$ was therefore taken to be zero.

The value of $L_\theta + C$ is to be obtained from experiment. The first attempt to find it seems to have been made by James Watt.[1] He discovered that the latent

---

[1] Browne's translation of Clausius on the "Mechanical Theory of Heat"

heat of steam diminishes as the temperature increases; and he supposed his experiments to show that the heat required to raise a given quantity of water from the freezing point to any temperature (θ), and then to convert it into steam at the constant temperature θ and the constant corresponding pressure, was independent of the temperature θ. This conclusion, known as Watt's law, was expressed by saying that "The sum of the free and latent heats is always constant."

Watt's law can easily be expressed in symbolical form. For if h be the heat required to raise one gramme of water at any ordinary pressure from 0° C. to the absolute temperature θ, the sum of the free and latent heats at the temperature θ is $h + L$. Again, the heat required to raise one gramme of water at any ordinary pressure from 0° C. to the absolute temperature $\theta + \tau$, where τ is very small, is $h + C\tau$; and the latent heat at the temperature $\theta + \tau$ is $L + \tau L_\theta$. Hence
$$h + L = (h + C\tau) + (L + \tau L_\theta),$$
or
$$\tau (C + L_\theta) = 0,$$
or
$$C + L_\theta = 0.$$

If we substitute this result in $L_\theta + C - C' = 0$, we obtain $C' = 0$. It was therefore supposed that when saturated steam was compressed or expanded in a vessel impermeable to heat, it neither began to condense nor became superheated.

In 1847, Regnault published his experiments from which it appeared that the sum of the free and latent heats of steam is not strictly constant, but increases with the temperature. If therefore the small quantity τ be positive, we have
$$(h + C\tau) + (L + \tau L_\theta) > h + L,$$
or
$$\tau (C + L_\theta) > 0,$$
and therefore since τ is positive,
$$L_\theta + C > 0.$$

Substituting in $L_\theta + C - C' = 0$, or $L_\theta + C = C'$, we get $C' > 0$—in words, C' is positive. From this it followed that when saturated steam was slowly compressed, it was necessary to impart heat to it to prevent condensation, and conversely, that when saturated steam was allowed to expand slowly, it was necessary to abstract heat from it to prevent superheating. Hence when saturated steam was compressed or expanded adiabatically, it was supposed to begin to condense in the first case and to become superheated in the second.

Art. 63. The experimental results obtained by Regnault for water and steam must now be substituted in equation (14).

If J be the number of ergs in a calorie (about 42 million) and θ' the temperature given by the ordinary centigrade thermometer (so that $\theta = \theta' + 273$), Regnault's results give

$$L_\theta + C = \cdot 305 \, J,$$

and

$$L = J\,(606 \cdot 5 - \cdot 695 \theta' - \cdot 000{,}02 \theta'^2 - \cdot 000{,}000{,}3 \theta'^3).$$

Hence if c' be the equivalent of C' in calories, so that $C' = Jc'$, we get

$$c' = \cdot 305 - \frac{606 \cdot 5 - \cdot 695 \theta' - \cdot 000{,}02 \theta'^2 - \cdot 000{,}000{,}3 \theta'^3}{\theta' + 273}.$$

From this formula the values of c' for saturated steam are easily calculated as in the table.

θ'	0	20	50	100	150	200
c'	$-1 \cdot 916$	$-1 \cdot 717$	$-1 \cdot 465$	$-1 \cdot 133$	$- \cdot 879$	$- \cdot 676$

The conclusion that the specific heat of saturated steam is negative, was arrived at by Clausius and Rankine, independently, early in 1850. The result has since been verified by the experiments of Hirn, and still later, by those of Cazin. They made use of a cylinder fitted at the ends with glass plates so that the behaviour of the vapour was visible to the eye. The steam was found to remain perfectly clear during compression, but to form a cloud during expansion. This showed that saturated steam condenses during adiabatic expansion, but not during adiabatic compression.

Hirn also made experiments with ether and bisulphide of carbon: Cazin with ether and chloroform. We will therefore calculate the values of the specific heats of the saturated vapours of these three liquids, making use of the experimental results obtained by Regnault.

Art. 64. For ether ($C_4H_{10}O$),

$$L_\theta + C = J(\cdot 45 - \cdot 001,111,12\theta'),$$
$$L = J(94 - \cdot 079\theta' - \cdot 000,851,43\theta'^2):$$

hence

$$c' = \frac{1}{J}\left(L_\theta + C - \frac{L}{\theta}\right)$$

$$= \cdot 45 - \cdot 001,111,12\theta' - \frac{94 - \cdot 079\theta' - \cdot 000,851,43\theta'^2}{\theta' + 273};$$

from which we find

θ'	0	50	100	150
c'	$+\cdot 105,7$	$+\cdot 122,2$	$+\cdot 130,9$	$+\cdot 134,4$

For Carbon Bisulphide (CS_2),

$$L_\theta + C = J(\cdot 146,01 - \cdot 000,824,6\theta'),$$
$$L = J(90 - \cdot 089,22\theta' - \cdot 000,493,8\theta'^2),$$

and therefore

$$c' = \cdot 146,01 - \cdot 000,824,6\theta' - \frac{90 - \cdot 089,22\theta' - \cdot 000,493,8\theta'^2}{\theta' + 273},$$

from which we find

θ'	0	50	100	150
c'	$-\cdot 183,7$	$-\cdot 160,01$	$-\cdot 140,6$	$-\cdot 132,5$

For Chloroform ($CHCl_3$),

$$L_\theta + C = \cdot 137,5 J,$$
$$L = J(67 - \cdot 094,85\theta' - \cdot 000,050,72\theta'^2),$$

from which

$$c' = \cdot 1375 - \frac{67 - \cdot 094,85\theta' - \cdot 000,050,72\theta'^2}{\theta' + 273},$$

and

θ'	0	50	100	150
c'	$-\cdot 107,9$	$-\cdot 054,9$	$-\cdot 015,3$	$+\cdot 015,5$

From this table we see that the specific heat of the saturated vapour of chloroform changes sign. The temperature at which it is zero is calculated by Cazin to be $123°\cdot 48$ C.

In the experiments on the saturated vapours of these three liquids, ether vapour formed a cloud during adiabatic compression, but remained perfectly clear during expansion; bisulphide of carbon, on the contrary, behaved like steam in remaining clear during compression and forming a cloud during expansion. In the case of chloroform, Cazin found by an apparatus that only allowed of expansion that clouds were formed during expansion up to 123° C., but that above 145° C. the vapour remained clear. Between 123° C. and 145° C., the result depended on the degree of expansion. With a small amount of expansion, there was no cloud; but with more expansion, a cloud appeared towards the end of the experiment. The explanation of this is simple. The expansion at first caused the vapour to become superheated; but when a considerable fall of temperature had taken place, it approached the saturated condition and finally began to condense. A second apparatus was afterwards used by Cazin which allowed of compression as well as expansion. It was found that up to 130° C., the saturated vapour formed a cloud during expansion but remained transparent during compression. Above 136° C., on the contrary, a cloud appeared during compression and none during expansion. The fact that the experiment showed the dividing temperature to be between 130° C. and 136° C. instead of between 123° C. and 124° C., is not a matter of great importance. Cazin states that the chloroform which he used was not chemically pure. In addition to this, some allowance must be made for the unavoidable defects of the experiment.

Art. 65. The important equation $L = u\theta p_\theta$, which has been obtained for evaporation, is also true for fusion.

In the case of water, the latent heat of fusion is 79·25 calories; and the volumes of one gramme of water and of one gramme of ice at 0° C. and atmospheric pressure, are respectively 1·000,116 and 1·087 cubic centimetres. Hence

$$p_\theta = \frac{L}{u\theta} = \frac{79\cdot 25 \times 41{,}540{,}000}{(-\cdot 086{,}884) \times 273}.$$

Now if τ' be the small (algebraic) rise of the temperature of fusion when the pressure rises from p to p', we have

$$p' - p = \tau' p_\theta,$$

or

$$\tau' = \frac{p' - p}{p_\theta}.$$

Denoting by P' and P the two pressures measured in atmos, we have

$$\left. \begin{array}{l} p' = 1{,}013{,}510 P'_1 \\ p = 1{,}013{,}510 P \end{array} \right\} :$$

hence

$$\tau' = \frac{1{,}013{,}510 \times (P' - P)}{p_\theta}$$

$$= -(P' - P) \times \frac{273 \times \cdot 086{,}884 \times 1{,}013{,}510}{79\cdot 25 \times 41{,}540{,}000}$$

$$= -(P' - P) \times \cdot 007{,}302$$

$$= -(P' - 1) \times \cdot 007{,}302,$$

since P denotes a pressure of one atmo.

Applications to an Unelectrified Body at Rest. 169

In this formula we have to put for P′ the new pressure in atmos. Thus P′ − 1 is the increase of pressure in atmos. The truth of the formula has been verified experimentally by Sir W. Thomson. In 1850, he placed a mixture of water and ice in an Oersted press which was fitted with an air-gauge to show the pressure. To measure the small differences of temperature correctly, he employed a thermometer filled with ether-sulphide and enclosed in a large glass tube to protect it from the pressure. On screwing down the press, the temperature was at once seen to fall, but returned to its original value when the additional pressure was removed. The table below gives the fall of temperature for two pressures, as shown in the experiment and as calculated by the formula.

Increase of Pressure.	Fall of temperature of melting point.	
	Observed.	Calculated.
8·1 atmos	·059° C.	·059° C.
16·8 ,,	·129° C.	·1227° C.

Art. 66. Bunsen was the first to make experiments on the behaviour of substances which expand during fusion. He took a thick glass tube, about the size of a straw. He drew one end out into a fine capillary tube A, 15 to 20 inches long, the other end into a somewhat larger tube C only 1½ inches long which he bent round as in the figure. The substance to be examined was placed in

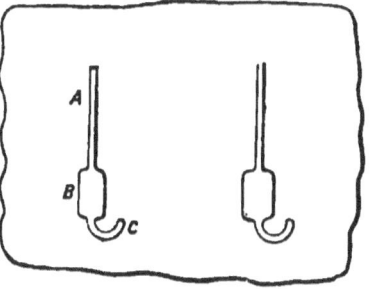

Fig. 35.

the bent part C, and the thick part B was filled with mercury. Both ends of the tube were sealed up and the tube fastened on a board. The temperature of the substance in C was varied at will by plunging the board into warm water of a known temperature. If the board was sunk so deep in the water that the part B was immersed, the expansion of the mercury compressed the substance in C and the air in A, and the pressure was measured by the volume of the air in the fine tube A. By sinking the board so that a greater or smaller part of B was immersed in the warm water, the pressure could be varied as was thought fit, and was capable of rising above 100 atmos.

Another tube, similar to the first except that the end A was left open, was fastened on the same board, in order to show the behaviour of the substance at atmospheric pressure and different temperatures.

In making an experiment, the board was plunged in water whose temperature was sufficient to melt the substance in C. Then on allowing the water to cool, the pressures were observed for which solidification took place at the successive known temperatures of the water.

Bunsen experimented with spermaceti and paraffin, and obtained the following results.

SPERMACETI.

Pressure.	Melting point.
1 atmo	$47°\cdot 7$ C.
29 atmos	$48°\cdot 3$ C.
96 ,,	$49°\cdot 7$ C.
141 ,,	$50°\cdot 5$ C.
156 ,,	$50°\cdot 9$ C.

PARAFFIN.

Pressure.	Melting point.
1 atmo	$46°\cdot 3$ C.
85 atmos	$48°\cdot 9$ C.
100 ,,	$49°\cdot 9$ C.

Applications to an Unelectrified Body at Rest. 171

More recently Hopkins experimented with spermaceti, wax, sulphur, and stearine, and showed that in all these substances, an increase of pressure raises the melting point.

The rise in the melting point cannot be calculated theoretically for these substances, because we do not possess sufficient experimental data to work with.

Art. 67. A last use of the equation $L = u\theta p'_\theta$ will be made in some calculations respecting the triple point of water.

At the triple point let L be the latent heat required to evaporate water and let L' be that required to convert ice into steam, both processes taking place at constant pressure and temperature. Then since ice may be evaporated at this particular point at constant pressure and temperature by first melting it and then evaporating the water, we evidently have $L' > L$.

The increase of volume is practically the same in the evaporation of ice as in the evaporation of water. Let u be this increase at the triple point, and let (p'_θ, p_θ) be the corresponding values of p_θ. Then if θ be the temperature of the triple point

$$\left. \begin{array}{l} L = u\theta p_\theta \\ L' = u\theta p'_\theta \end{array} \right\}$$

Hence $p'_\theta > p_\theta$

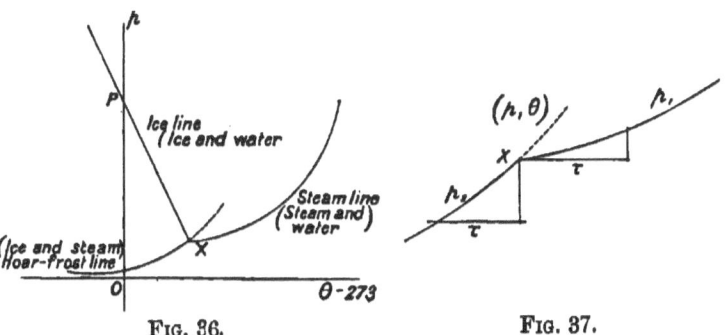

Fig. 36.   Fig. 37.

Now let X be the triple point. Take a point on the steam line whose temperature is higher than that of X by a small positive quantity τ; also take a point on the hoar-frost line whose temperature is lower than that of X by the same quantity τ. Let (p_1, p_2) be the pressures at these points. Then p being the pressure at X, we have

$$\left. \begin{array}{c} p_1 - p = \tau p_\theta \\ p - p_2 = \tau p'_\theta \end{array} \right\}.$$

Hence since p'_θ is greater than p_θ, it follows that $p - p_2 > p_1 - p$. In words, at the triple point X the hoar-frost line is more steeply inclined to the axis of temperature than the steam line, as shown in the figure by the dotted tangent at X to the hoar-frost line.

The temperature and pressure at the triple point can easily be calculated. Under a pressure of one atmo, the melting point of ice is 0° C. As the pressure diminishes, the melting point rises; but when the pressure has fallen nearly to zero, the melting point will only be just above 0° C. Now when the pressure is small and the temperature just above 0° C., steam can co-exist with water. Hence the temperature at the triple point is just above 0° C. and the pressure very small.

The equation which gives the effect of pressure on the ice line (the melting line of ice), is

$$\tau' = - (P'-1) \times \cdot 007{,}302,$$

where τ' is the (algebraic) rise in the melting point when the pressure is increased to P' atmos. Now at the point where the ice line meets the axis of p, the temperature is 0° C., and at the triple point the pressure is very small. Putting P' = 0, we get $\tau' = \cdot 007$. This therefore is the (centigrade) temperature of the triple point. Then by referring to the table of steam pressures, it will be seen that the pressure at the triple point is about the 1,500th of an atmo.

Art. 68. In the present article, two simple examples will be given of the application of thermo-dynamics to capillary phenomena.

If a fine glass tube be plunged into water, the liquid will be seen to rise in the tube; but if the tube be

Applications to an Unelectrified Body at Rest. 173

plunged into mercury, the mercury will stand at a lower level inside the tube than elsewhere.

It will be shown, in the first place, that if the tube has its bore in its lower part very fine and in the upper part considerable, it will be impossible for water to rise into the wide part of the tube so long as that part is above the proper level of the liquid. For if the liquid did ascend into the wide part of the tube, it might be let out (as in the figure) and employed to work a water-wheel in its descent. We should therefore be able to obtain a positive quantity of work, W say, during an operation which leaves the system in the same state as it found it. This can easily be shown to be impossible.

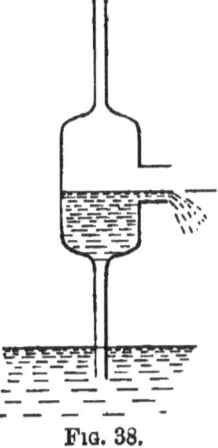

Fig. 38.

(a) If it can be assumed that no heat is absorbed or evolved during the operation, the supposition that work is obtained from the cyclical process is evidently at variance with the principle of energy.

(b) If the principle of energy be fulfilled, a positive quantity of heat must be absorbed during the cycle equal to the positive quantity of work obtained from it. But this is readily seen to involve a contradiction of Carnot's principle, because the system may be so arranged that the parts which gain and lose heat are all at one constant temperature.

(1) For by the fundamental axiom on which Carnot's principle is based, a positive quantity of work cannot be obtained from a cycle when all the parts of the system that gain or lose heat are at one constant temperature.

(2) Or we may make use of the equation $\frac{Q_1}{\theta_1} + \frac{Q_2}{\theta_2} + \ldots \leqslant 0$, which is a deduction from the fundamental axiom for any cycle whatever. When, as in our case, there is only one temperature (θ) at which heat is

absorbed or evolved, the equation becomes $\dfrac{Q}{\theta} \lesseqgtr 0$, where Q is the total quantity of heat absorbed during the cycle. Hence Q cannot be positive. This result is clearly but a slightly different form of the fundamental axiom, and it is opposed to the supposition that a positive quantity of heat is absorbed during the cycle.

Again, if the tube be plunged in a trough of mercury, it may be shown in like manner that the mercury cannot be depressed in the wide part of the tube below the level in the trough.

The second application we shall explain was given by Sir W. Thomson in 1870.

Let a fine glass tube A, open at both ends, be plunged in a vessel of water. Then the liquid will rise in the tube, and it is evident that when equilibrium is established, no change will be made by merely closing the lower end of the tube. Hence we infer that if a second tube B, equal and similar to A but with its lower end closed, be held parallel to and on the same level as A, equilibrium will not be established, if there is any aqueous vapour in the air, until the water is at the same level in B as in A. Now when given quantities of air and aqueous vapour (or of any two gases) exist together in the same vessel or space, the pressure and density of each is very nearly the same as if the other were altogether absent. The pressure and density of the aqueous vapour must therefore diminish as we ascend upwards. But if we suppose the temperature to be uniform and constant, the air can only be "saturated" with aqueous vapour at the level of the sheet of water. It therefore follows that a fine glass tube has

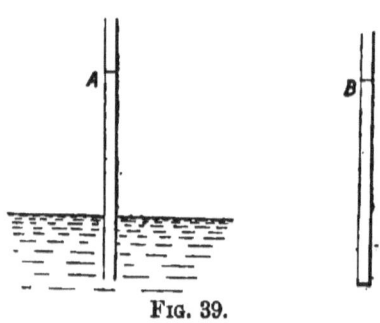

Fig. 39.

the power of condensing aqueous vapour from the air when the temperature is above the "dew-point"—that is, above the temperature at which dew begins to be deposited on ordinary objects.

The calculations will perhaps be simplified if we suppose the two glass tubes and the water contained in a closed vessel from which the air has been exhausted, so that there is nothing in the vessel but the two tubes, water, and aqueous vapour. Suppose that p is the pressure and ρ the density of the vapour just over the level sheet of water, and let p' be the pressure at the height h of the liquid column. Then $p' = p - g\rho h$, since the variations of the density in the small height h are negligible. If the temperature be 10° C., we have $p = 12{,}222$ and $\rho = \dfrac{1}{108{,}581}$. Putting $g = 981$, we obtain at 10° C.,

$$p' = p\left(1 - \frac{g\rho h}{p}\right)$$
$$= p\left(1 - \frac{h}{1{,}352{,}800}\right).$$

Now in a tube whose diameter is the 1,000th of a millimetre, it is calculated that water would ascend about 13 metres above its plane level. Substituting 1,300 for h in the above equation, we see that for such a tube, the pressure of the aqueous vapour at the top of the liquid column would be less than at the bottom by about the 1,000th part.

The foregoing calculation will hold if the water and tubes are situated in the open air instead of an exhausted vessel. Sir W. Thomson, to whom the calculation is due, thinks it probable that the moisture which vegetable substances, as flannel, cotton, etc., obtain from the air at temperatures far above the "dew-point," is due to the condensation of vapour in the minute cells of the substances.

Art. 69. We have seen that when a body is prevented

from gaining and losing heat, every spontaneous action[1] which takes place within the body causes the entropy to increase. If therefore we cause the entropy of the body to become a maximum (subject to the condition that heat is neither gained nor lost), no further spontaneous action can take place in the body, and the equilibrium will be stable.

The method of entropy is not of much practical use, owing to the difficulty of making the calculations. A very convenient method embodying the same principle has, however, been discovered for the case in which the volume is constant or in which the pressure is uniform and constant. Some of the properties of the functions (or expressions) employed in this method appear to have been first considered by M. Massieu in 1869, but he did not enquire whether the functions furnished a test of stability. This was done independently by two investigators—by Prof. W. Gibbs in 1875 and by Helmholtz in 1882.

If a system undergo any irreversible operation during which the entropy changes from ϕ' to $\phi' + \Delta\phi'$, we have $\dfrac{Q}{\theta} + \dfrac{Q'}{\theta'} + \dfrac{Q''}{\theta''} + \ldots < \Delta\phi'$. When heat is absorbed and evolved at one temperature θ only, the equation becomes $\dfrac{Q}{\theta} < \Delta\phi'$, or $Q < \theta\Delta\phi'$. This simplification is secured in the test of stability due to Gibbs and Helmholtz by enclosing the body or substance to be considered in a thick metallic vessel placed in a medium of uniform and constant temperature θ. When the thick metallic vessel and its contents are in a state of thermal and mechanical equilibrium, the temperature will everywhere be equal to θ. If any action then takes place in the body within the vessel, the temperature will generally cease to be uniform; but the temperature

---

[1] An instance of such an action occurs when explosive crystallization is produced by merely touching a saline solution with a crystal of the salt.

Applications to an Unelectrified Body at Rest.

of the exterior of the thick metallic vessel may always be supposed equal to θ. Thus if we take the metallic vessel and its contents as our system, heat will be gained and lost by the system at one temperature θ only.

Again, if sufficient time be allowed after the action has taken place in the contents of the vessel, a state of equilibrium will be attained in which the temperature is again everywhere equal to θ. The state of the metallic vessel will then be the same as before the action in its contents.[1] Let (U_0, ϕ_0) be its energy and entropy in this state. Also let $(U, \phi), (U+\Delta U, \phi + \Delta \phi)$ be the energy and entropy of the given body in the metallic vessel in the states of equilibrium in which the temperature is uniform and equal to θ before and after the action. Let $(U', \phi'), (U' + \Delta U', \phi' + \Delta \phi')$ be corresponding quantities for the compound system of the vessel and its contents. Then

$$\left. \begin{array}{l} U' = U + U_0 \\ U' + \Delta U' = (U + \Delta U) + U_0 \\ \phi' = \phi + \phi_0 \\ \phi' + \Delta \phi' = (\phi + \Delta \phi) + \phi_0 \end{array} \right\},$$

Hence

$$\left. \begin{array}{l} \Delta U' = \Delta U \\ \Delta \phi' = \Delta \phi \end{array} \right\}.$$

Again, let W be the work done on the compound system and Q the heat absorbed by it during the passage from one state of equilibrium to the other. Then $\Delta U' = W + Q$, and since no process which occurs in nature can be reversible, $Q < \theta \Delta \phi'$. Thus

$$\left. \begin{array}{l} \Delta U = W + Q \\ Q < \theta \Delta \phi \end{array} \right\}.$$

---

[1] The only use we make of the thick metallic vessel is to secure that the system absorbs and evolves heat at the temperature θ only. There is no particular restriction, however, as to the temperature at which the body within the vessel absorbs or loses heat. If we suppose heat conducted to and from it by the metallic vessel with considerable rapidity, the temperatures at which it gains or loses heat will seldom differ much from θ; but if the interior of the metallic vessel be lined with a substance which conducts heat slowly, the temperatures at which the body inside gains or loses heat may differ widely from θ.

By addition,
$$\Delta U + Q < W + Q + \theta \Delta \phi,$$
or
$$\Delta U < W + \theta \Delta \phi,$$
or
$$\Delta U - \theta \Delta \phi < W.$$

Now when the body within the vessel is in equilibrium, its temperature is uniform and the quantity $U - \theta \phi$ depends only on the state of the body. Let F and $F + \Delta F$ be the values of this quantity in the two states of equilibrium. Then since the body is at the same temperature θ in the two states, we have

$$\left. \begin{array}{l} F = U - \theta \phi \\ F + \Delta F = (U + \Delta U) - \theta(\phi + \Delta \phi) \end{array} \right\}.$$

Hence
$$\Delta F = \Delta U - \theta \Delta \phi,$$
and therefore
$$\Delta F < W.$$

By definition, W is the work done in the whole system; but since the parts of the metallic vessel may be supposed invariable in shape and size,[1] the work done on the whole system will be equal to the work done on the body within the vessel.

Let us first suppose that the volume of the body is kept constant. Then $W = 0$ and $\Delta F < 0$. Thus any change from one state of equilibrium to another at the same temperature causes F to decrease. The change of state will be originated by some spontaneous action in the body and completed by the conduction of heat through the metallic containing vessel in consequence of which the temperature becomes uniform and the same as before. If then we give to the body such a state of equilibrium that F has the least possible value for that temperature, no spontaneous action will be possible

---

[1] For example, the metallic vessel may be a cylinder fitted with a *smooth* air-tight piston. The cylinder and piston may change their relative positions, but they continue invariable in size and form.

Applications to an Unelectrified Body at Rest. 179

in the body, and the equilibrium will therefore be stable.

To find a state of stable equilibrium when the volume is kept constant, we have merely to write down the mathematical expression for F for the particular body considered and then to obtain the conditions that F should be a minimum *subject to the restriction that the temperature is constant*. The methods of doing this are somewhat too advanced to be given in the present work.

The quantity F, on account of the important property just explained, is called by Duhem the "Thermo-dynamic Potential at constant volume."

If the body is subjected to a uniform and constant pressure p, then if the volume change from v to $v + \Delta v$, during the change of state, we have $W = -p\Delta v$. Hence $\Delta F < -p\Delta v$, or $\Delta F + p\Delta v < 0$. From this it can easily be shown that $\Delta (F + pv) < 0$. Thus when the pressure is uniform and constant, the quantity $F + pv$, or $U - \theta\phi + pv$, possesses the same kind of property that F (or $U - \theta\phi$) does when the volume is constant.

Duhem writes Φ for $U - \theta\phi + pv$, and calls Φ the "Thermo-dynamics Potential at constant pressure."

Art. 70. We shall conclude this chapter with an account of Gibbs' geometrical test of stability which supposes the body subjected to a uniform and constant pressure.

We have first to explain Gibbs' geometrical method of representing the state of the body when in equilibrium, stable or unstable. In this will consist the chief difficulty of the article to non-mathematical readers.

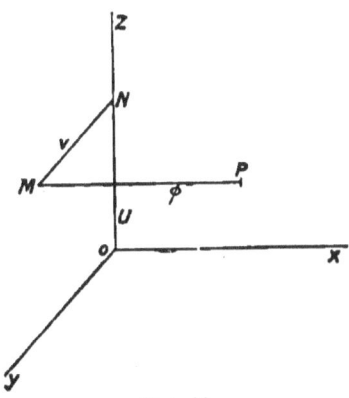

Fig. 40.

When the body or substance is in equilibrium, stable

or unstable, at a uniform temperature and pressure, there are five quantities (U, θ, ϕ, p, v), such that when we know any three of them we know the state of the body completely. Prof. Gibbs therefore represents the state of the body by the position of a point P such that three lengths depending on the position of P are respectively proportional to (ϕ, v, U). He does this by taking three rectangular axes Ox, Oy, Oz, and making ϕ proportional to the perpendicular distance PM of the point P from the plane yOz; v to the perpendicular distance MN of P from the plane zOx; and U to ON, the perpendicular distance of P from the plane xOy.

There are two points of fundamental importance in connection with this mode of representing the state of the body.

(1) The position of the point P, corresponding to all possible states of equilibrium, stable or unstable, of which the body is capable, lie on a surface.

(2) If at any two points the tangent planes to this surface are parallel, the temperature and pressure in the state represented by one of the two points are respectively equal to the temperature and pressure in the state represented by the other point.

The first of these propositions follows from the fact that when two of the three quantities (ϕ, v, U) are given, say v and U, there will be but one state of equilibrium, or only a limited number of states, which have the given value of v and U. The proof of the second proposition, though not difficult, is too advanced to be given here. The mathematical reader should have no difficulty in obtaining the proof for himself. Non-mathematical readers may take both propositions for granted. The rest of the article will be found very easy.

We can now determine whether a state of equilibrium is stable or unstable when the substance is subjected to a uniform and constant pressure. Let the substance be contained in a cylinder fitted with a smooth air-tight piston and placed within a closed vessel whose size is immense and form and volume invariable, and which

Applications to an Unelectrified Body at Rest. 181

is filled with air and covered with a perfect non-conductor of heat. Then since the changes in the volume of the comparatively small body in the cylinder and the heat absorbed and evolved by it, can have but little effect on the large quantity of air by which the cylinder is surrounded, it follows that the temperature and pressure of the air will practically remain constant, (θ, p) say. Thus the body in the cylinder is subjected to a uniform and constant pressure (p).

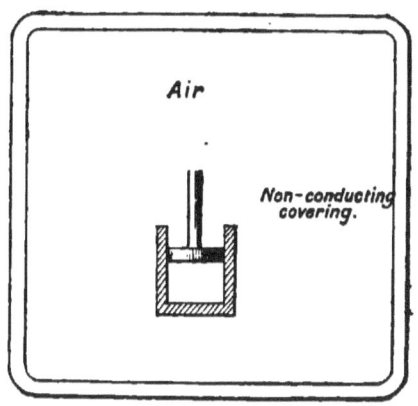

Fig. 41.

Again, if we take as our system the large closed vessel and all its contents, no work is done on the system and no heat is gained or lost by it. The energy of the whole system is therefore constant as well as the volume.[1] The entropy, however, increases if any change takes place within the system, because heat is neither gained nor lost, and every process in nature is irreversible. Now the only change (or changes) that can take place in the system will be originated by spontaneous actions occurring within the cylinder, and will not be completed

---

[1] There is a point in Maxwell's "Theory of Heat" which the author of the present work has never been able to understand. It is stated by Maxwell that the energy of the whole system does not remain constant. The same statement, the author also finds, is made by Prof. Gibbs. Now it would be absurd to suppose that these two authorities have made a fundamental mistake: the question is, What is the explanation of the statement?

It has been suggested that Maxwell does not use his terms with exactly the same meanings as the present writer assigns to them. A more probable explanation would be that Maxwell does not subject his system to the same restrictions that are imposed on the system in the present article, in which it is supposed that no work is done on the system and no heat gained or lost by it.

As it is very difficult to get an opinion on the point, perhaps some one will be good enough to criticize it.

until by the conduction of heat through the walls of the cylinder, the temperature becomes again everywhere equal to θ. In this process the temperature of the substance contained in the cylinder is not restricted to be constant. The cylinder may be a very good conductor of heat, in which case the temperature of the substance within will seldom differ much from θ: or the cylinder may be a very poor conductor of heat, and then the temperature of the substance may differ widely from θ during the change.

It will now be seen that if the entropy of the whole system when in a state of equilibrium A is less than the entropy when in another state of equilibrium B (the substance within the cylinder being at the same temperature in both states), a spontaneous action will be possible in this substance when this system is in the state A. The equilibrium of the substance will then be unstable. If, however, the entropy of the system be greater in the state A than in any other state B, no spontaneous action will be possible in the state A; and this state will therefore be stable.

Instances of unstable states occur " when a liquid not in presence of its vapour is heated above its boiling point, and also when a liquid is cooled below its freezing point, or when a solution of a salt or gas becomes supersaturated.

In the first of these cases, the contact of the smallest quantity of vapour will produce an explosive evaporation; in the second, the contact of ice will produce explosive freezing; in the third, a crystal of the salt will produce explosive crystallization; and in the fourth, a bubble of any gas will produce explosive effervescence."

Next, if we denote the state of the air within the large closed vessel by the position of a point with respect to three rectangular axes, the surface on which the point always lies will be such that its tangent planes at all points are parallel, because the air is always at the same temperature and pressure whatever its volume, etc., may be. The surface will therefore be a plane.

Applications to an Unelectrified Body at Rest. 183

Now let the axes $(O'x', O'y', O'z')$ used in representing the state of the air, be parallel to the axes (Ox, Oy, Oz) but in the opposite direction to them. Let A be the point representing the state of the substance within the cylinder and T the point representing the state of the air, when the whole system is in any state of equilibrium. Then since the sum of the energies corresponding to the points A and T has the same value for all states of equilibrium of the system, the distance of A above the plane xOy added to the distance of T below the plane $x'O'y'$, will be constant, h say. But this sum is clearly equal to the height of O' above xOy together with the height of A above T. The position of the point O' being at our choice, we will therefore choose it so that the height of O' above xOy is equal to h. The points A and T will then be at the same height above xOy.

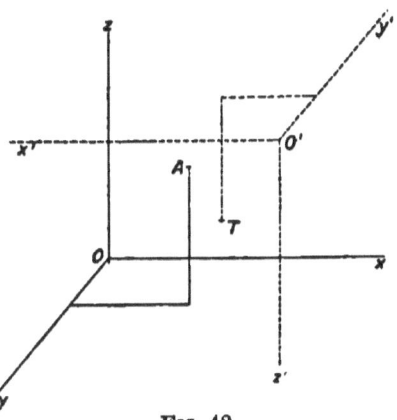

Fig. 42.

Similarly, since the sum of the volumes corresponding to the points A and T is constant, we can choose the position of the plane $xO'z'$ so that A and T are at the same perpendicular distance from xOz. The points A and T will then lie on a straight line parallel to Ox.

Suppose now that A and B are two points representing states of equilibrium of the substance within the cylinder when the temperature is θ and the pressure p. Then the tangent planes at A and B will be parallel to one another and to the plane which represents all the states of the air within the large vessel. Let parallels to Ox through A and B meet this plane in X and Y; and suppose the accompanying figure to be a view as seen by an observer at a great distance from O on Oy. Then the *increase* of the entropy of the substance in the cylinder in passing from A to B is proportional to BM, and the *decrease* in the entropy of the air in the large vessel is proportional to YN. The increase in the entropy of the whole system within the large vessel is therefore proportional to BM − YN, or (BL + LM) − YN, or (BL + LY) − MN, or simply BL. If then the point L be to the left of B, the system will be able to pass from the state A to the state B, and the equilibrium in the state A will be unstable. If, however, the surface be nowhere on the right of the tangent plane at A, the equilibrium in the state A will clearly be stable.

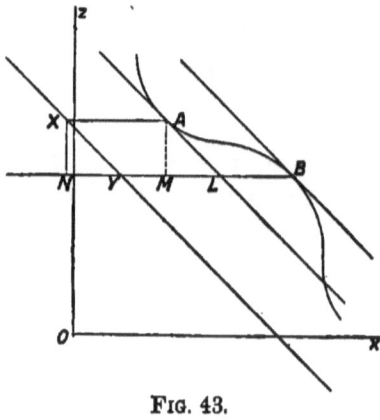

Fig. 43.

When a tangent plane touches the surface in two points A, B, and has the surface entirely on its left, portions of the substance can permanently co-exist at the same temperature and pressure in the two states represented by the points A and B. If a mass x co-exist in the state A with a mass $1 - x$ in the state B, the corresponding point in the diagram will be the centre of mass of a mass x placed at A and a mass $1 - x$ placed at B. Hence by giving to x all possible values between 0 and 1, we see that every point in the limited line AB represents a state of stable equilibrium, such that the substance is partly in one state and partly in the other at the same temperature and pressure.

If the tangent plane roll on the "primitive" surface, the locus of the limited line AB will be another surface which is called the "secondary" surface; and it is evident that the primitive surface represents the state of the substance when homogeneous, the secondary surface the state when heterogeneous.

Suppose the rolling plane AB belongs to the co-existence of the liquid and gaseous states. Then there will be two other rolling planes, one for the co-existence of the solid and liquid states, the other for that of the solid and gaseous states. If one of these rolling planes touch the surface in the same point A_0 as another of the three rolling planes, the two planes will then coincide, because (it is supposed) the primitive surface is continuous (or has no sharp points or edges) and therefore has only one tangent plane at a point. In this case the two rolling planes will each touch the surface in three points $A_0 B_0 C_0$, and will coincide with the third rolling plane. The physical interpretation of this result is that the substance can then co-exist at the same temperature and pressure in the three forms of solid, liquid, and gas.

The three surfaces obtained by means of the three rolling planes, together with the plain triangle $A_0 B_0 C_0$, which corresponds to the triple point, constitute what Prof. Gibbs calls the "Surface of Dissipated Energy."

CHAPTER V.

TIDAL FRICTION.

ART. 71. The theory of Tidal Friction is a deduction from Carnot's principle. It may also be obtained as a consequence of the law of gravitation alone without any particular reference to Carnot's principle. By deducing it from Carnot's principle we secure the great advantage that some of the chief results are apparent almost from the first.

The theory of tidal friction is a natural sequel to the theory of gravitation as laid down in the ordinary treatises. To those who are already acquainted with the principle of gravitation, the subject of tidal friction will appear very simple. The present chapter in which it is worked out by the help of Carnot's principle, is practically independent of the rest of the work and will present little difficulty to those who are sufficiently acquainted with dynamics even if their attainments in pure mathematics are of a very slender description. There are but few conclusions drawn in the chapter for which I have not been able to find authority in the standard works on the subject and the reader will be warned of such cases if it appears necessary.

Art. 72. The consequence of Carnot's principle which includes the theory of tidal friction will be readily understood even by those who have no previous acquaintance with Carnot's principle itself. If a body of any kind be protected from all external forces and from receiving or losing heat either by conduction or radiation, it follows as a corollary to Carnot's principle that the temperature of the body will ultimately be uniform and that it will ultimately move like a rigid body. If the body be protected from external forces and from receiving heat, and be merely at liberty to lose heat by radiation into infinite space, it may again be assumed from Carnot's principle that the temperature will ultimately be uniform and that the body will ultimately move as rigid. The final uniform temperature in this case will be very low, probably absolute zero itself, or that temperature at which a body has lost all the non-mechanical or invisible irregular motions which constitute "heat" and at which it therefore ceases to radiate heat. The body will not generally move *strictly* like a rigid body until the final uniform temperature is attained, owing to the motions which may be produced by the conduction of heat in its interior and by radiation into infinite space. It may, however, move *sensibly* like a rigid body long before the final state of uniform temperature is reached.

Except in special cases, a body will not be able to move strictly as rigid when it is not protected from external forces and from receiving or losing heat. The effect of the radiation of heat into infinite space in preventing a body behaving as rigid may be seen in the case of the Sun, the surface of which is thus kept in a state of violent agitation.

The proposition we have enunciated for a single body may be extended to the most general material system possible. Let any material system whatever be protected from all external forces and from receiving heat from external sources, and let its parts either be all protected from losing heat altogether, or let some or all of them be at liberty to lose heat by radiation into infinite space. Then it follows as a deduction from Carnot's principle that the whole system will ultimately move like a single rigid body, and this will be true whether all the parts of the system are ultimately in contact, or whether some of them are ultimately at a distance apart and move under the influence of gravitation, like the bodies of the solar system do at present. It also follows from Carnot's principle that the different parts of the system will ultimately be of uniform temperatures. All those which have been at liberty to radiate heat freely into infinite space will probably be reduced to the temperature of absolute zero. Of those which have been prevented from losing heat, the final temperatures will depend on various circumstances; and the ultimate uniform temperatures of two such parts will not necessarily or generally be equal.

The bodies or parts of the system may contain any number of hollow vacuous cavities of any kind, but radiation into these spaces is not to be considered as radiation into infinite space, because the radiation thus emitted returns to the system again after traversing the cavity. When we speak of radiation into infinite space, it is to be understood that the whole or part of the radiation is lost to the system for ever.

The proposition is applicable to electrified and magnetized bodies, but we do not propose to consider such cases. The existence of electric and magnetic forces will therefore be left out of account altogether. Gravitation will thus be the only mechanical force in our system in addition to the usual pressures, tensions, etc.

In this chapter we are only concerned with the tendency of the body or system to move ultimately as if rigid. The agencies which bring about this result are various, but in the solar system the principal agent is the friction of the tides. In the case of a planet on which there is no water or on which the water is all frozen, there will be no tides in the popular sense of the word; but we shall use the word "tide" to denote any kind of want of "perfect rigidity," and therefore every body will be liable to tidal friction.

Art. 73. The way in which tidal friction acts when two heavenly bodies attract one another can easily be explained by means of the law of gravitation. Neither body attracts all parts of the other body alike, and in consequence the bodies have not quite the same forms as if they were so far apart that their mutual attraction is insignificant; other things being equal. The deformations may be represented by supposing each body to consist of a rigid nucleus and a tide. For simplicity, each nucleus will be supposed to be spherical in form and either homogeneous throughout or made up of homogeneous shells bounded by concentric spherical surfaces. Such a sphere attracts and is attracted exactly as if it were condensed to a "particle" at its centre. The attraction of one of the bodies P on the other body Q will then consist of the following parts: (1) the attraction of the nucleus of P on that of Q, which is a force in the line joining the centres of the two nuclei, (2) the attraction of the nucleus of P on the tide of Q, (3) the attraction of the tide of P on the nucleus of Q, and (4) the attraction of the tide of P on that of Q. The forces (2) and (3) are generally very small in comparison with (1), on account of the small-

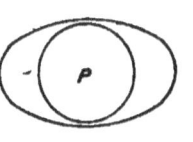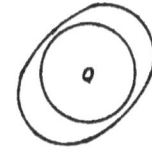

Fig. 44.

ness of the tides in comparison with the nuclei. The force (4) may be omitted altogether without any sensible error, because it is only the attraction between two small quantities of matter, whereas the forces (2) and (3) are each the attraction between a small quantity of matter and a large one. In the ordinary treatises on gravitation, these points are not considered, each body being treated as entirely "rigid." At present, the small forces (2) and (3) are what we are principally concerned with.

When P and Q move so as always to show the same faces to one another, it is obvious that the tides will be symmetrical

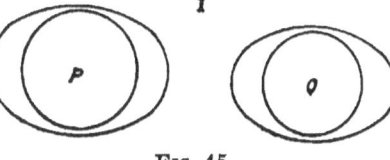

Fig. 45.

about the line PQ, joining the centres of the nuclei, as

in figure I, and the small forces (2) and (3) will therefore act along this line. If, however, Q does not always show the same face to P, the tide of Q will be carried to one side of the line PQ by the rotation of the nucleus of Q. The attraction of P on the tide of Q will then

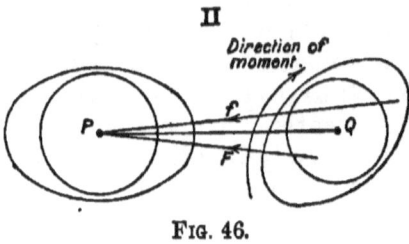

FIG. 46.

reduce to two forces, F, f, shown in figure II. Of these forces, F will be the greater, because the half of the tide corresponding to F is nearer to P than the other half. The attraction of P therefore produces a moment about the centre of Q (as shown in the fig.) which tends to make the tide of Q symmetrical about the line PQ. If we imagine the tide of Q to show always the same parts to P, it is obvious a considerable amount of friction

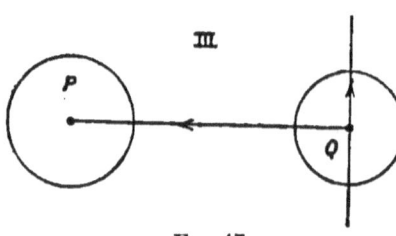

FIG. 47.

will be called into play to resist the motion of the nucleus under the tide. Again, the resolved parts of F and f perpendicular to PQ do not quite balance one another. Hence the attraction of P on Q gives a principal force along QP and a small force at right angles to QP, as in figure III.

The small force perpendicular to QP tends to alter the shape and size of the orbit of Q. The attraction of P on the tide of Q tends to make the tide symmetrical about the line QP, and acts indirectly by friction on the rotating nucleus of Q. The combined effects of these causes, as may be shown by calculations too complex to be given here, is to make Q show always the same face to P. We therefore see that if sufficient time be allowed, P and Q will move so as always to show the same faces

Tidal Friction. 191

to one another; and a similar result will follow in any other case.

The foregoing illustration will serve to show the difficulty of working out the theory by means of the law of gravitation alone. By making use of Carnot's principle, as in the text, many difficulties are avoided.

Art. 74. The body or system of bodies considered in Art. 72 ultimately moves as if rigid. But in such a case we know that the whole motion is equivalent to a motion of translation of the centre of mass combined with a motion of rotation of the whole body or system about a straight line drawn through the centre of mass in a fixed direction. The motions of the parts of the body or system with respect to the centre of mass, depend only on the rotation, and we shall therefore suppose the centre of mass to be at rest. In the final state, therefore, the motions of the parts of the body or system of bodies, will consist of circular orbits about the same straight line, to which the planes of the orbit are all perpendicular.

The effect of tidal friction in causing an orbit to become more circular may be expected to be most apparent in the case of very elongated elliptical orbits. In the solar system such orbits are described by comets; and as will be seen later on, the smallness of comets is also extremely favourable to tidal effects. Since by Kepler's second law, the line joining the comet to the centre of the sun describes equal areas in equal times, it is obvious that the comet's velocity will be so extremely small in the more distant parts of the orbit that nearly the whole of the periodic time will be spent in travelling

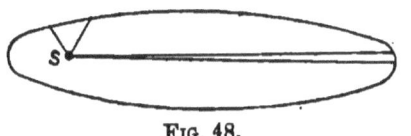

Fig. 48.

these parts. Hence tidal friction, by making the orbit less elongated (or more circular), and consequently diminishing the parts which are described most slowly, may produce a sensible shortening of the periodic time.[1]

A change in the periodic time has actually been observed in the case of one comet — that known as Encke's, and the observed fact is a shortening of the period. This fact, as we have just seen, may be accounted for tidal friction, at least in part. Another cause may be also assigned. When a comet approaches the sun, it is generally observed to throw off a tail, which points away from the sun, and is not merely left behind by the comet. The cause of the phenomenon is evidently equivalent to a repulsive force residing in the sun. And since the tail only appears when the comet is in that part of its orbit nearest the sun, it is clear that the forces which produce the tail will increase the perihelion distance of the comet (that is, the distance from the sun to the nearest point of the orbit). In consequence it may be shown that the orbit will become more circular and the periodic time diminish.[2]

A third explanation known as Encke's hypothesis is also given, and has, until recently, been generally accepted as a complete and satisfactory explanation of the phenomenon. It was supposed by Olbers and Encke[3] that the motion of the heavenly bodies is constantly but slightly resisted by the "ether," that subtle medium pervading space and in which radiant light and heat are transmitted. The effect of such a "resisting medium" on the solar system will be to cause all the other members of it to fall into the sun. For let S be the centre of the sun, P that of a body B revolving round it, PT the tangent to the orbit at P, and let the full line PM be the path that would be described if there was no "resisting medium." Then since the only force introduced by the resistance of the ether acts along the tangent, in a direction opposite to that of motion, it is

[1] It is one of the advantages of the method of treating the subject here adopted that it gives at once a simple explanation of the shortening of the period of the comet.
[2] This explanation must be taken for what it is worth.
[3] Newcomb's "Popular Astronomy," 2nd edition, page 394.

clear that the curve actually described by the centre of B will deviate further from PT than it otherwise would have done, like the dotted curve PN. Thus the effect of the resistance imagined by Encke will be to cause the orbit to grow continually smaller and smaller, and ultimately it will cause the body to fall into the sun. Now since by Kepler's third law the square of the periodic time varies as the cube of the mean distance, it follows that as the orbit gradually contracts, the periodic time will constantly diminish. Now in the case of bodies of the same density and moving with the same velocity, the effect of Encke's resisting medium will be greater for small bodies than large ones. For if d be the diameter of such a moving body, the area of the surface exposed to the resisting medium varies at d^2. Hence the resistance experienced varies as d^2. But the mass of the body varies at d^3. The amount of the resisting force per unit of mass of the moving body will vary as $\dfrac{d^2}{d^3}$, or as $\dfrac{1}{d}$, and this is greatest when d is least. Thus the effect of the resistance of the ether will be comparatively large and may be expected to produce sensible results in the case of small bodies like comets.

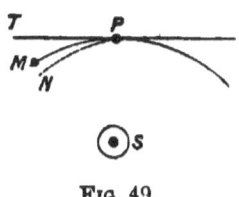

FIG. 49.

Encke's hypothesis, as usually understood, is entirely gratuitous. The mere fact that the periodic time of a single comet diminishes, is supposed to prove that the ether constantly offers a sensible resistance to motion; and the assumed resistance is supposed to account completely for the observed phenomenon. At one time this explanation seems to have been generally accepted, but at present there appears to be a disposition to reject it as insufficient.[1] It is not to be imagined, however, that in questioning Encke's hypothesis, the idea of an ethereal

[1] Newcomb's "Popular Astronomy," 2nd ed., p. 394.

resistance to the motion of bodies through space is altogether discarded. We know, in fact, from a theoretical argument due to the late Prof. Balfour Stewart, that in certain cases the ether does offer resistance to a body.

The theoretical argument respecting the resistance of the ether is considered in the chapter on Carnot's principle; and it is there inferred [1] that the amount of the resistance depends on the temperature, and practically disappears when the temperature is low enough. Now a comet being a small body is liable to very rapid cooling, and therefore the temperature may be supposed to be always very low except when near the sun. In consequence the resistance of the ether will be insignificant except near the perihelion. The effect of such a resisting force will evidently be to diminish the perihelion distance and may actually increase the periodic time.

If the resistance of the ether were strictly zero, the parts of a system would not necessarily be all in contact in the final state, but might remain at considerable distances apart and continue to describe orbits under the influence of gravitation alone. The same thing may happen when the ether resists provided the resistance dies away before the final state is reached. If the resistance continues until the end, the parts of the system will always be in contact in the final state.

In order to trace the history of a general system of bodies, it will be necessary to investigate the combined effects of tidal friction and the resistance of the ether. But in most cases the resistance of the ether is probably always extremely small. It is therefore permissible to treat it as altogether absent, so long as there is any appreciable amount of tidal friction at work. There are thus two distinct problems to be considered. We have first to investigate the effects of tidal friction alone, and then those of the resistance of the ether. In the rest of this chapter, we shall confine ourselves to the first

---

[1] This inference is merely a private speculation of the author's, which the reader may either accept or reject.

Tidal Friction. 195

problem and the final result obtained will be said to represent the final state of the system.

Art. 75. In what we have agreed to call the final state of the system, tidal friction will have come to an end and the whole system will move like a single rigid body. The motion of the system may then be represented as a translation of the centre of mass combined with a rotation of the whole system about a straight line drawn through the centre of mass and carried about with it. As there are no external forces acting on the system, the straight line about which it rotates will be fixed in direction, and the magnitudes of the translation and rotation will both be constant. The two motions are also independent, and it will simplify our task without detracting from its generality, to suppose that the motion of translation is zero, or the centre of mass at rest. We may then say that the system rotates like a single rigid body with constant angular velocity (ω say) about a fixed straight line drawn through the centre of mass which is a fixed point: and any part of the system will describe a circular orbit about this fixed straight line, the plane of the orbit being at right angles to the line and its centre on it.

The motion of any part of the system may in like manner be considered as a translation of *its* centre of mass together with a rotation of the said part about an axis passing through the centre of mass. For let O be the point where the axis about which the whole system rotates, meets the plane of the orbit described by the centre of mass G of any part of the system; and let G, G′ be the positions of G at any two instants. Then if OG meet the body in A and OG′ in A′, the line GA will have got to the new position G′A′ at the same instant that

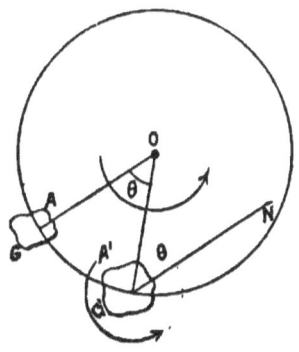

Fig. 50.

G has to G'. But if the body G had had no motion but the translation of its centre of mass G, the line GA would have moved into the new position G'N parallel to GA. It therefore follows that the displacement of the part of the system G is equivalent to a displacement of its centre of mass G together with a rotation about an axis through the moving point G parallel to the fixed axis through O, through the same angle θ and in the same direction as the rotation of the point G about the axis through O. We may therefore represent the motion of the body G either as a constant rotation (ω) of the whole body about the fixed axis through O or as a constant rotation (ω) of the centre of mass G about this axis together with a constant rotation (ω) in the same direction of the whole body about an axis through G parallel to the fixed axis through O.

This result leads at once to an important conclusion respecting the external forces[1] which act on any part of the system. For when any body is moving as rigid, the external forces which act on it are equivalent to a single force at the centre of mass together with a couple about some straight line. The motion of the centre of mass depends only on the force; and in discussing the motion of this point, the couple has not to be considered. Now in the case discussed in this Article, the centre of mass G (or any part of the system) describes a circular orbit with constant velocity. From this it follows that (1) because the orbit is plane, the force of G is in its plane, and (2) because the velocity is constant, the force is always at right angles to the path. It therefore tends to the centre of the circular orbit, *i.e.* it acts along the perpendicular from G on the fixed straight line through O about which the whole system revolves.

Now the heavenly bodies are generally roundish. Many of them are approximately spherical; and we may often treat them as if they attracted and were attracted

---

[1] The forces here spoken of are external as regards the part of the system considered, but internal as regards the whole system. When we consider the whole system, there are no external forces.

Tidal Friction.

as if they were concentrated into single particles at their centres, like solid homogeneous spheres (or solid spheres made up of homogeneous shells bounded by concentric spherical surfaces). In other words, the gravitational attractions of such bodies on each other reduce to a system of forces along the lines joining their centres. Suppose that, in our present case, the system consists entirely of such bodies, and, if possible, let them be so far apart that there is no material connection between any of them. Then the only external forces on any one of them will be the attractions of the rest; and if G be the centre (or centre of mass) of the body chosen, (G_1, G_2, G_3 . . .) of the others, the external forces on the body G will reduce to forces at G tending to the points (G_1, G_2, G_3, . . .). From this it easily follows that the centres all lie in one plane. For, if possible, let them not all lie in one plane. Then it will be possible to choose one of them, say G, such that all the other centres (G_1, G_2, G_3, . . .) lie in the orbit of G or on one side of it. From G draw GO perpendicular to ON, the fixed straight line about which the whole system rotates. Then it is clear that (with the supposition made) the forces which act at G cannot have a resultant along GO, as we know they always must when the whole system moves as rigid. Hence the centres all lie in one plane and the plane must evidently be at right angles to the fixed straight line ON.

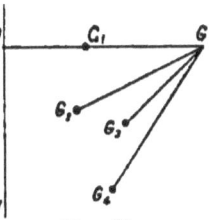

FIG. 51.

The question of the total number of bodies in the final state of the system will be considered later on (Art. 82) and it will be shown that in some cases there will be one, in others two. The proposition that the centres are all in one plane may therefore appear at first sight to be absurd. It may be pointed out, however, that tidal friction may have *practically* caused the system to move as rigid with the centres all in one plane long before the bodies begin to coalesce. This appears to be the case with the system of Saturn and his rings, where the numerous small bodies of which the rings are supposed to be made up, and the centre of Saturn himself, all lie in a plane perpendicular to the axis of rotation.

Art. 76. Take a system unacted on by external forces, but not in its final state. Let it consist of a number of small bodies and one large body, which last we shall treat as a solid homogeneous sphere. Suppose that tidal friction has already brought the small bodies into one plane and caused them to move as rigid, but let them be still so far apart that the mutual attractions are insignificant in comparison with the attraction of the large body. Then the only force on one of the small bodies G will be the attraction of the large one and will tend to its centre O. If S be the mass of the large body in grammes, r the distance in centimetres of the centre of the large body from the small body G, and λ the constant of gravitation, the external force on each gramme of the body G will be $\dfrac{\lambda S}{r^2}$ dynes acting along GO. Suppose now that the plane containing the small bodies and the straight line about which they rotate are both fixed. Then, as we have seen, the force on G acts along the perpendicular from G to the fixed axis of rotation,

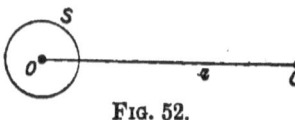

Fig. 52.

or from G to the point where the axis of rotation meets the plane containing the small bodies. This point is therefore the centre of the large body S, and each of the small bodies describes a circle round it. Denoting the common angular velocity of the small bodies by ω, the velocity of the body G will be

$$v = r\omega,$$

and its acceleration along the normal GO

$$\frac{v^2}{r} = r\omega^2.$$

But the force acting on each gramme of a body which can be treated as a point, is equal to the acceleration. Hence

$$\frac{\lambda S}{r^2} = r\omega,$$

or

$$r^3 = \frac{\lambda S}{\omega^2},$$

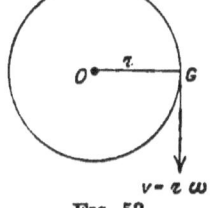

FIG. 53.

and therefore r is independent of the nature and size of the body. Thus all the small bodies lie at the same distance from the large body S, and therefore form a ring round it.

To complete this article it should be stated that in the case of ideal rigid bodies (that is, bodies not subject to tidal friction), we know that the motion of a system formed of one large body and a number of very small ones arranged in a ring round it, will be stable when there are no external forces. In other words, if the system be slightly disturbed from its state of steady motion, it will merely oscillate about it and not depart from it more and more. It will, however, be shown hereafter (Art. 82) that tidal friction may ultimately bring about the destruction of the ring by causing the small bodies which form it to coalesce.

The system considered in this Article is approximately that of Saturn and his rings according to Maxwell's generally accepted theory.[1] The rings of Saturn do not, however, form a simple circle like that we have just described, but a thin flat disc bounded by circles concentric with Saturn. But it is to be remarked that the small bodies which make up Saturn's rings are perhaps comparatively near together and the form of Saturn himself is certainly not spherical, as we have supposed that of our large body to be.

Art. 77. Several of the heavenly bodies consist chiefly of a hard solid nucleus and a liquid ocean of comparatively shallow water. In such cases tidal phenomena

[1] Newcomb's "Popular Astronomy."

will be due almost entirely to the agitation of the water; and so long as the water remains unfrozen, it will be permissible to treat the solid nucleus as perfectly rigid; only it must be remembered that tidal friction may still exist on a feeble scale after all the water is firmly frozen up. The external forces which act on the nucleus are therefore equivalent to a single force at its centre of mass and a couple. Now it appears that for many purposes the solid nucleus may be considered homogeneous and that the external forces which act on it are approximately of the same simple character as gravity at the surface of the earth—the forces on the different particles of the nucleus being practically all in parallel directions and proportional to the masses of the respective particles. The forces which act on the nucleus therefore reduce to a single force at its centre of mass, the couple being practically zero. But when the couple is zero, the motion of a rigid body about its centre of mass is the same as if there were no forces. Remembering that the mass of the ocean is comparatively small, it follows that there will be little error in treating the whole body as under the action of no forces.

Now we have seen that when a body whose centre of mass is at rest is under the action of no forces, tidal friction will sooner or later cause it to move as rigid with constant angular velocity about an axis fixed both in the body and in space. If the body be of the form generated by the revolution of an ellipse about its minor axis and can be treated as homogeneous, it may be shown that the axis of figure (or the minor axis of the revolving ellipse) will be the axis about which the body ultimately rotates.[1]

If there be a small couple, it will tend to separate the axis of rotation and the axis of figure; but owing to the fluidity of the ocean, this tendency will be strongly resisted by the tides. So long therefore as the couple remains small, the axis of figure will practically be the axis of rotation.

[1] Parker's "Elementary Thermo-dynamics," p. 191

If the axes of figure and rotation be separated to any considerable amount, and the body then left to itself in a state of steady motion, it can be proved [1] that the axis of rotation will describe a spiral in the body about and tending to the axis of figure. While the two axes are becoming coincident, the seas will be in a state of fluctuation, the mean level at any place being alternately above and below what it finally becomes. If the body considered in this Article be the earth, and the distance from the pole to the axis of rotation be a kilometre, the fluctuation of the sea on each side of its final level, may be 3 to 4 metres: if the distance be a mile, the fluctuation may be 15 to 20 feet.

Art. 78. In the rest of this chapter, we shall make some important limitations as to the kind of bodies forming our system, which will greatly simplify the questions with which we have to deal.

One of the chief difficulties in considering a general system of bodies arises from the fact that the centres of mass [2] may be moving about in different planes and the axes of rotation pointing in different directions. Now if at any instant the centres of mass are all lying in one plane with respect to which the masses and motions of each body are symmetrical (and the axes of rotation therefore all at right angles to the plane), it is obvious from symmetry that the centres of mass will all continue to lie in this plane and the axes of rotation all remain perpendicular to it. This will be supposed the case with our system; and in this it will bear a resemblance to the solar system and probably many other Astronomical systems of bodies. The forces exerted by one body of the system on any other will be symmetrical with respect to the fixed plane containing the centres of mass; and therefore when they can be reduced to a

---

[1] Routh's "Rigid Dynamics," or Parker's "Elementary Thermodynamics," p. 191.

[2] By the word "body" we mean a quantity of matter, the parts of which are in contact. When a quantity of matter consists of portions which, like the sun and earth, are not in contact with each other, each portion is called a "body."

force and a couple, the force will act in the fixed plane and the couple about the axis of rotation. The couple will be partly due to the permanent irregularities of form in the body and partly to those due to tidal causes. We shall, however, assume that the body when removed far enough from the action of the other bodies, takes up a form which is symmetrical about the axis of rotation. The couple will therefore be due entirely to the tides, and may be taken as a measure of their action.

The forces which act on a body cannot generally be replaced by a force and a couple, unless the body be perfectly rigid and therefore unlike any which occur in nature. We shall avoid the difficulty by supposing each body of our system to be very nearly rigid, or formed, like many of the heavenly bodies, of a hard central nucleus which may be treated as perfectly rigid, and a comparatively shallow ocean on the exterior of the nucleus.

The tides in the system will depend on the size, shape, density and fluidity of the bodies, on their rotations and orbits, and on their proximity to each other. The consideration of density and fluidity will be rendered unnecessary by assuming that the bodies are homogeneous and of the same density and fluidity: that of size and shape will be simplified as much as possible by supposing each body to be always spherical or very nearly so. We need therefore only concern ourselves with the size, rotations, orbits, and proximities of the bodies.

A last simplification will be effected later on by assuming that the system contains (like the solar system) one body which is overwhelmingly greater than the rest.

Art. 79. The tides are due to the fact that no body attracts all parts of any other body alike, and that the parts of two bodies do not always remain at the same distance from each other. In the case of our homogeneous spherical nearly rigid bodies, of the same

Tidal Friction.

density and fluidity, the tides will therefore depend on the sizes of the bodies, on their proximities to each other, and on their orbits and velocities of rotation. It will only be necessary for us, however, to consider the influence of size and proximity.

Let A, B be two such spherical bodies, (G, H) their centres, (M, m) their masses in grammes, D and d their diameters, and r the distance GH in centimetres. Then the attractions of B (in dynes) on particles whose masses are each a gramme, situated on the nearest and remotest parts of A, are

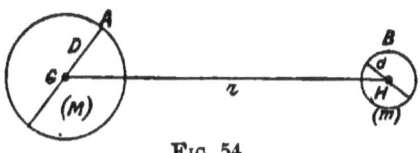

Fig. 54.

$$\frac{\lambda m}{\left(r - \dfrac{D}{2}\right)^2} \text{ and } \frac{\lambda m}{\left(r + \dfrac{D}{2}\right)^2}.$$

The difference between these forces may be taken as a measure of the tide-producing force of B on A, and is equal to

$$\lambda m \left\{ \frac{1}{\left(r - \dfrac{D}{2}\right)^2} - \frac{1}{\left(r + \dfrac{D}{2}\right)^2} \right\} = \lambda m \, \frac{\left(r + \dfrac{D}{2}\right)^2 - \left(r - \dfrac{D}{2}\right)^2}{\left(r - \dfrac{D}{2}\right)^2 \left(r + \dfrac{D}{2}\right)^2}$$

$$= \lambda m \cdot \frac{2rD}{\left(r^2 - \dfrac{D^2}{4}\right)^2}$$

$$= \lambda m \cdot \frac{2rD}{r^4} \cdot \frac{r^4}{\left(r^2 - \dfrac{D^2}{4}\right)^2}$$

$$= \frac{2\lambda m D}{r^3} \cdot \frac{1}{\left(1 - \dfrac{D^2}{4r^2}\right)^2}.$$

Now if, as usually happens, the bodies be small in comparison with the distance between them, $\dfrac{D}{r}$ will be small and therefore $\dfrac{D^2}{r^2}$ may be neglected. The tide-producing force of B on A therefore takes the simple form $\dfrac{2\lambda m D}{r^3}$.

Similarly, the tide-producing force (in dynes) of A on B is $\dfrac{2\lambda M d}{r^3}$.

The ratio of these is
$$\dfrac{2\lambda m D}{r^3} \cdot \dfrac{r^3}{2\lambda M d},$$
or
$$\dfrac{m}{M} \cdot \dfrac{D}{d}.$$

Now if ρ be the common density of the two spheres, we have
$$M = \tfrac{4}{3}\pi\rho \left(\dfrac{D}{2}\right)^3 = \dfrac{\pi}{6}\rho D^3,$$
$$m = \dfrac{\pi}{6}\rho d^3.$$

Hence
$$\dfrac{\text{tide-producing force on A(M) due to attraction of B}}{\text{\hphantom{tide-producing force on} B(m) \hphantom{due to attraction of} A}}$$
$$= \dfrac{m}{M} \cdot \dfrac{D}{d} = \dfrac{d^3}{D^3} \cdot \dfrac{D}{d} = \left(\dfrac{d}{D}\right)^2.$$

If A be very much greater than B, the ratio $\dfrac{d}{B}$ is very small and $\left(\dfrac{d}{B}\right)^2$, which is much smaller still, is altogether negligible. So long, therefore, as the rotations of the two bodies are comparable, the action (or couple) of the tides on the larger will be insignificant in comparison with the action on the smaller. In this case the effect of the tides in altering the angular velocity of rotation will be much greater for the small body than for the large one, for two reasons, (1) because the tides have a much more powerful action on the smaller body, and (2) because it requires a much smaller effort to produce a given change of angular velocity in the small body than in the large one. Hence it easily follows that the small body will be caused to turn always the same face to the large one long before the large one begins to turn always the same face to the small one. This is taken to be the explanation of the fact that the moon always shows the same face to the earth.

As the smaller body comes more and more nearly to show always the same face to the other, the tidal couple on it will gradually disappear. The very small tidal couple on the large body will, however, still continue; and though it is no greater than formerly, it will be comparatively important, because the tidal couple on the small body has disappeared altogether.

Tidal friction in causing the two bodies to show always the same faces to one another, may increase or diminish the rotations of the bodies about their axes. The question where this increase comes from or where the rotation lost goes to, can easily be answered if there be no external forces on the system of the two bodies. For since the attraction between them consists of a number of pairs of equal and opposite forces, the sum of the "moments" of these forces about any straight line is zero; and thence it follows from known dynamical principles that the "moment of the momentum" of the system about any fixed straight line remains constant. This quantity depends partly on the rotations of the bodies about their axes and partly on the orbits described by the centres of the bodies about their common centre of mass O, which may be treated as a fixed point. Thus whenever there is a change in the rotations, there is an accompanying change in the orbits. Now since

$M \times OG = m \times OH$, the ratio $\dfrac{OG}{OH}$ is constant $\left(\dfrac{m}{M}\right)$ and very small. The points G and O are therefore nearly coincident; and instead of speaking of two orbits we may speak simply of the orbit of B about A.

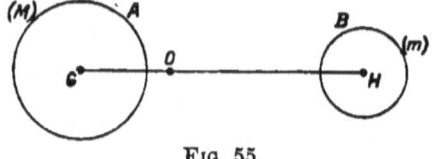

Fig. 55.

If the angular velocities of A and B are at first comparable, the large body A will start with a much larger wealth of "moment of momentum" of rotation than the small one. If we compare the large body to Dives[1] and the small one to Lazarus, the fundamental law of tidal friction may be expressed by saying that Dives and Lazarus are engaged in a common concern (altering the orbit of Lazarus about Dives) on the agreement that they contribute, not in proportion to their abilities (as the ethical notion requires they should), but inversely; Dives not having to pay a farthing for every pound extorted from the poor man's pocket. So long as Lazarus has anything to pay, the contributions of Dives will appear quite insignificant; but when the pockets of Lazarus are quite empty, the case will be different, because the mite of Dives is the only contribution then made.

Art. 80. The constant moment of momentum of our system about any fixed straight line is partly due, as we have already said, to the rotations of the bodies about their axes and partly to the translations of the bodies in their orbits about the common centre of mass O which is treated as a fixed point. The only straight line about which we shall suppose the moments to be taken will be a perpendicular through the fixed point O to the plane containing all the centres, and in comparing the values of the several parts of the total moment of momentum *we shall make a last assumption*,

---

[1] For this illustration we are indebted to Sir R. Ball.

Tidal Friction. 207

that the orbits about the fixed point O are always practically circular. Let us first of all compare the moment of momentum due to the translation of a large body with that due to the translation of a small one. To simplify the reasoning, imagine that these two bodies are the only bodies in the system: then at any instant the line joining their centres will pass through the fixed point O; and the two bodies will describe their orbits about O with the same angular velocity (ω say). Let A be the large body, B the small one, M and m their masses, G and H their centres at any instant, and let
OG $=$ R, OH $=$ r.
Then by a property of the centre of mass,

MR $= mr$.

Again, since the angular velocity of the point G in its orbit

Fig. 56.

is ω, its linear velocity will be Rω, and therefore the linear momentum of the translation of the body A will be MRω. To find the moment of momentum about the straight line through O perpendicular to the orbits (or, more shortly, about the point O), we multiply by the perpendicular from O on the tangent at G to the orbit of G, that is, by R. Thus the moment of momentum due to the translation of A is MR$^2\omega$. Similarly that due to the translation of B is $mr^2\omega$. The ratio of the first of these to the second is $\dfrac{MR^2\omega}{mr^2\omega}$, or $\dfrac{MR^2}{mr^2}$. Since MR $= mr$, this reduces to $\dfrac{R}{r}$, or $\dfrac{m}{M}$, and is therefore very small. Hence we shall introduce no sensible error by neglecting the translation of the large body and taking its centre G instead of the point O as the fixed point about which the orbits of the other centres are all described.

Let us now compare the moment of momentum due

to the rotation of a small body B about its axis with that due to its translation about the fixed centre G of a large body. Let H be the centre of the small body, m the mass, and r the distance GH. Then if ω be the angular velocity of the point H in its orbit, the moment of momentum of the translation is $mr^2\omega$. In considering the rotation of B about its axis, we have to take into account the fact that the parts of the body are not all at the same distance from the axis. Now the moment of momentum is the same

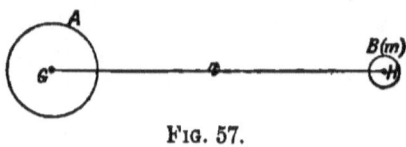

Fig. 57.

as if the whole body B were broken up into a number of small parts and every part placed at the same distance (k) from the axis of rotation and rotating about this axis with the same angular velocity (Ω say) as the whole body. Hence the moment of momentum of rotation is $mk^2\Omega$, and k is evidently less than the radius of the body (B).

The ratio of the moment of momentum of translation to that of rotation is $\dfrac{mr^2\omega}{mk^2\Omega}$, or $\dfrac{r^2\omega}{k^2\omega}$. If our system resembles the solar system in the magnitudes and distances of the bodies of which it is composed, Ω may be 10,000 times as great as ω, but r will be fully 10,000 times as great as k: hence $\dfrac{r^2\omega}{k^2\Omega}$ is equal to or greater than $\dfrac{(10,000)^2}{10,000}$ (or 10,000). But when the small body has been compelled by tidal friction to show always the same face to the large one, the inequality between Ω and ω will no longer hold, and the ratio $\dfrac{r^2\omega}{k^2\Omega}$ will then be simply enormous. In other words, the moment of momentum of rotation will be negligible in comparison with that of translation, the latter having practically absorbed the whole of the former. Since the former quantity was always small in comparison with the latter, it follows that its absorption will have made little difference in the distance GH (or r) of the small body from the fixed centre of the large one.

We shall therefore begin by supposing that all the small bodies of our system turn always the same faces to the large one, and the moment of momentum of the system will then be due entirely to the rotation of the large central body about its axis and to the translations of the small bodies in their orbits about the fixed centre of the large one.

If our system resembles the solar system, the values of the several parts of the total moment of momentum of the system can be roughly stated. Denoting the total value by 100, that due to the rotation of the central body is estimated to be 2, and the parts due to the translations of the two largest of the small bodies (Jupiter and Saturn) 60 and 24 respectively.

Art. 81. We have now to consider the effect of tidal friction on our system containing one very large body, when the small bodies always turn the same faces to the large one. Three cases may arise: (1) the small bodies may at first describe their orbits about the large one, some in one direction and some in the opposite direction, (2) the orbits may all be described in one direction and the rotation of the large body on its axis be in the opposite direction, and (3) the orbits may all be described in the same direction as the rotation of the large body on its axis. As we can only consider the problem briefly, it will be sufficient to take cases (2) and (3) alone.

In case (2) let us begin by imagining that the system contains only the large body and one small body. Let the small body be treated as a single particle at its centre H and the large one as a rigid homogeneous spherical nucleus (centre G) and two small symmetrical fluid tidal protuberances. Then it can be shown that so long as the small body continues to describe an orbit practically circular and the large body to rotate on its axis in the opposite direction, tidal friction will cause both the radius of the circular orbit and the rotation of the large body on its axis to diminish. For the tidal protuberance will not point along the line GH but will be carried by the rotation of the large body to one side of this line and will therefore point along some line tT (see fig.). The attraction of the large body on the small one is therefore equivalent to a force along HG due to the rigid nucleus, together with the forces P, p (shown in the fig.) due to the tidal protuberances.

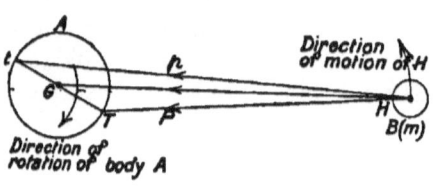

Fig. 58.

Now of these two protuberances, that of T is nearer to the body B than that of t. Hence the force P is greater than p. Again, the direction of P makes a greater angle with HG than that of p. For taking the points t and T to be at the opposite extremities of a diameter of the body A, the perpendiculars from these points on GH will be equal (n say), and the sines of the angles HT and Ht make with HG will be respectively $\dfrac{n}{\mathrm{HT}}$ and $\dfrac{n}{\mathrm{H}t}$. Of these fractions the former is the greater, because its denominator is less than that of the other. Thus of the two forces (P, p), the former is the greater and makes the greater angle with HG. The resolved part of P perpendicular to HG is therefore greater than that of p. Hence if we replace each of the forces (P, p) by two parts, one along HG and one perpendicular to HG, we see that the three forces at H which represent the attraction of the large body on the small one, are equivalent to two, one along HG, the other at right angles to HG in the opposite direction to the motion of H. The force perpendicular to GH impedes the motion of B and therefore causes it to fall

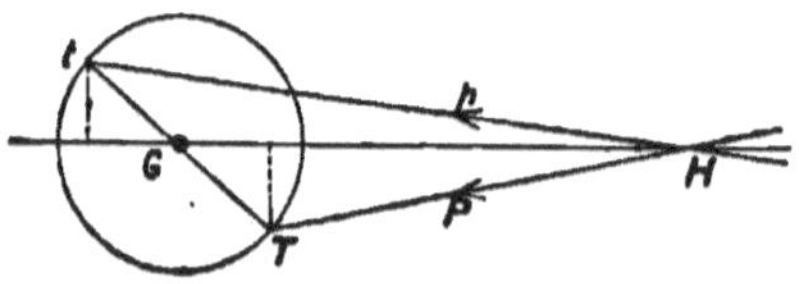

Fig. 59.

Tidal Friction. 211

inwards towards the large body A. In other words, tidal friction causes the practically circular orbit of H to become smaller and smaller.

FIG. 60.

The nature of the attraction of the small body on the large one may easily be seen in different elementary ways. First, because the large body is nearly rigid the attraction of the small one on it reduces to two forces at G equal and parallel to those at H, together with a couple about the axis of rotation through G. The mutual attrac-

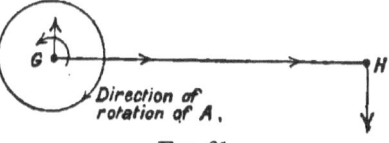

FIG. 61.

tions of the two bodies are therefore equivalent to four forces and a couple. Now the forces at G and H perpendicular to GH form a couple in the direction of the rotation of A. But since the sum of the moments about any straight line of the mutual attractions of any two bodies is zero, the couple at G must neutralize that formed of the forces at G and H by being equal and opposite to it. Hence the couple on the large body A acts in the opposite direction to its rotation and tends to stop it. We may also reason thus. The attraction of the small body on the large one is equivalent to three forces, one along GH due to the rigid nucleus, and forces (P, p) along TH and tH arising from the tidal protuberances. The forces (P, p) acting on

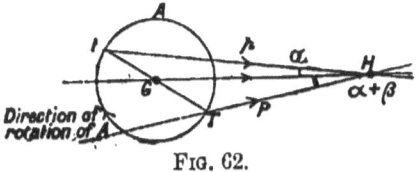

FIG. 62.

the tidal protuberances give a couple about G in the opposite direction to the rotation of the nucleus. For let the angle $t\hat{H}G$ be denoted by a and the larger angle $T\hat{H}G$ by $a + \beta$: then the moment about G of the force p acts in the direction of rotation and is equal to the product of p into the perpendicular from G on the direction of p, that is, is equal to $p \cdot \text{HG} \sin a$; the moment of P about G acts in the opposite direction and is equal to $P \cdot \text{HG} \sin(a + \beta)$. The latter moment is the greater, for P is greater than p. Hence it more than balances the former moment, and on the whole we have a moment about G in the opposite direction to the rotation of the nucleus. This moment will be transmitted by friction to the rotating nucleus of the body A. To prove this it is permissible to suppose the parts of the tides to occupy always the same positions with regard to the line GH, while the rigid nucleus rotates independently beneath, like a wheel under a friction brake. The tides not rotating about the axis through G, the friction of the rotating nucleus must produce a couple on it equal and opposite to that due to the attraction of the body B. Hence, since action and reaction are equal and opposite, the tides produce on the rotating nucleus a couple equal to and in the same direction as, the couple on the tides arising from the attraction of B, and therefore in the direction tending to stop the rotation of the nucleus.

So long as the small body continues to describe its circular orbit in one direction and the large body to rotate on its axis in the opposite direction, it is evident the two bodies cannot ever move as rigid. This state of motion must therefore come to an end. And clearly the change will take place in one of two ways: (*a*) the small body may continue in its circular orbit until the rotation of the large body on its axis ceases, or (*b*) it may cease to describe a circular orbit before the rotation of the large one is stopped.

Taking supposition (*a*), it is obvious that when the rotation of the large body ceases, the tidal protuberances will have the same general position with regard to the straight line GH as before. There will therefore be a force tending to impede the motion of B and a couple about the axis of A in the same direction as before. Hence as soon as the rotation of the large body on its axis stops, it begins to rotate in the opposite direction. The translation of B and the rotation of A will now be in the same direction, but the angular velocity of the former will be greater than that of the latter. So long as B continues to describe a practically circular orbit with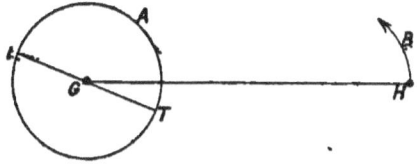

Fig. 63.

greater angular velocity than that of the rotation of A, it is easy to see B will continue to fall inwards towards G and the body A to rotate faster and faster. If while this state of motion lasts, the angular velocity of the rotation of A catches up to that of the translation of B, tidal action will be at an end. As we know that the orbit is stable when the bodies can be treated as perfectly rigid, there is clearly nothing to make the angular velocities of rotation and translation again unequal. Thus A will henceforth show always the same face to B and B will continue for ever to describe the same circular orbit about the point G. Before this event can occur, it may, however, happen that B either falls completely into A or ceases to describe a practically circular orbit. As we are limiting ourselves to the consideration of circular orbits, we cannot here investigate the last case or that in supposition (*b*) stated above. We shall merely observe that as the system must ultimately move as rigid, the small body must ultimately coalesce with the larger or describe a constant circular orbit about G and the large body A rotate on its axis in the same direction and with the same angular velocity.

Thus finally the two bodies may be reduced to one or may continue separate.

If in case (2) the system contains at first any number of small bodies, it is easy to see that tidal friction will either reduce the whole system to one body, or bring us to case (3).

In case (3) the small bodies all describe their orbits

in one direction and the large central body rotates on its axis in the same direction. Owing to the importance of this case, which is roughly that of the solar system, it will be considered in its generality, the number of small bodies being any whatever. It will be shown in the first place that the tidal effect of one small body on another is insignificant in comparison with its effect on the large one unless the two small bodies be comparatively close together. For let m be the mass of any small body, δ the diameter of any other small body and D that of the large central body. Also let r and ρ be the distances of m from the point G and the body δ respectively. Then the tide-producing forces of m on δ and D are respectively $\dfrac{2\lambda m \delta}{\rho^3}$ and $\dfrac{2\lambda m D}{r^3}$. The ratio of the former to the latter is $\dfrac{2\lambda m \delta}{\rho^3} \cdot \dfrac{r^3}{2\lambda m D}$, or $\dfrac{\delta}{D}\left(\dfrac{r}{\rho}\right)^3$. On account of the very small factor $\dfrac{\delta}{D}$, this ratio is very small except when $\dfrac{r}{\rho}$ is large, that is, except when $\dfrac{\rho}{r}$ is small, or the small bodies comparatively near together. Again, it will be shown that if B be any one of the small bodies which is always comparatively remote from all the other small bodies, the tidal action on B of the large central body is the same, on the whole, as if all

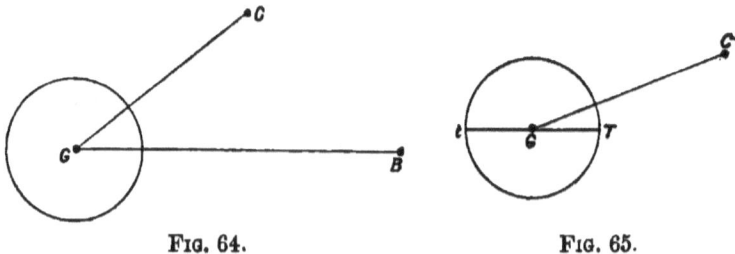

Fig. 64.   Fig. 65.

the other small bodies were absent, provided that the orbits are all circular. For if C be any one of these small bodies, the periodic times of B and C will be different owing to the unequal sizes of the orbits. Hence the angle between GB and GC alters at a constant rate and passes through all values. Now the line joining the summits (T, t) of the tidal protuberances on the large body caused by C makes a practically constant angle with GC. The angle it makes with GB therefore changes at a constant rate. Thence it follows that the force on B at right angles to GB is alternately accelerating and retarding; and the opposite effects thus produced will practically neutralize one another.

We reserve for another article the consideration of what happens when two or more of the small bodies come near together. For the present, we suppose them to be always so far apart that we may neglect their direct tidal action on each other and take the tidal action of the large body on a small one to be the same as if all the other small bodies were absent.

The tide-producing force of a small body B on the central body varies inversely as the cube of the distance of B from the point G. As the tides are small when B is sufficiently far from G, we shall suppose the orbit of B to be always practically circular except when it becomes small. The consideration of non-circular orbits will be avoided by assuming that when any orbit ceases to be circular, the small body is disposed by falling into the large one.

Now let there be several small bodies in the system of which B describes the smallest orbit, C the greatest. Then since the squares of the periodic times vary as the cubes of the distances from the central body, B will have the shortest periodic time and C the longest. In other words, B will have the greatest angular velocity of translation, and C the least. Hence in the initial state of the system, the angular velocity of the rotation of the central body will be either less than that of the translation of C, greater than that of the translation of B, or intermediate between them. In the first of these initial states, all the small bodies will tend to increase the rotation of the central body, and their orbits will all decrease. This will go on until all the small bodies fall into the large one, or until the angular velocity of the rotation catches up to that of the translation of the outermost small body C. If the body which is caught up to is the only small body left in the system, tidal friction will clearly be at an end and the distance of C from G will remain constant. If, however, there are other small bodies remaining in the system besides C, the tidal action of these bodies will still cause the rotation of the central body to go on increasing until it exceeds that of the translation of C. The angular velocity of the rotation will then be intermediate between the greatest and least angular velocities of translation, as in the third initial state. In the second initial state, all the small bodies will tend to diminish the rotation of the central body and their orbits will all increase. Now the orbits cannot increase indefinitely; for we know that the moment of momentum of the system about any fixed straight line remains constant; and if the fixed straight line about which moments are taken be chosen to be the axis of rotation through G, it is easy to show that when any orbit increases indefinitely, the corresponding moment of momentum becomes infinite. If m be the mass of the body, r its distance from G and ω its angular velocity of translation, its linear momentum is $mr\omega$ and its moment of momentum $mr^2\omega$. But by Kepler's third Law, the square of the periodic time varies as the cube of the distance r: hence since the periodic time is $\frac{2\pi}{\omega}$,

$$\left(\frac{2\pi}{\omega}\right)^2 \propto r^3 = \beta r^3,$$

where β is a quantity which remains constant as the orbit changes. Therefore

$$\omega^2 = \frac{4\pi^2}{\beta} \cdot \frac{1}{r^3} = \frac{\beta'}{r^3} \text{ (say)},$$

where $\beta' = \frac{4\pi^2}{\beta}$.

Hence
$$\omega = \frac{\sqrt{\beta'}}{\sqrt{r^3}} \text{ or } \frac{\sqrt{\beta'}}{r^{\frac{3}{2}}},$$
and
$$mr^2\omega = mr^{\frac{1}{2}}\sqrt{\beta'} \text{ or } m\sqrt{r}\sqrt{\beta'}.$$

Thus when r increases indefinitely, $mr^2\omega$ becomes infinite. Evidently then the orbits cannot increase indefinitely and hence there will be a limit to the angular velocities of translation. Consequently the incessant action of the tides in diminishing the rotation of the central body will sooner or later reduce the angular velocity of rotation to equality with the greatest angular velocity of translation—that of the body B. When this occurs, B will temporarily have no part in tidal action;[1] and therefore as the tidal influence of the other small bodies on the central body continues, the angular velocity of its rotation will become less than that of the translation of B. Thus we are again led to the third initial state.

When the angular velocity of rotation of the central body is intermediate between the greatest and least angular velocities of translation, tidal friction will drive the outermost small bodies further out and draw the others in. Meanwhile the rotation of the central body will probably vary; but so long as the orbits continue circular and none of the small bodies fall into the larger, the angular velocity of rotation can never be so great as the greatest angular velocity of translation (which is that of the innermost small body B). For whenever there is an approach to equality between them, the part played by B^1 in tidal action tends to disappear, while all the other small bodies still continue to diminish the rotation of the central body. Thus the innermost small body B is continually drawn in by tidal friction until at last it falls into the large body.

After this catastrophe, the angular velocity of the

[1] B will produce a tide in the central body, but this tide will neither have any effect on the motion of the central body nor on that of B.

rotation of the large body may possibly be greater than that of the translation of the innermost remaining small body. If so tidal friction will cause it to become less, and then another small body will be sucked into the large one. This will go on until there is only one small body left. If after the last but one of the small bodies falls in, the angular velocity of translation of the last small body is less than the angular velocity of rotation of the large body, the rotation will diminish and the small body be driven out until a permanent state is reached in which tides are absent. If, however, the angular velocity of translation be greater than that of rotation, the rotation will increase and the small body be drawn in until it either falls into the larger or until the two angular velocities become equal, after which there will be no further change.

Thus we see that in the third case, however many separate bodies the system may have once contained, there will ultimately be either one or two, provided no two of the small bodies ever come near together.

Art. 82. We have now to consider what happens in the third case when two or more of the small bodies come near together. Let us first suppose that only two small bodies come near together and let us determine their tidal action on one another. Each of the small bodies will consist of three parts, a nucleus which may be treated as rigid, and two tides, one due to the large central body, the other to the remaining small body. The first of these tides points always to the centre G of the large body. Hence if B be the inner of the two small bodies, C the outer, the

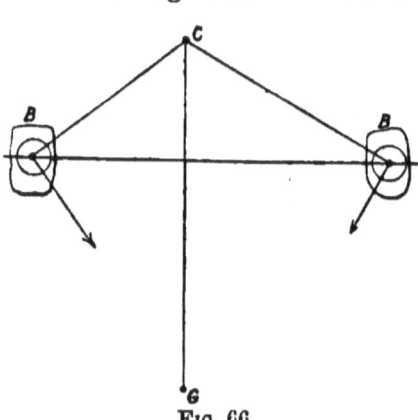

Fig. 66.

attraction of C on the tide raised on B by the central body G, produces a force at right angles to CB which tends to make B go faster when on one side of the line CG, slower when on the other. If the orbits continue circular, these opposite effects neutralize one another. Similarly, we may neglect the effect of B on the tide raised on C by the central body. Again, if (T, t) be the summits of the tide raised on B by C, the line Tt is always on one side of CB; and the tidal action of C on B produces a force at right angles to CB, which it is easy to see, tends to make B go slower and therefore approach the central body. Similarly, the direct tidal action of B on C tends to make C go faster and increase its orbit.

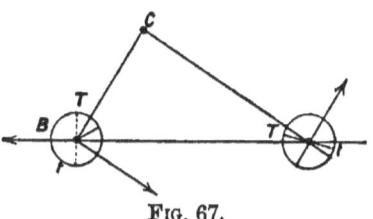

FIG. 67.

Thus the two small bodies may be prevented from coming near enough to collide by their direct tidal action.

If several orbits approach simultaneously to equality, the small bodies may possibly form a ring round the central one. If there were no tides, this state might be stable[1] and would continue for ever. It is easy to show, however, that the tides will tend to destroy it. For since the periodic times are equal, the tidal action of the central body on one of the small ones will no longer be the same as if all the other small bodies were absent. In consequence, the tidal action on one of the small bodies will not generally be proportional to its mass. In other words, some of the small bodies will tend to overtake those in front of them. If the stability of the system, considering all the bodies rigid, is sufficient to counteract this destructive tendency, the ring will continue: if not, the small bodies forming it will coalesce.

---

[1] The proposition that the motion in this state may be stable, has been proved by Maxwell in his Essay on Saturn's rings.

For the other statements in this article, I have little or no authority, and they are merely given as speculations.

It will now be seen that the tides will ultimately reduce our system of one large body and a number of small ones, to one body or to two, or perhaps to one large body and a ring.

Art. 83. The principles of tidal friction have an important application in the case of the sun, earth, and moon. We must first find the present relative values of the tides in the three bodies.

The tides produced in the sun by the earth and moon are comparatively unimportant, on account of the immense size and mass of the sun, and because the earth and moon do not yet turn always the same face to the sun.

The tides on the earth are partly due to the moon and partly to the sun. To compare the two causes, let M and S be the masses of the moon and sun, ρ and r the distances of their centres from that of the earth, and d the diameter of the earth. Then the tide-producing forces of the moon and sun are respectively $\dfrac{2\lambda M d}{\rho^3}$ and $\dfrac{2\lambda S d}{r^3}$. The ratio of these forces is $\dfrac{2\lambda M d}{\rho^3} \cdot \dfrac{r^3}{2\lambda S d}$, or $\dfrac{M}{S} \cdot \left(\dfrac{r}{\rho}\right)^3$. The mass of the sun being nearly 360,000 times that of the earth, and the mass of the earth 88 [1] times that of the moon, and the present value of $\dfrac{r}{\rho}$ about 385, we easily find by writing the ratio of the two tide-producing forces in the form $\dfrac{M}{E} \cdot \dfrac{E}{S} \cdot \left(\dfrac{r}{\rho}\right)^3$, that it is equal to about $\tfrac{16}{9}$.[2]

---

[1] These numbers are taken from Herschell's "Outlines of Astronomy."

[2] Although the moon is a more efficient tide-producing agent than the sun, it can easily be shown that the attraction of the moon on the earth is far smaller than that of the sun. For if E be the mass of the earth, these forces are respectively $\dfrac{\lambda E M}{\rho^2}$ and $\dfrac{\lambda E S}{r^2}$, and therefore their ratio is $\dfrac{\lambda E M}{\rho^2} \cdot \dfrac{r^2}{\lambda E S}$, or $\dfrac{M}{S}\left(\dfrac{r}{\rho}\right)^2$, which is easily calculated to be about $\tfrac{2}{15}$.

The moon is hardly susceptible to tidal influence at present, owing to its want of fluidity, there being neither water nor air on the moon. Even if the moon were as fluid as the earth, the tidal effect of the earth on it would be zero, because the moon always turns the same face to the earth. The tidal effect of the sun on the moon would not, however, be zero. For since the moon rotates on its axis in a month, and describes in company with the earth an orbit round the sun in a year, it is clear that the moon does not at present show always the same face to the sun. Indeed, it may be seen that the sun's tidal effect on the moon's rotation would be comparable with the sun's tidal effect on the earth's rotation. The sun would produce a considerably smaller tidal couple on the moon than on the earth; but a given couple acting for a given time will produce about 1,200 times as great a change of rotation in the case of the moon as in that of the earth, owing to the smaller mass and size of the moon. The tidal couple on the moon would be less than that on the earth for two reasons:—
(1) because the moon rotates slower than the earth, and (2) because the sun has a smaller tide-producing force on the moon than on the earth; for since the distance from the sun to the moon may be taken to be the same as that from the sun to the earth, if δ be the diameter of the moon, the ratio of the tide-producing forces of the sun on the moon and earth is $\dfrac{2\lambda S\delta}{r^3} \cdot \dfrac{r^3}{2\lambda S d}$, or $\dfrac{\delta}{d}$, which is a little greater than $\frac{1}{4}$. If then we take into account the fact that the moon at present is vastly less fluid than the earth, it will follow that the effect of the tides on the moon's rotation is negligible in comparison with the effect on the rotation of the earth.

Thus of the three bodies, sun, earth and moon, the earth is the only one which is not free from tides at the present time.

As the earth rotates on its axis too fast to show always the same face either to the moon or sun, the lunar and solar tides will concur to diminish the rotation, that is, to increase the length of the day. This effect is too feeble for instrumental verification; but it has been detected by the astronomical calculations of Adams and Delaunay, which could not be made to agree with observation except on the supposition that the day is slowly lengthening.

It may be noticed that as the spheroidal form of the earth is a consequence of the axial rotation, the slackening of the rotation will give the earth a tendency to become spherical. If we suppose the solid parts of the earth to be indifferent to this tendency, the effect will be confined entirely to the seas, which will slowly leave the equatorial for the polar regions.

Art. 84. The action of external bodies on our system of the three bodies is small and not very important to us. If we neglect it altogether, it will follow that the moment of momentum of our system about any fixed straight line remains constant. Hence the change in the rotation of the earth on its axis will be accompanied by changes in other parts of the system.

The moment of momentum of the system depends on the rotations of the three bodies on their axes, and on the orbits of the moon about the earth and of the earth about the sun.[1] The part due to the rotation of the moon on her axis is very small in comparison with that due to the rotation of the earth, partly on account of her small mass and size, and also on account of the slowness of her rotation on her axis. We shall therefore neglect it. The part due to the rotation of the sun is very large, but the tides have no influence on it, comparatively speaking, so long as there are tides on the earth. In what follows we confine ourselves to the case

---

[1] Strictly speaking, not the orbit of the earth about the sun, but the orbit about the sun of a mass $E + M$ placed at the centre of the earth. The moment of the momentum of this particle about any straight line is greater than that (about the same straight line) of the earth's translation in the constant ratio $\dfrac{E + M}{E}$, or $1\frac{1}{88}$.

in which there are tides on the earth. The moment of momentum due to the sun's axial rotation will therefore be constant and need not be taken into account. The moment of momentum of the system depends therefore only on the rotation of the earth on its axis and on the two orbits.[1] We shall now assume that the lunar tides affect only the orbit of the moon round the earth, and the solar tides only the orbit of the earth about the sun. It will also be supposed, for simplicity, that the two orbits are circular and in one plane, and that the axis of the earth is at right angles to this plane.

Now let ω be the angular velocity of the moon about the earth, ω' of the earth about the sun, and Ω that of the rotation of the earth on its axis. Then the moment about the earth's axis of the momentum of the translation of the moon is $M\rho^2\omega$, and of the rotation of the earth $Ek^2\Omega$, where k is less than the earth's radius (see Art. 81). The ratio of the former quantity to the latter is $\dfrac{M\rho^2\omega}{Ek^2\Omega}$, or $\dfrac{1}{88}\left(\dfrac{\rho}{k}\right)^2\dfrac{\omega}{\Omega}$. The present value of this ratio is $\dfrac{1}{88}\left(\dfrac{\rho}{k}\right)^2\dfrac{3}{82}$; and therefore even if k were equal to the earth's radius, the ratio would be greater than $1\frac{1}{2}$. The moment of momentum of the moon's translation is consequently several times greater than that of the earth's rotation. Again, the moment of momentum about the sun's axis of the translation of the earth is $Er^2\omega'$. This is evidently very great in comparison with the present values of $M\rho^2\omega$ and $Ek^2\Omega$.

It is now easy to predict part of the moon's future history. So long as the lunar and solar tides both continue to check the earth's axial rotation, the moment of momentum lost by the rotation goes to increase the orbits of the moon about the earth, and of the earth about the sun. Thus only part of the moment of momentum lost by the earth's rotation goes to increase that of the moon's translation about the earth, a part

[1] See previous footnote.

going to increase that of the earth's translation about the sun. The distance of the moon from the earth cannot, however, increase from its present value in a ratio much greater than one before the earth will have been brought to show always the same face to the moon and the moment of momentum of its rotation be practically exhausted. For since at present the moment of momentum of the moon's translation is several times greater than that of the earth's rotation, it is clear that even if the latter were transferred wholly to the former, the former would not increase in a ratio much greater than unity. When the earth begins to turn always the same face to the moon, the day and the month will then be of the same length. By calculations which cannot be given here, because we cannot here find the value of k, it may be shown that the length of the day will then be nearly 60 times as long and the distance of the moon from the earth nearly $1\frac{1}{2}$ times as great, as at present. The lunar tides will then be at an end, but only for an instant. Owing to the action of the solar tides, the rotation of the earth will continue to diminish, and the angular velocity of the rotation will become less than that of the moon's translation. In other words, the day will become longer than the month. Lunar tides will then be called into existence; but this time they will cause the moon to approach the earth and tend to increase the rotation of the earth on its axis.

It is interesting to notice that there is a known instance in the solar system at present of the day being longer than the month. We refer to the case of the inner satellite of Mars, discovered by Prof. Hall at Washington in 1877. The day of Mars, or the period in which he completes a rotation on his axis, is about $24\frac{1}{2}$ hours, while the revolution of the satellite about the primary is accomplished in less than 8 hours.

As the moon is at present receding from the earth and the earth's axial rotation decreasing, it is evident that in the past the moon must have been nearer the earth, and the earth's rotation greater than at present. If we go back far enough, we can imagine the moon close to the earth and the earth rotating very rapidly. Indeed, it is the accepted theory that the earth and moon were originally one body which broke in two owing to some instability, and the parts of which were then separated by the action of the tides.

In the early stages of the moon's history, the moon was probably at a high temperature and in a molten state, and therefore much more susceptible to tidal influences than at present. Now at present the tide-producing force of the earth on the moon exceeds that of the sun on the moon in the ratio $\dfrac{2\lambda E}{\rho^3} \cdot \dfrac{r^3}{2\lambda S}$, or $\dfrac{E}{S}\left(\dfrac{r}{\rho}\right)^3$, or about 150. When the moon was much nearer the earth than at present, the ratio would be much greater still. The fact that the moon turns always the same face to the earth at present therefore seems to be a consequence (as Helmholtz originally pointed out) of the tidal superiority of the earth over the sun on the moon while the moon was still in a molten condition.

Art. 85. Whenever friction is at work, we know that there is a production of heat, and that the bodies subject to the frictional action tend to rise in temperature. In the case of the earth, the tides will evidently cause the general temperature to rise unless there is a sufficient radiation of heat into space. Suppose that q is the positive quantity of heat lost from the earth in a given time during which the general temperature remains constant. Then q may be called the heat developed by tidal friction. The question now arises, Is this quantity exactly equal to the kinetic energy lost in the same time from the earth's rotation? The problem is somewhat complex when there are three bodies, the sun, earth and moon, to be taken into account. We shall therefore imagine the sun to be absent, and consequently that there are no solar tides.

It will then be easy to show that at present the heat developed by tidal friction is less than the kinetic energy lost from the earth's rotation.

As the system of the earth and moon is imagined to be unacted on by external forces, it will be able to lose or gain energy only in the form of heat. Also as the moon is free from tides, there will be no heat lost or gained except in case of the earth. Thus the energy lost by the system in the given time will simply be q.

Again, since the moon is free from tidal action, its axial rotation (which is small) will be constant. The part of the energy of the system that can vary, will therefore consist of the following items:—

(1) The kinetic energy due to the rotation of the earth on its axis.

(2) The kinetic energy due to the translations of the two bodies about their common centre of mass, which is a fixed point.

(3) The gravitational potential energy due to the separation of the two bodies from one another.

In calculating (2), each body is to be supposed concentrated into a particle at its centre of mass. The same supposition may also be made without much error in calculating (3).

In (2), let E be the centre of the earth, M that of the moon, and G their common centre of mass. Also let $GE = x$, $GM = y$, and let ω be the common angular velocity of the two points (or of the line EM) about G. Then when the points are describing circular orbits about G,

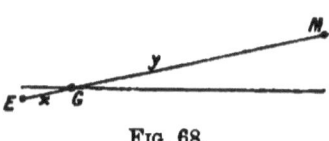

FIG. 68.

their velocities will be respectively $x\omega$ and $y\omega$. Hence the sum of the kinetic energies of the two particles at E and M is $\tfrac{1}{2}\mathrm{E}(x\omega)^2 + \tfrac{1}{2}\mathrm{M}(y\omega)^2$. Now if a particle be describing a circular orbit of radius R with linear velocity v and angular velocity ω, the acceleration along the normal is $\dfrac{v^2}{R}$, or $R\omega^2$, since $v = R\omega$. But this is equal

to the normal force on the particle per unit of mass. Hence
$$\frac{\lambda E}{(x+y)^2} = y\omega^2 \text{ and } \frac{\lambda M}{(x+y)^2} = x\omega^2.$$

Thus the sum of the kinetic energies of the two particles is $\frac{1}{2}\frac{\lambda EM}{(x+y)^2}(x+y)$, or $\frac{1}{2}\frac{\lambda EM}{\rho}$, since $x+y=\rho$.

Next, the potential energy of the two bodies (or particles) increases with the distance and may be written
$$-\frac{\lambda EM}{\rho}.$$

Hence the sum of (2) and (3) is $-\frac{1}{2}\frac{\lambda EM}{\rho}$, and increases with ρ.

If during the time that the heat q is given out, the decrease in (1) be d, and i the increase in (2) and (3), the total decrease in the energy of the system may either be written q or $d - i$. Hence $q = d - i$, or $q + i = d$. Thus at present q is less than d; or only a part of the energy d lost by the earth's rotation appears as heat, a certain amount of d being required to supply the increase in sum of (2) & (3).

CHAPTER VI.

APPLICATIONS TO ELECTRICITY.

ART. 86. In this chapter some applications of the principles of thermo-dynamics will be made to electrical questions. It was not à priori obvious that the principles of thermo-dynamics would be true in such cases; they have been proved to be so (1) by giving results which had already been obtained by experiment and observation, and (2) by leading to predictions which have since been verified by experiment. Among the predictions we may mention Sir W. Thomson's discovery of the "specific heat of electricity" and the formula of Helmholtz for the electromotive force of a galvanic battery in terms of the strength of the aqueous solution forming the electrolyte.

The principles of thermo-dynamics were first applied to electrical problems by Sir W. Thomson in 1852. The questions he discussed refer to a thermo-electric circuit, and the results he obtained are universally accepted.

In addition to homogeneous pieces of metal, each of which is at one temperature, a thermo-electric circuit contains two kinds of junctions, one kind of which is the junction of two pieces of different metal at the same temperature, and the other the junction of two pieces of the same metal at different temperatures. If we know all the properties of the two kinds of junctions, we can immediately obtain all the properties of a thermo-electric

Applications to Electricity. 229

circuit. Now before we can investigate all the properties of a junction, it is necessary to have an accurate expression for the energy of a system electrified statically. This expression has been given by Helmholtz and it has recently been deduced from thermo-dynamical considerations by Duhem. But in 1852 the expression was not known. In his paper of 1852 Sir W. Thomson therefore adopted a very simple method which does not require it. In this way he obtained several of the properties of a thermo-electric circuit, but there are some important questions in connection with the circuit which the method is insufficient to solve. However, it is easy, by means of Helmholtz' expression referred to above, to supply the defects of Thomson's method. In fact, the principle result relating to a junction of two pieces of different metal at the same temperature has been obtained independently by several investigators, among whom we may mention Maxwell, and perhaps also Gibbs and Helmholtz. At any rate, the result in question is almost included in a similar result obtained by Gibbs and Helmholtz for the galvanic battery.

We shall therefore begin the chapter with some investigations which do not require a knowledge of the energy of an electrified system. Then Helmholtz' expression will be assumed and some deductions made from it. For a junction of two pieces of different metal at the same temperature, we easily deduce the result given by so many independent investigators. In the case of a junction of two pieces of the same metal at different temperatures, a very simple and striking result is obtained which seems to have as firm a foundation as the result mentioned just previously: for (1) the two results are obtained by exactly the same kind of reasoning, (2) if we take both of the results for granted, we can deduce all Thomson's formulæ from them, and (3) if we assume Thomson's formulæ, we can all but deduce the second result from them.

After the discussion of the properties of a thermo-electric circuit, the fundamental formula of Gibbs and

Helmholtz for the electromotive force of a galvanic battery will be given. The chapter will then be practically completed by some discussions of experiments and by a comparison of some of the methods of obtaining work from the consumption of food or fuel.

Art. 87. In this chapter, a previous knowledge of electricity is required as well as a knowledge of the principles of thermo-dynamics. The latter may be obtained from Chaps. I. and III., and the chief results are briefly re-stated at the beginning of Chap. IV. The former must be obtained from works on electricity: a few definitions and results, however, will be given that will be found useful for reference.

Throughout the chapter we work in the electro-magnetic C.G.S. absolute system of electrical units, which are obtained as follows:—

Let two equal magnetic poles, both of which are positive, be placed in a vacuum at a distance of one centimetre apart. Then if the mutual repulsion between the two poles be a dyne, each pole is defined to be of unit strength.

Next, let a portion of a wire conveying a current be in the form of an arc of a circle whose radius is one centimetre, and let a unit magnet-pole be situated at the centre. Then if a force of one dyne be exerted on the pole by each centimetre of the circular arc of the conducting wire when the whole is in a vacuum, the current which the wire conveys is defined to be the electro-magnetic C.G.S. unit current,[1] and the quantity of electricity it conveys in one second the electro-magnetic C.G.S. unit of quantity.

Again, let two conductors whose charges are respectively q and q' electro-magnetic C.G.S. units, be situated

---

[1] The circumference of the circle is 2π centimetres. Hence if the circular part of the wire form a complete circle, the force which this circle exerts on the pole will be 2π dynes when the current is unity. If there be n turns of wire all in the same direction and the thickness of the wire be so small that all the turns may be supposed to lie in one plane and be everywhere at a distance of one centimetre from the pole, the force on the pole will be $2\pi n$ dynes for unit current, and $2\pi n I$ dynes for a current I.

Applications to Electricity. 231

in a vacuum at so great a distance apart that the conductors may be treated as mere points. Then if this distance be r centimetres, the mutual force between the conductors is found to be $\varepsilon\frac{qq'}{r^2}$ dynes, where ε is a large positive constant the value of which experiment gives as 9×10^{20}.

If the charges of an electrified system in electric equilibrium be supposed divided into a number of parts each of which is so small that it may be considered a point in comparison with its distance from a point P, the value of $\frac{q}{r} + \frac{q'}{r'} + \frac{q''}{r''} + \ldots$, or $\Sigma\frac{q}{r}$, where ($q, q', q'',$ ...) are these small charges and ($r, r', r'',$ ...) their respective distances in centimetres from the point P, is defined to be the Potential of the electrified system at the point P.

If a steady current be flowing in a homogeneous wire of uniform temperature at rest, and all other bodies near it be at rest in electric and magnetic equilibrium, or at rest merely conveying steady currents, there is found to be a steady fall of potential in the conducting wire in the direction of the current. The electro-magnetic C.G.S. electromotive force in any part PQ of the wire is defined to be ε times the difference of the electro-magnetic C.G.S. potentials of the points P, Q. If the length of PQ be such that this electromotive force in it is unity, and the current be also unity, the resistance of PQ is defined to be the electro-magnetic C.G.S. unit of resistance.

In addition to the electro-magnetic C.G.S. system of units, there are two other systems of electrical units in use which it is necessary to understand and to be able to compare with the electro-magnetic C.G.S. system. These two additional systems are the electro-static C.G.S. absolute system and the practical system.

In the electro static C.G.S. system, the unit of quantity is such that if two small conductors which may be treated as points be each charged with unit charge and placed in a vacuum at a distance of one centimetre apart, the repulsion between them is one dyne. Hence if q be the value of this unit measured in the electro-magnetic C.G.S. system, we have $\epsilon q^2 = 1$, and therefore $q^2 = \dfrac{1}{\epsilon}$ or $q = \dfrac{1}{3 \times 10^{10}}$. Thus the electro-magnetic C.G.S. unit of quantity is 3×10^{10} times as great as the electro-static C.G.S. unit.

In the electro-static C.G.S. system, electromotive force is the same as difference of potential. Hence the electro-magnetic C.G.S. units of potential, electromotive force, and resistance, are respectively 3×10^{10} times as great, 3×10^{10} times as small, and 9×10^{20} times as small, as the corresponding electro-static C.G.S. units.

The practical units form an electro-magnetic absolute system in which the fundamental units of length, mass, and time, are respectively 10^7 metres, 10^{-11} gramme, and a second. The practical system of units derives its name from the fact that several of its units are comparable with quantities that occur in the working of galvanic batteries.

The practical unit of work or energy, called the Joule, is equal to 10^7 dynes.

The practical unit of current, called the Ampere, is equal to $\tfrac{1}{10}$ of the electro-magnetic C.G.S. unit current.

The practical unit of electromotive force, called the Volt, is equal to 10^8 of the electro-magnetic C.G.S. units. It is about equal to the electromotive force of a Daniell's battery. The practical unit of quantity, called the Coulomb, is the quantity conveyed in one second by a current of one ampere.

The practical unit of resistance, called the Ohm, is equal to 10^9 electro-magnetic C.G.S. units. The resistance of 100 metres of copper wire 1 millimetre in diameter, is about 2 ohms.[1]

---

[1] For further information see Mascart's "Electricity and Magnetism."

Applications to Electricity.

The following table gives the values of the Ampere, Volt, and Ohm, in terms of the corresponding units of the electro-magnetic C.G.S. and electro-static C.G.S. systems.

Practical Units.	Value in the Electro-Magnetic C.G.S. Units.	Value in the Electro-Static C.G.S. Units.
Ampere	$\frac{1}{10}$	3×10^9
Volt	10^8	$\frac{1}{300}$
Ohm	10^9	$\frac{1}{9 \times 10^{11}}$

Art. 88. When an electrified system is at rest in electric and magnetic equilibrium, it is known that the electric potential is constant throughout any homogeneous conductor whose temperature is uniform. From this it is deduced that there is no distribution of electricity in the substance of the conductor: the charge is therefore confined to the surface or surfaces.[1] Again, if a small charge which may be treated as a point is placed anywhere in the substance of the conductor without disturbing the electric state of the system, the electric force exerted on the small charge by the given system is known to be zero. We therefore see that the conductor forms a gap or neutral region in the electric field.

If the poles of a battery are joined by a conducting wire, a current is produced which quickly becomes uniform. When this is the case, there is a continual fall of potential in the direction of the current in those parts of the circuit which are homogeneous and of uniform temperature. But in travelling round a closed circuit, the total change of potential is evidently zero. There must therefore be abrupt rises of potential at the junc-

---

Also Everett's "Units and Physical Constants." Also the Appendix to this work.

[1] If there is a hollow within the conductor, its surface may be electrified. If, however, the conductor is solid throughout, there will be no charge at all within the conductor.

tions whose sum is equal to the total fall in the rest of the circuit.

Hence we see that when two homogeneous conductors [1] of uniform temperatures are put in contact, there is an abrupt change of potential at the surface of contact. This change appears to be the same whatever current be crossing the junction, and is found to be independent of :—

(1) The size and form of the two conductors.
(2) The extent and form of the surfaces of contact.
(3) The charges of the two conductors.

In short, it is found to depend only on the natures, temperatures, and physical and chemical states of the two conductors.

The theory of sudden changes of potential at junctions which we have just explained, is fully adopted by Mascart, Tait, Maxwell, and other leading writers on electricity and thermo-dynamics. It is known as the "contact theory" of electricity. Formerly a different theory, which is now universally rejected, was known as the "contact theory." It was supposed that the abrupt differences of potential at the junctions of a galvanic battery could maintain a current *without chemical action*. This supposition has been found to be completely at variance with experiment, by which it has been shown that no current, great or small, can traverse an electrolyte without causing chemical action.[2] In the case of a

---

[1] The word "conductors" is here supposed to include electrolytes.
In the case of non-conductors, a difference of potential has also been shown to exist.

[2] The experimental discoveries of Faraday in electro-chemistry were the chief cause of the disappearance of the original contact theory. These experiments were so decisive that many persons hastily concluded not only that a current cannot traverse an electrolyte without causing a chemical action, but also that there is no sudden change of potential at a junction. It is therefore interesting to find that Faraday expressly said he had no objection to the theory of a sudden change of potential at a junction, but only to the theory that such sudden changes could maintain a current in a galvanic battery *without chemical action*.

We may add that at one time it was supposed that Faraday's result was not true when the current which traverses the electrolyte is very feeble. It has, however, been shown by Helmholtz, one of the founders

Applications to Electricity.

thermo-electric circuit of metals, there is no liquid for chemical action to take place in; but it is found, both by experiment and theory, that when a current flows in such a circuit, heat must be absorbed or evolved at some of the junctions or the temperatures of the junctions will change.

When a current traverses a conductor, it tends to heat or cool the parts of the conductor, and the thermal phenomena at a junction are different from what they are elsewhere.

When a current traverses a homogeneous conductor of uniform temperature, the temperature tends to rise, or heat must be abstracted to keep the temperature constant. If R be the resistance of the conductor and I the intensity of the current, it was found by Joule that the heat that must be abstracted per second to keep the temperature constant is RI^2 ergs. This expression remains unchanged when the sign of I is changed. The current therefore tends to heat the conductor whether it flows in one direction or the opposite; in other words, the heating effect of the current is irreversible. Again, if q be the quantity of electricity conveyed by the current, supposed constant, in t seconds, so that $q = It$, the heat that must be abstracted in accordance with Joule's law during this time, or RI^2t, will be equal to RIq. But RIq can be made as small as we please by taking I small enough. Hence if a given quantity of electricity is to traverse the conductor, the heat developed by Joule's law can be made to vanish by making the current slow enough or allowing plenty of time.

When any given quantity of electricity crosses a junction, it is found that the thermal effect at the junction is simply proportional to the quantity of electricity that passes and independent of the strength of the current (or of the time taken by the given quantity of electricity to cross). It cannot therefore be made to

of the modern contact theory, that even when the current is very feeble there is no exception to Faraday's law that conduction takes place in electrolytes only by electrolysis.

disappear by diminishing the strength of the current. It is also found that the thermal effect is reversible; so that if it is necessary to impart heat to keep the temperature of the junction constant when the charge crosses in one direction, it is necessary to abstract an equal quantity when the charge crosses in the other direction. Lastly, it is found that the thermal effect is independent of :—

(1) The size and form of the two conductors.
(2) The extent and form of the surfaces of contact.
(3) The charges of the two conductors.

Art. 89. All the properties of the difference of potential at a junction can be recovered theoretically by means of thermo-dynamics with the exception of the fact that the difference of potential is independent of the current that happens to be crossing the junction. Several of these properties are the same whether the junction is formed by the contact of two conductors which a current traverses without electrolysis or whether one or both of the substances in contact is an electrolyte. The theoretical treatment of the two cases must, however, be different, because we cannot cause a current to traverse an electrolyte without electrolysis. In the present work we are not much concerned with the properties of electrolytes: we shall therefore restrict the word "conductor" in what follows to mean a substance which a current traverses without electrolysis.[1]

The theoretical discussion of the difference of potential at a junction depends on the following proposition:—

Let any conductors A, B, C, . . . be put in contact so as to form a closed chain at rest, and let the temperature of the chain be kept uniform and constant. Then there cannot be a current in the chain if there are no external objects acting on it.

The passage of a current in the chain is not attended with electrolysis. Hence if the current be the same at

[1] The word "metal" will sometimes be used as equivalent to "conductor" in this sense, and will include liquid conductors, like mercury, as well as solid substances.

Applications to Electricity.

two instants, the states of the chain will be identical at those instants. Suppose therefore, if possible, that there is a permanent current in the chain. Since we may suppose the current to have the same value at two instants, the states of the chain will be identical at these two instants. Let W be the work done on the chain and Q the total quantity of heat absorbed by it during the interval. Then the principle of energy gives for the cycle $W + Q = 0$. But clearly $W = 0$: hence $Q = 0$. Now since the passage of the current in the chain can nowhere be reversible except at the junctions, it is evident that the cycle under discussion is irreversible. Hence if θ be the uniform temperature of the chain, Carnot's principle gives $\dfrac{Q}{\theta} < 0$, or $Q < 0$. Thus the supposition that there is a permanent current is not compatible both with the principle of energy and with Carnot's principle. It is therefore absurd.[1]

Another proof of a less elementary kind may also be given. Let the chain be broken and suppose the two ends of the chain consist of wire. Let these wires be made to reach as far as a fixed straight magnet and suppose the end of one wire made to touch the magnet at about its middle point P while the other wire terminates at a point Q in the axis of the magnet. Let the points P, Q be connected, as in the figure, by a bent wire PXQ which is pivoted at P and Q in such a way that it completes the circuit of the chain and yet is free to rotate about the magnet-pole M. Then we know that if a current

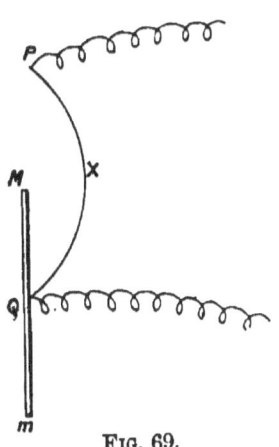

FIG. 69.

---

[1] If there be an electrolyte in the chain, as in the circuit of a galvanic battery, the passage of a current will produce a chemical change and a constant current may therefore coexist with a continual change in the state of the chain. For such a chain, the proposition given in the text evidently does not hold.

traverse the movable piece of wire PXQ, the wire will tend to rotate round the magnet-pole M, and work may be obtained from the rotation. This rotation, it is evident from symmetry, makes no change in the position of the wire relative to the magnet-pole. Suppose therefore that the system of the conducting circuit and magnet is so arranged that there can be no current in the chain which does not also traverse PXQ; and let the temperature of the system be kept uniform and constant. Then if there could be a permanent current in the chain, we could obtain a positive quantity of work from a cycle during which the temperature is uniform and constant, contrary to Carnot's principle. The supposition of a permanent current is therefore absurd.

Thus when the chain is formed, it assumes a state in which there is no current. When this is the case, the potential is constant throughout each conductor. If a point be then supposed to travel round the chain in the interior of the conductors, the sum of the abrupt rises[1] of potential at the junctions in its path will be zero.

Suppose, in the first place, that the chain consists of three conductors, A, B, B', two of which B, B' are of the same kind so that there is no difference of potential between them. Let a point travel in the interior of the conductors from a point P in the interior of A to a point Q in the interior of B, (1) through the junction of A and B, and (2) through the junction of A and B'. Let a be the abrupt rise of potential at the junction of A and B in passing from A to B, and let a' be the abrupt rise of potential at the junction of A and B' in passing from A to B'. Then since these are the only changes of potential in the paths, we obtain $a = a'$. Noticing therefore that

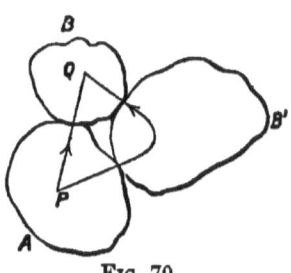

Fig. 70.

[1] Abrupt falls of potential being considered negative abrupt rises.

Applications to Electricity.

B and B' may be chosen as that their size, form, charges, and surfaces of contact with A, differ as we please, we see that the difference of potential at the junction of two conductors of the same temperature is independent of:—

(1) The size and form of the two conductors.
(2) The extent and form of the surfaces of contact.
(3) The charges of the two conductors.

Now let there be any number of conductors A, B, C, ... M, N in the chain; and suppose that A and N are directly in contact as well as connected by the intermediate conductors B, C, ... M. Also let (V_A, V_B, V_C, ... V_M, V_N) be the constant potentials of the conductors, and write V_{BA} for $V_B - V_A$, the abrupt rise of potential at the junction of any two conductors A, B in contact in passing from A to B.

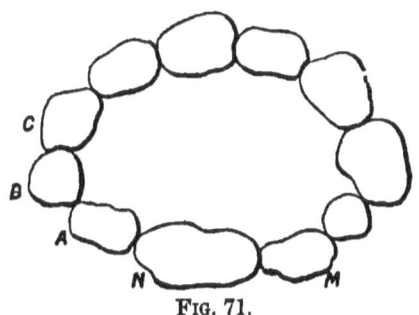

FIG. 71.

Then $V_{AB} = V_A - V_B = -V_{BA}$; and since the sum of the abrupt rises of potential is zero in travelling round the chain,

$$V_{BA} + V_{CB} + \ldots + V_{NM} + V_{AN} = 0.$$

Hence

$$V_{BA} + V_{CB} + \ldots + V_{NM} = -V_{AN}$$
$$= V_{NA}.$$

Thus the sum of the abrupt rises of potential in passing from one conductor A to any other conductor N through any number of intermediate conductors, *all at the same temperature*, is independent of the number of intermediate conductors and the same as if A and N were directly in contact.

This law is attributed to Becquerel.[1]

In what follows we shall often write D_{BA} for the product of V_{BA} and the positive constant ϵ ($9 \times 10_{20}$),

[1] Mascart's "Electricity and Magnetism," Volume I.

and the result will be called the "electromotive force of contact."

Art. 90. In this article will be given some properties of the thermal effects of a current at a junction corresponding to the properties of the difference of potential given in the last article.

For this purpose we require a machine which shall give an electric current in exchange for work. The simplest form of such a machine is a metallic disc touched at its centre and circumference by fixed contact-pieces of the same metal as itself and capable of rotating freely between the poles of a permanent horse-shoe magnet. On completing the circuit of the machine by connecting the contact-pieces by a conductor of any kind, a current may be produced by forcibly rotating the disc; and, conversely, on passing a current through the machine, the disc will tend to rotate and work may be obtained from the motion.

When the machine is used for producing a current, it is called a "dynamo": when used to obtain work from a current, it is called an "electro-motor," or simply, a "motor." The reason why two names are given to the same machine, is that in the more complicated machines of commerce it is necessary to alter the positions of the fixed contact-pieces of a machine used to produce a current from work before it can be used to obtain work from a current. The term "electro-magnetic engine" may be used as the name of both dynamos and motors.

If the temperature of the machine be kept uniform and constant, there will only be two thermal effects in it when at work—one in accordance with Joule's law, the other due to the rubbing of the fixed contact-pieces. The heat developed according to Joule's law may be made to disappear by increasing the size of the machine and the frictional rubbing may be obviated by substituting rolling for rubbing contact, etc. The temperature of the machine will then remain uniform and constant when at work, and the operation will be reversible whatever the current.

Applications to Electricity.

Now let the contact-pieces be joined by a chain of any number of conductors A, B, C, ... M, N; and suppose the temperature of the system kept uniform and constant. Then if every part of the system be at rest, there will be no current, and a feeble current may be produced by slowly turning the disc. But we know that a reversibly feeble current produces no thermal effects while traversing a homogeneous conductor of uniform temperature: the only thermal effects in our system are therefore at the junctions. If then θ be the uniform and constant temperature of the system and Q the sum of the quantities of heat that must be absorbed during a cycle at the junctions to keep the temperature constant, Carnot's principle gives for the reversible cycle, $\frac{Q}{\theta} = 0$, or $Q = 0$.

We assume as a result of experiment that the thermal effect at a junction is proportional to the quantity of electricity that crosses it. If therefore q be the quantity of electricity that passes in the cycle just described, the quantities of heat absorbed at the junctions may be written $(qP_1, qP_2, qP_3, \ldots)$. Hence $q(P_1 + P_2 + P_3 + \ldots) = 0$ and $P_1 + P_2 + P_3 + \ldots = 0$.

In the equation $P_1 + P_2 + P_3 + \ldots = 0$, let P_1 refer to the junction of the bodies B, C; and suppose B connected with A by any number of intermediate conductors $(B_{\prime}, B_{\prime\prime}, B_{\prime\prime\prime}, \ldots)$ of the same kind as B, and in like manner let C be connected with D by any number of intermediate conductors $(C_{\prime}, C_{\prime\prime}, C_{\prime\prime\prime}, \ldots)$ of the same kind as C. Then the current will produce no thermal effects anywhere between the junction

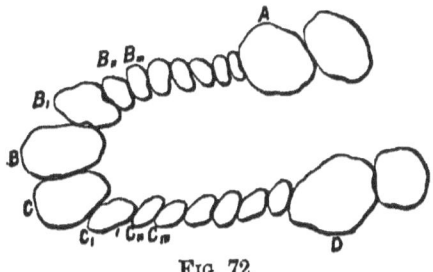

Fig. 72.

BC and the body A nor between the same junction BC and the body D. Now it is evident the intermediate bodies

($B_{\prime\prime}$, $B_{\prime\prime\prime}$, $B_{\prime\prime\prime\prime}$, ...), ($C_{\prime\prime}$, $C_{\prime\prime\prime}$, $C_{\prime\prime\prime\prime}$, ...) may be so chosen that we can replace the bodies B, C by two other bodies of the same kind but differing from them as much as we please in size, form, surface of contact, and charge, without making any change near any important junction with the sole exception of the junction B, C. If therefore we denote the new value of P_1 by P_1', we get $P_1' + P_2 + P_3 + \ldots = 0$. Hence $P_1 = P_1'$. Thus the value of P_1 (like the difference of potential) is independent of:—

    (1) The size and form of the two conductors.
    (2) The extent and form of the surfaces of contact.
    (3) The charges of the two conductors.

The reversible thermal effect of a current at a junction was discovered by Peltier, and the words "Peltier effect" will therefore be used as the name of the quantity P—the heat that must be absorbed at the junction on the passage of unit charge to keep the temperature of the junction constant.

The Peltier effect for the passage of a current across a junction from A to B will be written P_{BA} by analogy with the expression V_{BA} for the rise of potential in the direction of the current.

Suppose the system of machine and chain contains only two kinds of conductors A, B. Then there will only be two Peltier effects, and $P_{AB} + P_{BA} = 0$. Hence $P_{AB} = -P_{BA}$, which is simply the expression of the fact that the Peltier thermal effect is reversible.

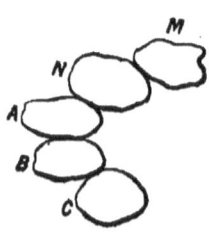

The temperature at which the Peltier effect vanishes is called the Neutral Point of the two metals or conductors.

Fig. 73.

Lastly, let there be any number of conductors in the system. Then

$$P_{BA} + P_{CB} + P_{DC} + \ldots + P_{NM} + P_{AN} = 0.$$

Hence

$$P_{BA} + P_{CB} + \ldots + P_{MN} = -P_{AN} = P_{NA}.$$

Thus the sum of the Peltier effects in passing from one conductor A to any other conductor N through any number of intermediate conductors, *all at the same temperature*, is independent of the number of intermediate conductors and the same as if A and N were directly in contact.

Art. 91. We have seen that there is no current in a circuit of conductors kept at a uniform and constant temperature. By making experiment, it is found that there is no current if the junctions of the different conductors be all at the same temperature whatever values the temperature may have in the rest of the circuit, but that if the junctions of the different conductors be raised to different temperatures, a current will generally be produced.

In a thermo-electric circuit there are two kinds of junctions, (1) the junction of two pieces of different metal at the same temperature, and (2) the junction of two pieces of the same metal at different temperatures. A peculiar thermal effect was discovered by the experiments of Peltier to be produced at the first kind of junction, but it was not suspected that a current produces any particular thermal effect at the second kind of junction until the fact was discovered by Sir W. Thomson[1] in 1852.

The following is the reasoning by which Sir W. Thomson discovered the existence of the thermal effect of a current at the junction of two unequally heated pieces of the same metal. Take the simplest thermo-electric circuit possible, consisting of two metals, and let the two junctions be raised to different temperatures. Let the hotter of these two junctions be at the neutral point θ, or at the temperature at which the Peltier effect is zero for the two metals: also let no part of the circuit be at a temperature below θ_0, the temperature of the colder junction of the two metals. Then experiment shows that there is a current in the circuit, and that the current is steady if the temperatures of all the

[1] Now Lord Kelvin.

different parts of the circuit be kept constant. Consider the circuit for any interval during which it is in such a state. Since the temperatures and current are invariable, the state of the circuit is invariable. But evidently no work is done on the system during the cycle: the total quantity of heat absorbed is therefore zero. If therefore we assume that a current produces no thermal effect in crossing the junction of two unequally heated pieces of the same metal, and denote by h and h_0 the positive quantities of heat given out during the cycle according to Joule's law, there will be a positive quantity of heat $h + h_0$ absorbed at the colder junction of the two metals. Hence the value of $\Sigma\left(\dfrac{Q}{\theta}\right)$ for the cycle is $-\dfrac{h}{\theta} - \dfrac{h_0}{\theta_0} + \dfrac{h+h_0}{\theta_0}$, or $h\left(\dfrac{1}{\theta_0} - \dfrac{1}{\theta}\right)$ or $h\dfrac{\theta - \theta_0}{\theta_0\theta}$. Since θ_0 is less than θ, the value of $\Sigma\left(\dfrac{Q}{\theta}\right)$ for the cycle is positive, contrary to Carnot's principle. The supposition that there is no thermal effect produced by a current in crossing the junction of two unequally heated pieces of the same metal is therefore absurd.

The thermal effect just discovered is known as the 'Thomson effect"; and it can be shown, both by experiment and theory, that it is reversible and proportional to the quantity of electricity that passes the junction. It may also be proved that the effect is proportional to the difference of the two temperatures when that difference is small.

When the temperatures are nearly equal, their positive difference is written τ and $\sigma\tau$ is used to denote the quantity of heat that must be absorbed at the junction to keep the temperatures constant on the passage of unit charge from the cooler to the warmer piece of metal. Hence $\sigma\tau$ is the quantity of heat that must be evolved at the junction to keep the temperatures constant on the passage of unit charge from the warmer to the cooler piece of metal.

The quantity σ is called by Sir W. Thomson the "specific heat of electricity" in the metal considered.

Art. 92. The following are some of the properties discovered by Sir W. Thomson in 1852 for a thermo-electric circuit of two metals. Their discussion is not quite so elementary as the propositions of the preceding articles.

Take an open chain of three pieces of metal, of which the two end pieces are of the same kind, as in the figure. Let the two end pieces be at the same temperature θ, and suppose the temperatures of the junctions are respectively θ and $\theta - \tau$, where τ is very small.

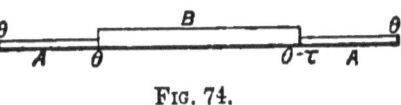

FIG. 74.

Lastly, let every part of the chain be either at the temperature θ or the temperature $\theta - \tau$.

Next, take a rotating-disc dynamo of the same metal as A and at the uniform temperature θ. Complete its circuit by connecting the two contact-pieces with the ends of the chain just described. Then if everything be at rest, there will be a current in the system; but if the disc be turned, the current may be prevented. Suppose therefore that the disc is rotated at such a speed that there is no current. Then by turning the disc at a slightly different speed, let a unit charge be made to pass slowly

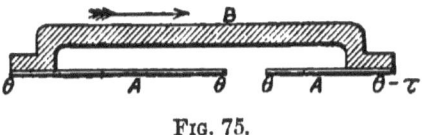

FIG. 75.

round the circuit in one direction, say from A to B through the θ junction; and keep the temperature of every part of the system constant during the process. The process forms a reversible cycle, and we intend to apply the principle of energy and Carnot's principle to it. It will be best to begin with Carnot's principle, because this principle only considers the quantities of heat absorbed and given out during the cycle, while the principle of energy requires us to know also the work done in turning the disc.

There is no heat absorbed or evolved during the

reversible cycle except at the four junctions. The heat absorbed at the junction of the unequally heated pieces in the metal A is $\sigma_A \tau$. The heat absorbed at the similar junction in the metal B is $-\sigma_B \tau$, because in this metal the current passes from the piece at temperature θ to the piece at temperature $\theta - \tau$. The heat absorbed at the θ junction of the two metals is P_{BA}, and that absorbed at the $\theta - \tau$ junction may be written $-P'_{BA}$, where P'_{BA} is the value that P_{BA} assumes when θ is changed to $\theta - \tau$.

In forming $\Sigma \left(\dfrac{Q}{\theta} \right)$ for the cycle, it will perhaps appear doubtful whether we should divide $\sigma_A \tau$ and $-\sigma_B \tau$ by θ or $\theta - \tau$. It is easily shown, however, that it is immaterial which divisor we choose. For example, the difference between $\dfrac{\sigma_A \tau}{\theta - \tau}$ and $\dfrac{\sigma_A \tau}{\theta}$ is $\sigma_A \tau \left(\dfrac{1}{\theta - \tau} - \dfrac{1}{\theta} \right)$, or $\sigma_A \tau \times \dfrac{\tau}{(\theta - \tau)\theta}$. Since τ is small, it is easily seen that this difference is small in comparison with either quotient. We will therefore take θ as the divisor. Hence Carnot's principle gives for the reversible cycle

$$\frac{P_{BA}}{\theta} - \frac{P'_{BA}}{\theta - \tau} + \frac{\sigma_A \tau - \sigma_B \tau}{\theta} = 0.$$

The second term $\dfrac{P'_{BA}}{\theta - \tau}$ is evidently what $\dfrac{P_{BA}}{\theta}$ becomes when θ is changed into $\theta - \tau$. On referring to Art. 59, Chap. IV., it will be found that $\dfrac{P'_{BA}}{\theta - \tau} = \dfrac{P_{BA}}{\theta} - \tau \left(\dfrac{P_{BA}}{\theta} \right)_\theta$, where $\left(\dfrac{P_{BA}}{\theta} \right)_\theta$ is the rate at which $\dfrac{P_{BA}}{\theta}$ increases with respect to θ.

Hence

or
$$\left(\frac{P_{BA}}{\theta}\right)_\theta = \frac{\sigma_B - \sigma_A}{\theta},$$
a result given by Thomson.

We have next to apply the principle of energy to the cycle. If W be the work done on the system in turning the disc, we have, since the energy of the system in the final state is the same as in the first,
$$0 = W + (P_{BA} - P'_{BA}) + (\sigma_A \tau - \sigma_B \tau),$$
or
$$0 = W + \tau(P_{BA})_\theta + \tau(\sigma_A - \sigma_B),$$
since
$$P'_{BA} = P_{BA} - \tau(P_{BA})_\theta.$$

It remains to express W in terms of quantities relating only to the metallic chain which connects the contact-pieces of the dynamo. Let E be ε times the sum of the abrupt rises of potential at the four junctions in the chain in the direction in which the unit charge is sent through the chain by the dynamo[1]—that is, from A to B through the θ junction. Then in the system of the dynamo and chain, it is evident that the fall of potential in the dynamo in the direction of the current during the reversible cycle is $\dfrac{E}{\varepsilon}$. Hence E is the electromotive force lost by the current in passing through the dynamo.

Now consider the dynamo as a system of itself. It undergoes a cyclical process during which no heat is absorbed or evolved by it. Hence, by Art. 8, if e be the "electric energy" absorbed by it, we have $W + e = 0$, or $W = -e$. But since E is the electromotive force lost by the current in passing through the dynamo and the quantity of electricity that passes is unity, we have, as in Art. 8, $e = +E$. Thus $W = -E$.

[1] To an observer who travels in the opposite direction, E will be ε times the sum of the abrupt *falls* of potential at the four junctions.

The equation given by the principle of energy may therefore be written

$$E = \tau\{(P_{BA})_\theta + \sigma_A - \sigma_B\}.$$

This result may be much simplified be means of the previous result $\left(\dfrac{P_{BA}}{\theta}\right)_\theta = \dfrac{\sigma_B - \sigma_A}{\theta}$. To do this we require a little knowledge of pure mathematics.

Let x and y be any two quantities or expressions depending on θ, and let us find the rates of increase of xy and $\dfrac{x}{y}$ with respect to θ in terms of those of x and y themselves. When θ increases by the small amount τ let x and y increase by the small amounts Δx and Δy. Then the increase of xy is $(x + \Delta x)(y + \Delta y) - xy$, or $x\Delta y + y\Delta x + \Delta x \Delta y$. If this be divided by τ we get $x\dfrac{\Delta y}{\tau} + y\dfrac{\Delta x}{\tau} + \dfrac{\Delta x}{\tau}\Delta y$. But to find the rate of increase of xy with respect to θ, we divide by τ the increase of xy corresponding to the small increase τ of θ, and then make τ, Δx, and Δy all gradually vanish. Hence $(xy)_\theta = xy_\theta + yx_\theta$. Again, the increase of $\dfrac{x}{y}$ corresponding to τ is $\dfrac{x + \Delta x}{y + \Delta y} - \dfrac{x}{y}$, or $\dfrac{y\Delta x - x\Delta y}{(y + \Delta y)y}$. Hence $\left(\dfrac{x}{y}\right)_\theta = \dfrac{yx_\theta - xy_\theta}{\theta_2}$.

Thus the result $\left(\dfrac{P_{BA}}{\theta}\right)_\theta = \dfrac{\sigma_B - \sigma_A}{\theta}$ may be written

$$\dfrac{\theta(P_{BA})_\theta - P_{BA}}{\theta^2} = \dfrac{\sigma_B - \sigma_A}{\theta},$$

or

$$\theta(P_{BA})_\theta = P_{BA} + \theta(\sigma_B - \sigma_A).$$

Hence

$$E = \tau\dfrac{P}{\theta},$$

another result due to Thomson.

The interpretation of the simple result $E = \dfrac{\tau}{\theta} P_{BA}$ is important. If a steady current be flowing in any homogeneous conductor of uniform temperature, we know that there is a gradual fall of potential in the direction of the current. Hence if we choose one direction as the positive direction and consider currents positive when in this direction, negative when in the opposite, the fall of potential in the positive direction will be positive when the current is positive, negative when the current is negative. Now let a closed circuit be formed of the metallic chain by taking it away from the dynamo and joining its ends together. Let the positive direction be from A to B through the θ junction. Then if R be the resistance of the circuit and I the current (considered positive when in the positive direction), RI will, by Ohm's law, be ϵ times the gradual fall (positive or negative) of potential in this direction. But E is ϵ times the sum of the abrupt rises of potential at the four junctions in the same direction. Hence since the total change of potential in travelling round the circuit must be zero, $E = RI$

Thus $RI = \dfrac{\tau}{\theta} P_{BA}$; and from this result we may obtain two propositions which are known to be in accordance with experiment.

(1) Suppose the temperature θ fixed. Then P_{BA} will also be fixed; and since R hardly varies with temperature, we see that I changes sign with τ. Hence when one junction is kept at a constant temperature θ, the current will gradually diminish, disappear, and flow in the opposite direction, as the temperature of the other junction rise up to and above θ. (2) Suppose the θ junction the hotter of the two, so that τ is positive. Then P_{BA} and I are of the same sign. If P_{BA} be positive, the current flows from A to B through the θ junction; if P_{BA} be negative, the current flows from B to A through the θ junction. In words, the current flows in such a direction that it absorbs heat at the hotter junction when

the temperature of the junction is kept constant or cools the junction when no heat is supplied to it from without.

The quantity E, which is ϵ times the sum of the abrupt rises of potential at the four junctions in the direction chosen as the positive direction, may be called the "electromotive force of the circuit." It may obviously be positive or negative. In common parlance, however, the "electromotive force of the circuit" is ϵ times the sum of the abrupt rises of potential in the direction of the current—not necessarily in the direction which is arbitrarily chosen as positive. These two definitions are identical, arithmetically speaking; but the algebraic definition possesses some advantages and will be adopted in what follows.

The use of the rotating disc in theoretical investigations, as in the last few articles, is due to Sir W. Thomson. It does not require a knowledge of the energy of an electrified system; but it does not tell us what is the abrupt change of potential at any particular junction nor what is the relation of that abrupt change to the corresponding thermal effect. To answer these points we make use of Helmholtz' expression for the energy of an electrified system.

Art. 93. The electrified system at rest of which we wish to know the energy U will have all its electric properties internal. The principle of energy willl therefore apply to it in the usual form $\Delta U = W + Q$. The system will be supposed to consist of two parts between which there is no connection except that a current flows from one to the other. Hence (it will be seen from Art. 8) the energy of the whole system is equal to the sum of the energies of the two parts.

All the thermal effects which occur in the system will be confined to a particular one of the two parts, and all the work done on the system will be done on the other part. The first of these parts will either be in an invariable state or all the operations it undergoes will be cyclical. It is therefore unnecessary to have an expression for its energy U'. The second part will be supposed

Applications to Electricity. 251

to consist entirely of metallic or other conductors situated in a vacuum. There may, however, be gases or any other non-conductors in the first part of the system.

The second part will be supposed to be constantly in a state of electric equilibrium.[1] Then Helmholtz' expression for its energy U'' is

$$U'' = U''_0 + \varepsilon \Sigma \frac{qq'}{r} + q_A F_A + q_B F_B + \dots,$$

where A, B, C, ... are the different homogeneous conductors of uniform temperatures; $q_A, q_B, q_C, \dots$ their charges; $F_A, F_B, F_C, \dots$ quantities which depend on the temperatures of these conductors respectively; and $\varepsilon \Sigma \frac{qq'}{r}$ has the following signification:—Imagine the charges divided into elements so small that each elementary charge may be treated as a point in comparison with its distance from any other element. Then if (q, q') be any two of these elementary charges and r their distance apart, $\Sigma \frac{qq'}{r}$ denotes the sum of all such quantities as $\frac{qq'}{r}$ that can be formed by taking the elementary charges two at a time in every possible way. Lastly, U''_0 is what U'' becomes when the part of the system under consideration is deprived of its electrification but otherwise unchanged.

If $(V_A, V_B, V_C, \dots)$ be the potentials of the conductors, it is known that $\Sigma \frac{qq'}{r} = \frac{1}{2} \{q_A V_A + q_B V_B + q_C V_C + \dots\}$.

Writing U_0 for $U' + U''_0$, we finally get for the energy of the system

$$U = U_0 + \varepsilon \Sigma \frac{qq'}{r} + q_A F_A + q_B F_B + \dots$$
$$= U_0 + \varepsilon \{\tfrac{1}{2} q_A V_A + \tfrac{1}{2} q_B V_B + \dots\} + q_A F_A + q_B F_B + \dots \quad (15),$$

where U_0 is a quantity the changes of which are zero in the operations to be described.

[1] And therefore to contain no appreciable electric currents.

Before the investigations of Helmholtz on the subject, the terms $q_A F_A + q_B F_B + \ldots$ were omitted from the expression for U. As these terms are of vital importance to our discussions, it will be well to give reasons for believing Helmholtz' expression for U to be the true one.

(1) Helmholtz' expression has been universally accepted for many years.

(2) Duhem has recently obtained the same expression from thermo-dynamical considerations alone.

(3) The expression leads to the accepted fundamental formula of Gibbs and Helmholtz for the galvanic battery.

(4) It also leads to several results in thermo-electricity which are either identical with those of Thomson or completely in accordance with them. One of these results, connecting the Peltier effect with the corresponding electromotive force of contact, has been given by several independent investigators, and is so similar to the fundamental formula of Gibbs and Helmholtz for the galvanic battery that it may almost be said to be contained in that formula.

(5) Lastly, if we assume Joule's law, which is a result of experiment, we can obtain an expression for the entropy of the given system; and it contains terms exactly similar to the terms $q_A F_A + q_B F_B + \ldots$ given by Helmholtz in the expression for the energy.

Art. 94. The first application we shall make of Helmholtz' expression for the energy will be to deduce Joule's law from Ohm's law.

Take two large plates M, N of any of the same metal and connect them by a wire of this metal. Parallel and opposite to these two plates place equal plates of any the same metal, as iron, and connect the iron plates by very stout iron wires with each other, or with a large distant mass of iron in the neutral state, so that the two iron plates are always at potential zero. Let the air be exhausted from between the parallel plates, but let there be air or gas or any non-conductor around the wire which connects M and N. Also let the temperature of the system be kept uniform and constant; and suppose that all parts of the system with the exception of the wire joining M, N are so massive that they may be considered to be in a state of electric equilibrium.

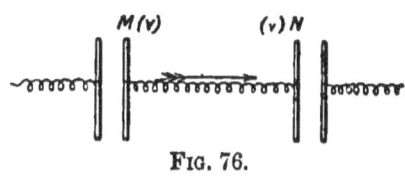

Fig. 76.

Now let the plate M be raised to a higher potential than N. Then there will be a current in the connecting wire from M to N. In consequence, the potential of M will tend to fall and that of N to rise. But if the plates N be moved nearer together and the plates M separated, the potential of N will tend to fall and that of M to rise. Let us therefore suppose the motions of the plates to be such that the current is steady and the potentials of M and N constant, (V, v) say. Then since the wire from M to N is in an invariable state and the rest of the system in electric equilibrium, the energy is given in the last article. If q_M be the charge of the plate M and q_N the charge of N, the expression for the energy of the system is

$$U = U_1 + \epsilon(\tfrac{1}{2}q_M V + \tfrac{1}{2}q_N v),$$

where U_1 is a quantity which remains constant.

Now let a quantity of electricity q be conveyed along the wire from M to N in t seconds during which the current is constantly equal to I. Then $q = It$; and at the end of the time the energy of the system will be

$$U + \Delta U = U_1 + \epsilon\{\tfrac{1}{2}(q_M - q)V + \tfrac{1}{2}(q_N + q)v\}.$$

Hence

$$\Delta U = -\tfrac{1}{2}\epsilon q(V - v).$$

If W be the work done on the system during the interval and Q the heat absorbed by it, then $\Delta U = W + Q$. But with the exception of the wire joining M, N, the system is so massive that there is nowhere any appreciable current in it. There is therefore no thermal effect in the system except in this wire. Again, there is no work done on the system except in moving the plates. The work done in forcing the plates M further apart will be shown presently to be $\frac{1}{2}\epsilon qV$, and the work obtained from the attraction of the plates N in their approach, $\frac{1}{2}\epsilon qv$. Thus $W = \frac{1}{2}\epsilon q(V - v)$. Substituting in $\Delta U = W + Q$, we get

$$-\tfrac{1}{2}\epsilon q(V - v) = \tfrac{1}{2}\epsilon q(V - v) + Q,$$

or

$$Q = -\epsilon q(V - v).$$

But by Ohm's law, if R be the resistance of the wire which conveys the current,

$$\epsilon(V - v) = RI.$$

Hence

$$Q = -RIq = -RI^2 t.$$

Thus heat is evolved in the wire and RI^2 is the rate of evolution per second. This result was discovered experimentally by Joule, and is known as Joule's law. It has been obtained here as a consequence of Ohm's law; and, conversely, if we take Joule's law for granted, we can deduce Ohm's law from it.

Ohm's law is seldom true except in the case given above. It is not generally true when the current is variable, or when the wire is in motion, or when there is a magnet acting on the current and there is motion in the wire or magnet. But Joule's law appears to be true in all cases. It would therefore seem to be desirable to give Joule's law first and then deduce Ohm's law from it. We may observe, however, that Ohm's law was discovered long before Joule's.

It is clear that Joule's law could have been deduced from Ohm's if the expression for U had been taken to be $U = U_0 + \epsilon\Sigma\dfrac{qq'}{r}$; and it is easily shown that this would not have been the case if the expression had

Applications to Electricity. 255

contained any additional terms except of the form $q_A F_A + q_B F_B + \ldots$

It remains to show that the expression given for the work W is the true one. As the question is of great importance to us in the following articles, and is also very interesting, none but easy arguments will be made use of.

(1) It may perhaps be thought that we ought to put $W = \varepsilon q(V - v)$, by analogy with qE in the equation $\Delta U = W + Q + qE$. If this be taken as the value of W, the equation $\Delta U = W + Q$ in the foregoing reasoning gives $-\tfrac{1}{2}\varepsilon q(V - v) = \varepsilon q(V - v) + Q$. Hence this expression for W makes the heat evolved by the wire to be $\tfrac{3}{2}RI^2$ per second, contrary to experiment.

(2) The complete proof that $W = \tfrac{1}{2}\varepsilon q(V - v)$ is rather difficult; but a somewhat similar case may be given in which the calculation is easy.

Take a system consisting of two distant conducting spheres of the same metal and let them be connected by a long wire of the same substance as themselves. Let the temperature of the system be kept uniform and constant; and denote the common potential of the two spheres by V. Then if (q', q'') be the charges of the spheres (r', r'') their radii, we have

$$U = U' + \tfrac{1}{2}\varepsilon(q' + q'')V,$$

and

$$V = \frac{q'}{r'} = \frac{q''}{r''}.$$

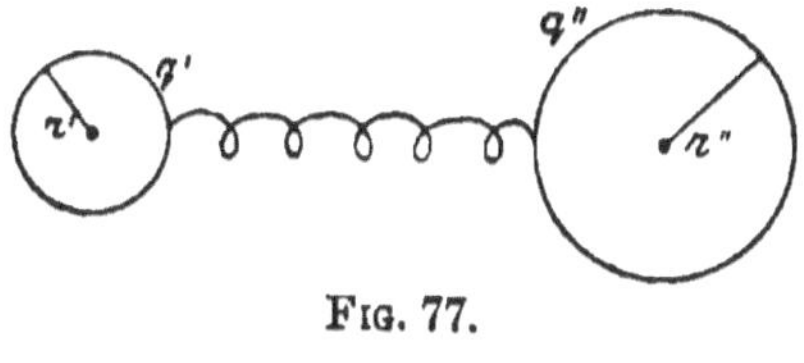

Fig. 77.

Now suppose the sphere (r') capable of expanding or contracting without any change in its non-electric state, and let it expand slowly till a charge q passes in a reversible manner from the sphere (r'') to the sphere (r'). There will be no change in U', and we shall have

$$V + \Delta V = \frac{q'' - q}{r''},$$

$$U + \Delta U = U' + \tfrac{1}{2}\varepsilon(q' + q'')(V + \Delta V).$$

Hence

$$\Delta V = - \frac{q}{r''},$$

$$\Delta U = \tfrac{1}{2}\epsilon(q' + q'')\Delta V = - \tfrac{1}{2}\epsilon q \frac{q' + q''}{r''}.$$

But since the operation is reversible, there is no thermal effect in the system. If therefore W be the work done on it, we have $\Delta U = W$, or

$$W = - \tfrac{1}{2}\epsilon q \frac{q' + q''}{r''} = - \tfrac{1}{2}\epsilon q \frac{q''\left(1 + \dfrac{q'}{q''}\right)}{r''}..$$

If the sphere (r') be small in comparison with the other, or $\dfrac{r'}{r''}$ small, $\dfrac{q'}{q''}$ will also be small. We have then $W = - \tfrac{1}{2}\epsilon q \dfrac{q''}{r''}.$ Again, ΔV will be small, or the potential practically constant, and W may be written $- \tfrac{1}{2}\epsilon q V$.

In the case of the parallel plates, a charge q enters the plate N during the operation to which it is subjected and a charge q leaves the plate M. Hence, since the potentials are constant, it is seen the work done on the plates N is $- \tfrac{1}{2}\epsilon q v$, and the work done on the plates M, $\tfrac{1}{2}\epsilon q V$. The total work done on the plates is therefore $\tfrac{1}{2}\epsilon q v(V - v)$, as previously stated.

Art. 95. In this article we take an electrified system at rest consisting of conductors situated in a vacuum in a state of electric equilibrium (and therefore containing no appreciable currents); and Joule's law will be used to find an expression for its entropy ϕ.

It is easy to prove that if a charge be transferred across a homogeneous conductor of uniform temperature, there will be no thermal effect produced and the transference will be reversible, if only it be slow enough. Suppose therefore that our system is made to undergo a reversible operation of any kind in which no part of it is compressed or distorted, and no charge made to pass from one body to a different body or to a body of the same kind but at a different temperature. Then there will be no thermal effect produced in the system and consequently no change of entropy. So long, therefore, as the charges are not made to leave the bodies on which they were at first, the entropy of the system is unaltered by *any* change in the distribution of the charges or in the relative positions of the bodies. We may therefore put

$$\phi = \psi_A(q_A) + \psi_B(q_B) + \psi_C(q_C) + \ldots,$$

where $\psi_A(q_A)$ only depends on the body A and its charge, $\psi_B(q_B)$ only on B and its charge, ...

Now take a second system identical with the first, forming with it a compound system whose entropy is 2ϕ. By a reversible process, such as we have already described, let a charge be made to pass from one metallic body A to the other body A, and suppose that no other charge passes from one body to another. Then, since by what precedes, the entropy of the compound system is unchanged, it follows that, if q be the final charge of one body A and $2q_A - q$ of the other, $\psi_A(q) + \psi_A(2q_A - q)$ is independent of q, *whatever* q may be. We therefore infer that $\psi_A(q_A) = \psi_A(0) + q_A H_A$, where H_A is independent of q_A.

Next, let us take a system formed of the original system ϕ and of a second system which also contains a metal body A at the same temperature as the body A in the first system but different in form and size and with any charge q'_A. Let H'_A be the quantity corresponding to H_A. Then by supposing any charge to pass from one metal A to the other metal A, without the passage of a charge between any other bodies, we find $H'_A = H_A$. Hence H_A is independent of the size and form of A as well as of its electric state.

Thus finally

$$\phi = \psi_A(0) + \psi_B(0) + \ldots + q_A H_A + q_B H_B + \ldots$$
$$= \phi_0 + q_A H_A + q_B H_B + \ldots \ldots \ldots \ldots (16),$$

where ϕ_0 is the entropy of the system when deprived of its electrification but otherwise unchanged.

Art. 96. There is an important identical relation between the quantities F and H. Take an electrified system consisting entirely of conducting bodies and in electric equilibrium. Then

$$\left.\begin{aligned} U &= U_0 + \epsilon\Sigma\frac{qq'}{r} + q_A F_A + q_B F_B + \cdots \\ \phi &= \phi_0 + q_A H_A + q_B H_B + \cdots \end{aligned}\right\}.$$

Let the body A, which is homogeneous and of uniform temperature, be slowly heated from temperature θ_A to temperature $\theta_A + \tau$, where τ is small; and let the parts of the system be at the same time slowly moved about so that no charge passes from one body to a different body or to a body of the same kind but at a different temperature. Then there will be no thermal effect except in the body A. Since therefore the temperature of A never differs much from θ_A, we have for the reversible operation, if $\Delta\phi$ be the small increase of entropy and Q the heat absorbed, $Q = \theta_A \Delta\phi$. But if W be the work done on the system, $\Delta U = W + Q$. Therefore $\Delta U - W - \theta_A \Delta\phi = 0$. Also

$$\left.\begin{aligned} \Delta U &= \Delta U_0 + \epsilon\Delta\Sigma\frac{qq'}{r} + q_A \Delta F_A \\ \Delta\phi &= \Delta\phi_0 + q_A \Delta H_A \end{aligned}\right\}.$$

Hence

$$\Delta U_0 - \theta_A \Delta\phi_0 - W + \epsilon\Delta\Sigma\frac{qq'}{r} + q_A \Delta F_A - q_A \theta_A \Delta H_A = 0.$$

Now take a second system identical with the first except that every charge is reversed in sign, and let it undergo a reversible operation similar to that just described. Then since ΔU_0, $\Delta\phi_0$, W, and $\epsilon\Delta\Sigma\frac{qq'}{r}$ are the same as before, we obtain

$$\Delta U_0 - \theta_A \Delta\phi_0 - W + \epsilon\Delta\Sigma\frac{qq'}{r} - q_A \Delta F_A + q_A \theta_A \Delta H_A = 0$$

By subtraction,
$$q_A \Delta F - q_A \theta_A \Delta H_A = 0,$$
or
$$\Delta F_A = \theta_A \Delta H_A.$$
Remembering that $\Delta F_A = \tau (F_A)_\theta$, $\Delta H_A = \tau (H_A)_\theta$, where $(F_A)_\theta$ and $(H_A)_\theta$ are the rates at which F_A and H_A increase with respect to θ, we get
$$(F_A)_\theta = \theta_A (H_A)_\theta \qquad \ldots \ldots \ldots (17).$$

Art. 97. To obtain a relation between the Peltier effect and the corresponding change of potential, let two long wires A, B, of different metals, be joined together at J, and also connected, as in the figure, with two plates A, B, which are respectively of the same metals as the two wires. Parallel and opposite to these two plates place equal plates of any the same metal, as iron, and connect the iron plates by iron wires with each other, or with a large distant mass of iron in the neutral state, so that the two iron plates are always at potential zero.

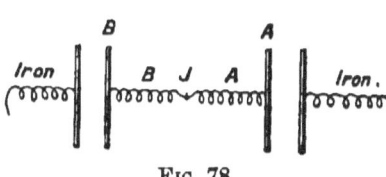

Fig. 78.

Also, to make the calculations simple, let us suppose the air exhausted about the plates; but, to make the results general, the junction J must be surrounded by air, and to prevent the air coming near the plates, it must be enclosed in a bag and the junction J and the bag kept at a great distance from the plates. Then when the system is at a uniform temperature θ the potential (V_A, V_B) of the plates A and B will differ by an amount equal to the abrupt change of potential at J.

Now let the system be made to undergo the following reversible operation during which the temperature of the system and the potentials of A and B are kept constant. By slowly moving the plates B nearer together and slowly separating the plates A, let a quantity q of electricity be made to pass slowly from A to B against the abrupt rise of potential $V_B - V_A$ at J.

In this reversible operation, there is no thermal effect produced in the system except at J, and no work done except on the plates.

If (q_A, q_B) be the initial charges of the plates A, B, the expressions for the energy and entropy of the system at the beginning of the operation are easily seen to be

$$\left. \begin{array}{l} U = U_1 + \epsilon\{\tfrac{1}{2}q_B V_B + \tfrac{1}{2}q_A V_A\} + q_B F_B + q_A F_A \\ \phi = \phi_1 + q_B H_B + q_A H_A \end{array} \right\},$$

where U_1 and ϕ_1 remain constant during the operation. Hence

$$\left. \begin{array}{l} \Delta U = \tfrac{1}{2}\epsilon q(V_B - V_A) + q(F_B - F_A) \\ \Delta \phi = q(H_B - H_A) \end{array} \right\}.$$

But if W be the work done on the plates and Q the heat absorbed at the junction, the principle of energy gives

$$\Delta U = W + Q,$$

and Carnot's principle gives

$$Q = \theta \Delta \phi.$$

Thus

$$\Delta U = W + \theta \Delta \phi,$$

or

$$\tfrac{1}{2}\epsilon q(V_B - V_A) + q(F_B - F_A) = W + \theta q(H_B - H_A).$$

Now the work done on the plates B is $-\tfrac{1}{2}\epsilon q V_B$, and the work done on the plates A, $\tfrac{1}{2}\epsilon q V_A$. The total work W done on the plates is therefore $-\tfrac{1}{2}\epsilon q(V_B - V_A)$. Consequently

$$\epsilon q(V_B - V_A) + q(F_B - F_A) = \theta q(H_B - H_A).$$

If we divide out by q and write D_{BA} for $\epsilon(V_B - V_A)$, the electromotive force of contact, we get

$$D_{BA} + F_B - F_A = \theta(H_B - H_A).$$

Making use of the property $(xy)_\theta = xy_\theta + yx_\theta$, we find

$$\begin{aligned}(D_{BA})_\theta + (F_B)_\theta - (F_A)_\theta &= \theta(H_B - H_A)_\theta + (H_B - H_A) \\ &= \theta(H_B)_\theta - \theta(H_A)_\theta + H_B - H_A.\end{aligned}$$

Now the bodies both being at temperature θ, we have
$$(F_B)_\theta = \theta(H_B)_\theta, (F_A)_\theta = \theta(H_A)_\theta:$$
hence
$$(D_{BA})_\theta = H_B - H_A.$$

To complete the investigation, we must return to the thermal effect. The heat absorbed at the junction has been shown to be $\theta q(H_B - H_A)$. We have thus a theoretical proof that it is proportional to the quantity of electricity that crosses. Denoting the heat by qP_{BA}, we have $P_{BA} = \theta(H_B - H_A)$; and we find $P_{AB} = -P_{BA}$, $P_{CA} = P_{CB} + P_{BA}$. Lastly, putting $(D_{BA})_\theta$ for $H_B - H_A$, we obtain

$$P_{BA} = \theta(D_{BA})_\theta \quad \ldots \ldots \ldots \quad (18),$$

which is a relation between the Peltier effect P_{BA} and the electromotive force of contact D_{BA}.

If we had omitted the terms $q_B F_B + q_A F_A$ from the expression of energy, we should have found

$$D_{BA} = \theta(H_B - H_A) = P_{BA}.$$

The result $P = \theta D_\theta$ has been given on four independent occasions:—by Prof. J. J. Thomson in his Application of Dynamics to Chemistry and Physics; by Maxwell in his small Treatise on Electricity, where he has abandoned the older result that $P = D$; by Duhem; and, lastly, by the present writer. Later on it will be shown that the result $P = \theta D_\theta$ is in agreement with and almost included in, the accepted fundamental formula of Gibbs and Helmholtz for the galvanic battery. It may therefore be considered to be fully established.

Art. 98. To investigate the properties of the junction of two pieces of the same metal in the same molecular state but at different temperatures, take a system of plates and wires as in the last article, but let the plate B be of the same metal as A and in the same molecular state but at a slightly different temperature. Suppose the plate A and wire A to be at the uniform temperature θ, and the plate B and wire B at the uniform temperature $\theta + \tau$, where τ is small. Also let the other plates and wire which the system contains be at a uniform temperature.

By slowly moving the plates, let a charge q be made to pass slowly from A to B; and let the temperatures and potentials of all parts of the system be kept constant during the operation. There will be no thermal effect in the system during the operation except at the junction J. Now if the change of temperature at J be gradual, the heat conducted across the junction in a finite time, say a second, will be exceedingly small. We may therefore suppose that the amount conducted while the charge q passes from one plate to the other, can be neglected. The operation will then be reversible. Hence if Q be the heat absorbed at the junction and $\Delta\phi$ the increase of the entropy of the system during the operation, Q will lie between $\theta\Delta\phi$ and $(\theta + \tau)\Delta\phi$, heat being absorbed at the junction at all temperatures between θ and $\theta + \tau$. But since τ is small, the difference between $\theta\Delta\phi$ and $(\theta + \tau)\Delta\phi$ will be insignificant in comparison with $\theta\Delta\phi$. It will therefore be sufficient to put $Q = \theta\Delta\phi$.

If V and V' be the constant potentials of the plates A and B, (q_A, q_B) their charges at the beginning of the operation, we have

$$\left. \begin{array}{l} U = U_1 + \epsilon(\tfrac{1}{2}q_A V + \tfrac{1}{2}q_B V') \\ + q_A F + q_B F' \\ \phi = \phi_1 + q_A H + q_B H' \end{array} \right\},$$

where U_1 and ϕ_1 remain constant during the operation, and F' and H' are the values that F and H assume when θ is changed to $\theta + \tau$. Hence

$$\left. \begin{array}{l} \Delta U = \tfrac{1}{2}\epsilon q(V' - V) + q(F' - F) \\ \Delta\phi = q(H' - H) \end{array} \right\}.$$

But if W be the work done on the plates,

$$\Delta U = W + Q:$$

also

$$Q = \theta\Delta\phi.$$

Therefore
$$\Delta U = W + \theta \Delta\phi,$$
or
$$\tfrac{1}{2}\epsilon q(V' - V) + q(F' - F) = W + \theta q(H' - H).$$
Now since the charge q passes from the plate A to the plate B, $W = -\tfrac{1}{2}\epsilon q(V' - V)$. Substituting this value for W and dividing out by q, we get
$$\epsilon(V' - V) + F' - F = \theta\,(H' - H).$$
On reference to Art. 59, it will be seen that
$$F' - F = \tau F_\theta,\ H' - H = \tau H_\theta,$$
where F_θ and H_θ are the rates of increase of F and H with respect to θ. Hence we have
$$\epsilon(V' - V) + \tau F_\theta = \theta\tau H_\theta,$$
or
$$V' - V = 0 \quad \ldots \ldots \ldots \quad (19),$$
since F_θ is known to be equal to θH_θ.

For the result $V' - V = 0$,[1] which asserts that there is *no* electromotive force of contact between two pieces of the same metal in the same molecular state at different temperatures, the author has little or no independent authority to give. He has merely been informed that a similar result is given by a continental investigator. However, there appears to be strong internal evidence for the truth of the result.

(1) It is obtained by the same kind of reasoning as the result $P = \theta D_\theta$, and later on the same kind of reasoning will be found to give an accepted formula due to Gibbs and Helmholtz.

(2) If we take the two results $P = \theta D_\theta$, $V' - V = 0$,

---

[1] The various stages through which the author has been led to this result may be mentioned.

First, without using the formula for the entropy of the electrified system of plates, etc., he proved that $V' - V = k(\theta_1^2 - \theta_2^2)$, where θ_1 and θ_2 are the two temperatures and k is a constant which can only depend on the nature of the metal.

Next, he proved that k has the same value for all metals.

Lastly, by making use of the formula for the entropy, he proved, as in the text, that the constant value of k is zero.

for granted, we can deduce the results of Thomson for a thermo-electric circuit from them; and if we assume Thomson's results and $P = \theta D_\theta$, we can all but deduce the result $V' - V = 0$ from them.

Returning to the thermal effect, we have proved $Q = \theta \Delta \phi = \theta q(H' - H)$. Thus $Q = \theta q \tau H_\theta$. We have therefore a theoretical proof that the thermal effect is proportional to the charge which crosses the junction and to the difference of temperature at the junction.

We may obviously put $Q = q\sigma\tau$, where the quantity σ, which is called the "specific heat of electricity," is equal to θH_θ or F_θ.

It should be noticed that $q\sigma\tau$ is the heat absorbed when the charge passes across the junction against the rise of temperature τ. Hence $q\sigma\tau$ will be the heat *evolved* when the charge crosses in the *opposite* direction.

Art. 99. From the results relating to the two kinds of metallic junctions—the junction of two pieces of different metal at the same temperature and the junction of two pieces of the same metal at different temperatures, we can easily obtain the properties of a closed metallic circuit.

The total change of potential in travelling round a closed circuit is zero. Hence the sum of the abrupt rises of potential at the junctions is equal to the gradual fall of potential in the homogeneous parts of the circuit.

First, let the circuit be of one kind of metal in the same molecular state throughout, and let the temperature vary from point to point in any gradual way we please. Then there will be no abrupt change of potential at any of the junctions, and therefore there can be no current to produce a gradual fall of potential in the rest of the circuit. This is the result obtained experimentally by Magnus, who found it impossible to obtain a current by unequal heating in a homogeneous circuit.[1]

---

[1] An exception to the experimental result of Magnus has been discovered by Maxwell. He found that if we take a homogeneous circuit and by filing make a junction of a very thick piece and a very

Now let a circuit be formed of two different metals A, B, as in the figure, and let the temperature of every part of the circuit be kept constant; θ and θ_0 being the temperatures of the junctions. Then the "electromotive force of the circuit" is defined to be ε times the sum of the abrupt rises of potential as we travel round the circuit in the direction chosen as positive. Suppose the positive direction to be from A to B through the θ junction, as shown by the arrow; and let D be ε times the abrupt rise of potential encountered in passing through the junction in this direction. Then since the electromotive force of contact of two pieces of the same metal at different temperatures is zero, the electromotive force E of the circuit will be

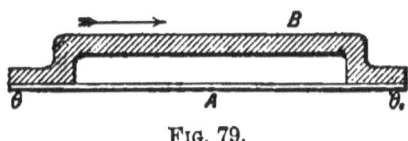

Fig. 79.

$$E = D - D_0,$$

where D_0 is what D becomes when θ_0 is put for θ.

For a circuit of several metals, we shall have

$$E = D_1 + D_2 + D_3 + \ldots,$$

where $(D_1, D_2, D_3, \ldots)$ are respectively ε times the abrupt rises of potential met with at the various junctions of two different metals in travelling round the circuit in the positive direction.

If I be the steady current of a circuit and R the resistance, we must have

$$E = RI,$$

however many metals the circuit may contain, since the sum of the abrupt rises of potential at the various junctions must be exactly balanced by the gradual fall of potential in the other parts of the circuit.

Again, when θ and θ_0 are so nearly equal that we

thin piece, a current is produced in the circuit on applying a flame to the junction.

There are one or two possible explanations of this fact in accordance with the results of the preceding articles.

may put $\theta_0 = \theta - \tau$, where τ is small, we have (by Art. 60)

$$D_0 = D - \tau D_\theta = D - \tau \frac{P}{\theta},$$

and therefore for a circuit of two metals

$$E = D - D_0 = \frac{\tau}{\theta} P,$$

a result due to Sir W. Thomson.[1]

For simplicity taking τ to be positive, we see that E and P are of the same sign. The current therefore travels in such a direction as to absorb heat at the hotter junction of the two metals.

Next, suppose the temperature of the θ junction raised from θ to $\theta + \tau'$, where τ' is small, and let the temperature of the θ_0 junction be kept constant. Then if E' and D' be the new values of E and D, we have

$$E' = D' - D_0.$$

Combining this equation with $E = D - D_0$, we get

$$E' - E = D' - D;$$

and therefore

$$\tau' E_\theta = \tau' D_\theta,$$

or

$$E_\theta = D_\theta.$$

Hence $\quad P = \theta D_\theta = \theta E_\theta,$

another result of Thomson's.

A very simple formula, in accordance with experiment, for the electromotive force of a circuit of two metals is obtained if we assume $D = a + b\theta + c\theta^2$, where a, b, c are constants. With this value of D we have

$$E = D - D_0 = b(\theta - \theta_0) + c(\theta^2 - \theta_0^2)$$
$$= (\theta - \theta_0)\{b + c(\theta + \theta_0)\}.$$

Now $D_\theta = b + c(\theta^2)$. To find $(\theta^2)_\theta$, put x and y each

---

[1] When $P = 0$ the formula becomes $E = 0$. Thus the value of E to a first approximation is zero. A nearer approximation will be given presently.

Applications to Electricity. 267

equal to θ in $(xy)_\theta = xy_\theta + yx_\theta$. Thus $(\theta^2)_\theta = 2\theta$, and $D_\theta = b + 2c\theta$. Hence $P = \theta D_\theta = \theta(b + 2c\theta)$. If therefore T be the "neutral point" of the two metals, or the temperature at which the Peltier effect vanishes, we find $b + 2cT = 0$, or $b = -2cT$. Substituting for b in E, we obtain

$$E = (\theta - \theta_0)\{-2cT + c(\theta + \theta_0)\}$$
$$= -2c(\theta - \theta_0)\left\{T - \frac{\theta + \theta_0}{2}\right\} \quad . \quad (20),$$

the formula of Avenarius and Tait.

We see from this formula that the electromotive force of a circuit of two metals vanishes in two cases, (1) when the two junctions are at the same temperature, and (2) when the mean of their temperatures is T, that is, when one is as much above the "neutral point" as the other is below.

If we put $\theta_0 + \tau$ for θ, the formula of Avenarius and Tait for the electromotive force becomes

$$E = -2c\tau\left\{T - \frac{2\theta_0 + \tau}{2}\right\}$$
$$= -2c\tau\left(T - \theta_0 - \frac{\tau}{2}\right).$$

If the temperature θ_0 be kept constant and τ be small enough, we have $E = A\tau$, where A is sensibly constant. But $E = RI$: hence $RI = A\tau$. Since R varies little with temperature, the current is practically proportional to the difference of temperature τ. This is the principle of Melloni's thermo-pile—an instrument used to measure small differences of temperature in connection with experiments on radiant heat.

There is one case in which we cannot put $E = A\tau$, and that is when the quantity within brackets in the expression for E is not practically constant. Let the θ junction be at the "neutral point" T: then $T = \theta = \theta_0 + \tau$ and $T - \theta_0 - \frac{\tau}{2} = \frac{\tau}{2}$. Hence $E = -c\tau^2$. Thus when

τ is small, E is exceedingly small (for example, if $\tau = \dfrac{1}{1,000}$, $\tau^2 = \dfrac{1}{1,000,000}$); and moreover the value does not change sign with τ. In words, when one junction is at the "neutral point," the current flows in the same direction whether the other junction is hotter or colder.

Combining the theoretical result $(D_{BA})_\theta = H_B - H_A$ with the above assumption for the value of D, we find
$$H_B - H_A = b + 2c\theta.$$
This is satisfied by taking
$$H = k\theta + k',$$
where k and k' are constants;
and we deduce $H_\theta = k$.
Now the specific heat of electricity is equal to θH_θ. Thus $\sigma = k\theta$, or the specific heat of electricity is proportional to the absolute temperature. In lead, the specific heat of electricity is zero: in this metal, k is therefore zero, and $H = k'$, a constant.

Prof. Tait has taken advantage of the fact that the specific heat of electricity is zero in lead by choosing lead as the standard metal in his "thermo-electric diagram." The "thermo-electric power" of any metal M with respect to the standard metal (ω) is taken to be $\dfrac{P_{M\omega}}{\theta}$ or $(D_{M\omega})_\theta$, which is equal to $H_M - k'$, where k' is constant when the standard metal is lead. Take two rectangular axes and find a point P in the plane of these axes such that the abscissa is proportional to θ and the ordinate to the thermo-electric power $H_M - k'$. Then the positions of P for different temperatures will lie on a curve by means of which the thermo-electric properties of the metal M may be studied.

Lastly, from the equations
$$\left.\begin{array}{r}P_{BA} = \theta(H_B - H_A)\\(D_{BA})_\theta = H_B - H_A\\\sigma = \theta H_\theta\\E = D - D_0\end{array}\right\}$$

Applications to Electricity.

we immediately obtain two formulæ due to Thomson connecting the specific heats of electricity in the two metals.

Firstly,
$$\left(\frac{P_{BA}}{\theta}\right)_\theta = (H_B - H_A)_\theta$$
$$= (H_B)_\theta - (H_A)_\theta$$
$$= \frac{\sigma_B - \sigma_A}{\theta}.$$

Next,
$$E_\theta = D_\theta = H_B^{\cdot} - H_A.$$
Now $(P_{BA})_\theta = (H_B - H_A) + \theta (H_B - H_A)_\theta$
$$= H_B - H_A + \theta \{ (H_B)_\theta - H_A)_\theta\}$$
$$= H_B - H_A + \sigma_B - \sigma_A.$$
Hence
$$E_\theta = P_\theta - (\sigma_B - \sigma_A).$$

Art. 100. Let two equal metallic plates A, B, be placed near together in a vacuum and parallel to one another. Then if they be at different potentials, the parallel opposed surfaces will be equally and oppositely electrified. If the potential of B be higher than that of A, the charge of B will be positive and that of A negative. If (V_B, V_A) be the potentials of the two plates, d the distance between them, and (ρ, $-\rho$) the electric densities of the parallel opposed surfaces of the plates at any point sufficiently far from the edges, we have

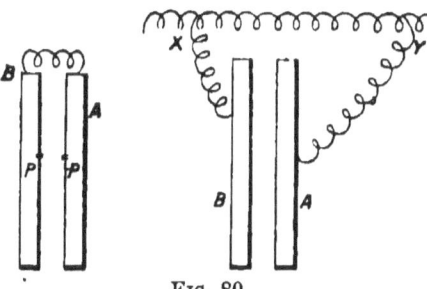

Fig. 80.

$$\frac{V_B - V_A}{d} = 4\pi\rho.$$

If the plates A, B, which are supposed to be at

different potentials, be of the same metal, they must either be insulated from one another or connected with two different points X, Y, of a wire in which a current is flowing: if they be of different metal, they may be in electric communication by means of a wire in which there is no current. In the latter case, the difference of the potentials $V_B - V_A$ will remain constant however the distance of the plates is varied; and the densities of the parallel electric layers will increase continually as the distance of the plates decreases. The smallest possible value of d, the distance of the parallel electric layers, occurs when the plates are actually in contact; and this value of d may be taken to be the distance of the atoms, which Sir W. Thomson calculates to be $\frac{1}{3 \times 10^8}$ centimetre. We thus see that when two different conductors are in contact, (1) the surfaces of contact are equally and oppositely electrified, and (2) the change of potential in passing from one body to the other is not abrupt, in the strict sense of the word, but takes place within an exceedingly small distance.

Suppose the electromotive force of contact of the two metals, or $\epsilon(V_B - V_A)$, denoted by D. Then $\frac{D}{d} = 4\pi\epsilon\rho$. At the surfaces of contact, we may put $d = \frac{1}{3 \times 10^8}$. For simplicity, suppose the electromotive force of contact to be a volt, or $D = 10^8$. Then $4\pi\epsilon\rho = 3 \times 10^{16}$. Hence $\rho = \frac{3 \times 10^{16}}{4\pi \times 9 \times 10^{20}} = \frac{1}{4\pi \times 3 \times 10^4} = \frac{1}{38 \times 10^4}$, nearly. If ρ' be the density in electro-static C.G.S. units, $\rho' = \rho \times 3 \times 10^{10} = \frac{10^6}{4\pi} = 80{,}000$ nearly.

Let metallic filings of the same metal as B be allowed to slide over that surface of B which is parallel to A, and suppose they are not allowed to come in contact with A. Then they will become electrified and will carry their electrification away with them as they fall to the ground.

This enables us to understand the following passage in Tait's "Sketch of Thermo-dynamics":—"By means of a very simple arrangement, not involving electrolysis, Thomson has shown how to collect the electricity developed in either of two metals in contact; but, as the principle of energy requires, mechanical energy has to be expended. He allows water, or copper filings, to drop from a copper can, the drops falling (without touching it) through a vertical zinc cylinder which is in metallic connection with the can. Each drop carries with it electricity induced by electrostatic induction in the air between the zinc and copper; and, if they be collected in an insulated dish, the latter may be charged to any extent. The apparatus is, in fact, an electrical machine worked by gravity; and the energy of the charge acquired by the insulated body on which the drops fall is accounted for by a deficiency in the heat produced by their impacts."

If the potentials of the two parallel plates be kept constant by a connecting wire, the formula $\dfrac{V_B - V_A}{d} = 4\pi\rho$ shows that a decrease in the distance of the parallel plates will increase the electrification of their opposed surfaces by causing a charge to pass from one plate to the other along the connecting wire. Conversely, an increase in the distance of the plates will cause the electrification of the opposed surfaces to decrease. If, however, the plates be insulated, the electrification of the parallel surfaces can only increase at the expense of the electrification of the other surfaces of the plates; and it is easy to see that a decrease in the distance of the insulated plates will diminish their potentials and simultaneously increase the electrification of the parallel surfaces. Conversely, if the distance of the insulated plates be increased, the potentials will also increase. Now when two plates are of different metal and in contact, the opposite layers of electricity are very large but very near together. Hence "if we could separate the plates without a partial recombination of the opposite layers of electricity, each of them would be found at an extremely high potential." "Assuming that the distance of the electric layers is originally $\frac{1}{10^6}$ of a millimetre, Helmholtz has shown that if a disc of zinc 10 centimetres in radius were in contact with a disc of copper in connection with the earth, the potential of the zinc when carried to a great distance would be 39×10^6 times as great as the original potential due to contact. With Sir W. Thomson's numbers, the final potential would be 30 times as great."[1]

When the distance of the plates is changing, the motion of the charges will give rise to an absorption or evolution of heat. If the plates are of the same metal at the same temperature, the only thermal effect will be in accordance with Joule's law, and will be negligibly small when the motion of the plates is slow enough. If, however, the plates are of different metal and connected by a wire, there will also be a reversible thermal effect at the junction; and it is easily seen that whether the plates are insulated or kept at a constant difference of potential by a connecting wire, the work done in slowly separating them will be exactly equal to the work obtained on allowing them slowly to approach to their former distance provided that the temperature remains constant during the motions.

We can readily calculate the electric attractions of the parallel surfaces of the plates (supposed to be in a vacuum) per square centimetre at any point sufficiently far from the edges of the plates. For the "electric force" F at any point between the plates at a sufficient distance from the edges is $\frac{\epsilon(V_B - V_A)}{d}$, or $\frac{D}{d}$; and the attraction on unit area of the plates is $\frac{1}{2}F\rho$. Substituting $\frac{D}{4\pi\epsilon d}$ for ρ, we find the attraction per square centimetre to be $\frac{1}{8\pi\epsilon}\left(\frac{D}{d}\right)^2$. Now when the two

---

[1] Mascart's "Electricity and Magnetism," vol. i., p. 181.

Applications to Electricity.

plates are solid, they can only be made to touch in a few isolated points. To obtain continuous contact, one of the plates must be liquid and able to adapt itself to the inequalities in the surface of the other. Taking $D = 10^8$ and $d = \dfrac{1}{3 \times 10^8}$ for this case, we find the attraction per square centimentre in dynes to be $\dfrac{(3 \times 10^{16})^2}{8\pi \times 9 \times 10^{20}}$, or $\dfrac{10^{12}}{8\pi}$, or $40{,}000 \times 10^6$, nearly. Hence since the pressure of one atmo is only about 10^6 dynes per square centimetre, the electric attraction on each plate is about 40,000 times the pressure of the atmosphere on an equal area. In the case when both plates are solid, the unavoidable imperfections of workmanship will impose a limit on the distance to which the parallel surfaces can be brought without touching. If this distance be taken to be $\tfrac{1}{30}$ centimetre, the corresponding attraction per square centimetre for an electromotive force of one volt will be $\dfrac{(3 \times 10^9)^2}{8\pi \times 9 \times 10^{20}}$, or $\dfrac{1}{2{,}500}$ dyne. This is about the 2,500 millionth of the pressure of the atmosphere, or less than the millionth of a pound per square foot.

One other point must be mentioned in connection with the electromotive force of contact before concluding this article. It is sometimes said that a sharp point has the power of discharging a conductor perfectly or of equalizing the potentials of two neighbouring conductors. It can be easily shown, however, that in the case of two neighbouring conductors situated in a vacuum, the action of a point will be, not to reduce the difference of potential to zero, but to make it equal to the difference of contact. If the conductors be surrounded by air, the result will probably be different, because, as will appear more clearly a little later, there is an abrupt change of potential at the surface of contact of a conductor and air.

Art. 101. We can now obtain the fundamental formula

of Gibbs and Helmholtz for the electromotive force of a galvanic battery. Throughout the article, the temperature of the system will be supposed to be uniform at every instant.

If two conductors of the same metal be connected by an electrolyte, the potentials of the two conductors will evidently be equal. There will therefore be no current produced on joining the conductors by a wire. To obtain a current, the conductors which are plunged in the electrolyte must be of different metal. If the two conductors be capped with pieces of any the same metal, the arrangement constitutes a galvanic battery, and the caps are called its poles. When the poles of the battery are not joined by a wire, the difference of their potentials is evidently the same whatever kind of metal they are of. We therefore define the electromotive force E of the battery to be ε times the sum of the abrupt rises of potential that we meet with in travelling from one pole to the other through the electrolyte in the direction in which the battery gives a current when the circuit is closed. If R be the resistance of the wire which completes the circuit, r the internal resistance of the battery, and I the steady current, it follows from Ohm's law that $E = (R + r) I$, and that the difference of the potentials of the poles multiplied by ε is then equal to $E - rI$.

To obtain the formula of Gibbs and Helmholtz for E, let us suppose that when the circuit of the battery is open, the poles are connected by means of wires with two plates A, B; both wires and plates being of the same metal as the poles. Parallel and opposite to these two plates place equal plates of any the same metal, as iron, and connect the iron plates by iron wires with each other, or with a large distant mass of iron in the neutral state, so that the two iron plates are always at potential zero. Also let the air be exhausted about the plates. Lastly, let the temperature of the system be uniform and equal to θ.

The potentials of B and A will be different. The

Applications to Electricity. 275

parallel surfaces of the plates will therefore be electrified; and it will be possible, by moving the plates, to cause any quantity of electricity to pass through the battery in a reversible manner. Suppose the potential of B to be higher than that of A. Then if a charge be made to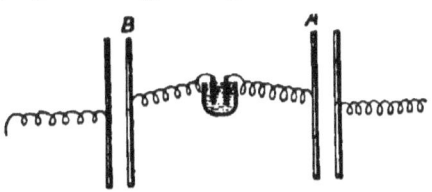

Fig. 81.

pass from A to B through the battery, its direction will be the same as that of the current the battery gives when its poles are connected by a wire.

Now let the system be made to undergo the following cycle of reversible operations at every instant of which the temperature of the system is uniform throughout. During the cycle, there may be considerable changes produced in the volume of the system by chemical action due to the passage of charges through the battery. We will therefore suppose all parts of the system at liberty to expand acted on by a uniform and constant normal pressure. The work done by this pressure on the system during the cycle being zero, the pressure need not be taken into account.

(1) By slowly moving the plates B nearer together and slowly separating the plates A, let a charge q be made to pass slowly through the battery from A to B; and let the uniform temperature of the system and the potentials of the plates be kept constant during the operation. The work done on the plates will be $-\frac{1}{2}q\mathrm{E}$, and there will be no thermal effect except at the junctions within the battery. The heat absorbed there may be written $q\psi$.

(2) We are next going to raise the temperature of the system from θ to $\theta + \tau$, where τ is small. The electromotive force will consequently change from E to $\mathrm{E} + \tau \mathrm{E}_\theta$, where E_θ is the rate at which E increases with respect to θ. Now if the plates are not moved during the rise of temperature, an unknown charge will pass

through the battery from one plate to the other. To prevent this undesirable occurrence, we will suppose the plates moved slowly about in such a way that their charges remain constant and the difference of their potentials gradually changes from $\dfrac{E}{\epsilon}$ to $\dfrac{E + \tau E_\theta}{\epsilon}$. Denote by β the increase of the potential of the plate B and by a the increase of the potential of the plate A, so that $\epsilon(\beta - a) = \tau E_\theta$. Also let q_B be the charge of B during the operation and q_A that of A. Then the work done on the plates during the operation will be $\tfrac{1}{2}\epsilon q_B \beta + \tfrac{1}{2}\epsilon q_A a$. Lastly, let the heat imparted to the system in raising the temperature be written $G\tau$.

(3) Then by slowly moving the plates, let the charge q be made to return from B to A without altering the potentials of the plates; and suppose the temperature of every part of the system kept equal to $\theta + \tau$ during the operation. The work done on the plates will be $\tfrac{1}{2}q(E + \tau E_\theta)$. Let the heat evolved be written $q\psi_1$.

(4) Lastly, let the temperature be slowly reduced from $\theta + \tau$ to θ, and suppose the plates moved about during the process in such a way that their charges remain constant. The work done on the plates in the operation will be $-\{\tfrac{1}{2}\epsilon(q_B - q)\beta + \tfrac{1}{2}\epsilon(q_A + q)a\}$, and the heat abstracted from the system may be written $H\tau$.

The total work done on the plates during the cycle is $\tfrac{1}{2}q\tau E_\theta + \{\tfrac{1}{2}\epsilon q_B \beta + \tfrac{1}{2}\epsilon q_A a\} - \{\tfrac{1}{2}\epsilon(q_B - q)\beta + \tfrac{1}{2}\epsilon(q_A + q)a\}$ or $\tfrac{1}{2}q\tau E_\theta + \tfrac{1}{2}\epsilon q(\beta - a)$, which is equal to $q\tau E_\theta$, since $\epsilon(\beta - a) = \tau E_\theta$.

Again, since the operations (2) and (4) are not the reverse of one another, we cannot take $G - H$ to be zero. If we put C for $G - H$, the principle of energy gives for the cycle

$$q\tau E_\theta + q(\psi - \psi_1) + C\tau = 0.$$

The heat $C\tau$ is absorbed at temperatures varying from θ to $\theta + \tau$; but since $C\tau$ is small, we get practically the same result whether we divide it by θ or $\theta + \tau$. Carnot's principle therefore gives for the cycle

$$q\left(\frac{\psi}{\theta} - \frac{\psi_1}{\theta_1}\right) + \frac{C\tau}{\theta_1} = 0,$$

where θ_1 stands for $\theta + \tau$.

[It is not immaterial whether we divide ψ_1 by θ or θ_1, because ψ_1 is not small.]

Multiplying the equation of entropy by θ_1 and subtracting from the equation of energy, we get

$$q\tau E_\theta + q\psi\left(1 - \frac{\theta_1}{\theta}\right) = 0.$$

Hence

$$\tau E_\theta = \psi\left(\frac{\theta_1}{\theta} - 1\right) = \psi\frac{\tau}{\theta},$$

or

$$\psi = \theta E_\theta.$$

We have thus found the value of the heat ψ absorbed at the various junctions in the battery on the passage of unit charge in the direction in which the battery tends to give a current; the temperature being uniform and kept constantly equal to θ during the process.

To apply the result, we call to mind that when the only external force acting on a system is a uniform and constant normal pressure, the heat evolved in any change of state is the same however the change is effected. Let the system be the battery with its poles connected by a wire of the same metal as themselves; and let the temperature be kept uniform and constantly equal to θ. Then if I be the current, r the internal resistance of the battery, and R the resistance of the wire which joins the poles, the heat *evolved* per second in the homogeneous parts of the circuit will be $(R + r)I^2$ ergs, in accordance with Joule's law: again, the heat *absorbed* at the junctions is $I\psi$ ergs per second. We will now assume the perfect accuracy of Faraday's laws, that conduction takes place in an electrolyte only by electrolysis and that "the chemical power of a current of electricity is in direct proportion to the absolute quantity of electricity which passes." If therefore we denote by L the heat *evolved* on effecting in any way at constant pressure the same change as is produced by the passage of unit charge, we have

$$(R + r)I^2 - I\psi = IL.$$

Hence

$$(R + r)I - \psi = L,$$

or

$$E - \theta E_\theta = L \quad \ldots \quad \ldots \quad (21),$$

which is the fundamental formula of Gibbs and Helmholtz.

At one time it was usual to put $E = L$: in other words, the thermal effects of a current at the junctions were assumed to be zero. In some cases of common occurrence, it actually happens that the thermal effects at the junctions are practically zero. For such batteries we have $E = L$.

Faraday's laws of electrolysis, assumed in establishing the equation $E - \theta E_\theta = L$, are completely at variance with the old contact theory of electricity, which supposed it possible for a current to be produced by contact alone without chemical action, and therefore possible for a current to traverse an electrolyte without electrolysis. If there be any doubt in connection with Faraday's laws, there will also be some uncertainty attaching to the equation $E - \theta E_\theta = L$. Fortunately, the last element of doubt respecting these laws has been dispelled by some remarkable experiments of Helmholtz. The point cannot be better explained than in the following words taken from Maxwell's "Elementary Treatise on Electricity."

"If an electrolytic cell, consisting of a vessel of acidulated water, in which two platinum plates are placed as electrodes, is inserted in the circuit of a single Daniell's cell, along with a galvanometer to measure the current, it will be found that though there is a transient current at the instant the circuit is closed, this current rapidly diminishes in intensity, so as to become in a very short time too weak to be measured except by a very sensitive galvanometer." "The current, however, never entirely vanishes, so that if the electromotive force is maintained long enough, a very considerable quantity of electricity may be passed through the electrolyte without any visible decomposition.

Hence it was argued that electrolytes conduct electricity in two different ways, by electrolysis in a very conspicuous manner and also, but in a very slight degree, in the manner of metals, without decomposition. Helmholtz has recently shown that the feeble permanent current can be explained in a different manner, and that we have no evidence that an electrolyte can conduct electricity without electrolysis.

In the case of platinum plates immersed in dilute sulphuric acid, if the liquid is carefully freed from all trace of oxygen or of hydrogen in solution, and if the surfaces of the platinum plates are also freed from adhering oxygen or hydrogen, the current continues only till the platinum plates have become polarized and no permanent current can be detected, even by means of a sensitive galvanometer. When the experiment is made without these precautions, there is generally a certain amount of oxygen or of hydrogen in solution in the liquid, and this, when it comes in contact with the hydrogen or the oxygen adhering to the platinum surfaces, combines slowly with it, as even the free gases do in presence of platinum. The polarization is thus diminished, and the electromotive force is consequently enabled to keep up a permanent current, by what Helmholtz has called electrolytic convection. Besides this, it is probable that the molecular motion of the liquid may be able occasionally to dislodge molecules of oxygen or of hydrogen adhering to the platinum plates. These molecules when thus absorbed into the liquid will travel according to the ordinary laws of diffusion, for it is only when in chemical combination that their motions are governed by the electromotive force. They will therefore tend to diffuse themselves uniformly through the liquid, and will thus in time reach the opposite electrode, where, in contact with a platinum surface, they combine with and neutralize part of the other constituent adhering to that surface. In this way a constant circulation is kept up, each of the constituents travelling in one direction by electrolysis, and back again by diffusion, so that a permanent current may exist without any visible accumulation of the products of decomposition. We may therefore conclude that the supposed inaccuracy of Faraday's law has not yet been confirmed by experiment."

Art. 102. The relation $\psi = \theta E_\theta$, obtained in the last article for the galvanic battery, is strictly analogous, both in form and meaning, with the relation $P = \theta D_\theta$ between the Peltier effect and the electromotive force of contact in the case of two different metals at the same temperature. To show the close connection between the two relations, take a galvanic battery with its poles X, Y, of the same metal, and let the temperature be uniform and equal to θ. Suppose that the direction in which we must travel through the battery from one pole to the other that ϵ times the sum of the abrupt rises of potential at the various junctions met with may be $+ E$, is from X to Y. To the metal Y solder a piece of metal Z, of a different kind from Y but at the same temperature θ. Let D be ϵ times the abrupt rise of potential in passing from Y to Z. Then if the temperature be kept constant, the sum of the quantities of heat absorbed at the junctions on the passage of unit charge from X to Z is $\theta E_\theta + \theta D_\theta$. This is equal to $\theta(E_\theta + D_\theta)$, or $\theta(E + D)_\theta$. If we write E' for $E + D$, and ψ' for $\psi + \theta D_\theta$, we have $\psi' = \theta E'_\theta$. Conversely, if we assume $\psi' = \theta E'_\theta$, *whether the terminal pieces X, Z, are of the same or different metal*, we can deduce from it the relation $P = \theta D_\theta$.

The terminal pieces X, Z, which are of different metal, are not to be considered the poles of the battery.

For let them be joined by a homogeneous wire of the same metal as X and at the temperature θ. Then the abrupt rise of potential in passing from Y to Z is equal to the abrupt fall of potential in passing from Z to the wire which joins Z to X. Hence ϵ times the sum of the abrupt rises of potential met with in travelling round the circuit, is equal to E and not to E'. It is therefore agreed that the poles of a battery are to be of the same metal; and it is evidently immaterial what kind of metal is chosen for this purpose.

Art. 103. The present article will be devoted to the consideration of a few very interesting and important experiments that have been made to test the existence of a difference of potential at a junction and to measure its value.

Take two equal metal plates A, B, each of which is homogeneous and of a uniform temperature; and let them either be of different metal and at the same temperature, or of the same metal and at different temperatures. Insulate the plates and let them be placed parallel and very near together. Then remove the air so completely from the plates that not a particle of air remains either adhering to the plates or existing near them. Lastly, let the plates be put in electric communication for a short time by means of a wire, so that the difference of their potentials assumes the value corresponding to the junction of the plates. When the plates are of different metal at the same temperature, the connecting wire may be of any metal but must be at the same uniform temperature as the plates. When the plates are of the same metal at different temperatures, the connecting wire is to be of the same metal as the plates but may be at the temperature of either of them.

If the plates be at different potentials, the parallel surfaces will be equally and oppositely electrified. At a point sufficiently far from the influence of the edges, the densities of the parallel electric layers will be uniform. If ρ be the charge per square centimetre at such a point of the surface of B; (V_B, V_A) the potentials of the two plates; and d the distance between them, we have

$$\frac{V_B - V_A}{d} = 4\pi\rho.$$

If we knew the values of ρ and d, we could easily deduce the value of $V_B - V_A$, the difference of potential at the junction in question. To find ρ, we may proceed as follows. Take two equal and insulated conductors X, Y, whose capacities are so great that their potentials do not change much even when they receive charges equal to those of the plates. Let these conductors be permanently connected with the plates, by means of wires, X with A and Y with B. Then when the distance of the plates is d, let the plates be put in communication for a short time by means of a wire, so as to make the difference of their potentials that which corresponds to their junction, which we have written $V_B - V_A$.

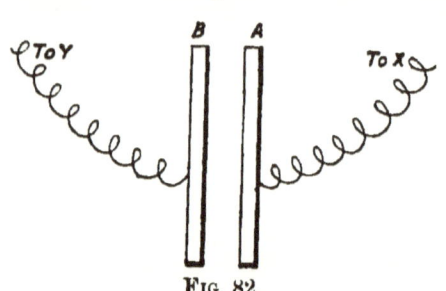

Fig. 82.

On separating the plates, the electrification of their opposed surfaces will evidently be diminished by the flow of electrification along the wires into the conductors X, Y. If these conductors be far enough from the plates, there will be no electrification at the back of the plates to take account of. Hence on separating the plates far enough, the whole of the electrification on their opposed surfaces may be sent into the conductors X, Y.

The conductors X, Y, are chosen to form an electrometer, the indications of which tell us the total charge received from the plates. We therefore know the total charge of each plate; and if it was not for the irregular distribution of electrification near the edges of the plates, we should know the value of ρ.

To avoid the influence of the edges, we may make

use of the principle of the "guard-ring." Each plate forms the central part of a larger disc of the same metal and at the same temperature as itself; and there is only just sufficient room allowed between a plate and its surrounding ring to allow freedom of movement without the risk of touching. Let the rings be insulated and fixed at the distance d. Then when the plates are placed in the centres of the rings at the distance d apart, let the potential of each ring be made the same as that of the corresponding plate, and let the difference of the potentials of the plates at the same time be made $V_B - V_A$, as before. Then if S be the area of each plate, the charges of the plates are respectively $S\rho$ and $-S\rho$; and these quantities are given by the deflection of the needle of the electrometer on separating the plates. In this way we find ρ and thence $V_B - V_A$.

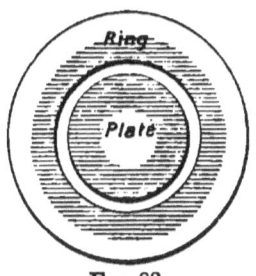

FIG. 83.

The foregoing is practically the method of Prof. Clifton, except that he does not remove the air from the apparatus. A slightly different method is used by M. Pellat, who also works in the open air. In the experiments of M. Pellat the plates A, B, when at the distance d, are connected for a short time, not with each other, as in the method of Prof. Clifton, but with two different points M, N, of a wire in which a steady current is flowing. If the plates are of different metal at the same temperature, the wire in which the steady current is flowing and the wires used to connect A with M and B with N, may be of any metal but must be at the same temperature as the two plates. If the plates are of the same metal at different temperatures, the three wires AM, MN, NB may be of the same temperature as one of the plates, say A. The wires AM, MN may then be of any metal; and if this metal is not of the same kind as the plates, the wire NB must be of two kinds of metal— a part NT, next to MN, of the same kind as MN, and the other part TB of the same kind as the plates. Thus if the plates be iron and the wire MN copper, the wires AM and NT are to be copper and TB iron, all at the temperature of the plate A. Then if I be the steady current flowing in the homogeneous wire MN from M to

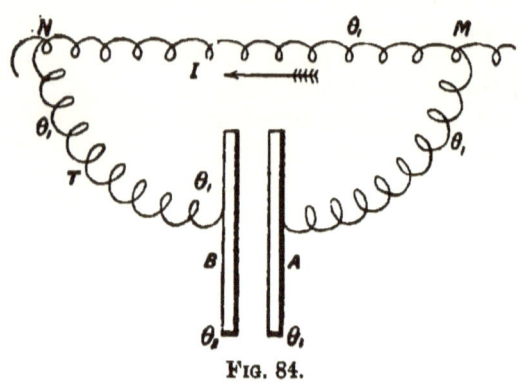

Fig. 84.

N, R the resistance of MN, the potentials of the plates when at the distance d will differ by $V_B - V_A - \frac{1}{\varepsilon}RI$, where $V_B - V_A$, as before, is the difference of potential corresponding to the junction of the two plates. The quantities R and I are now chosen so that $V_B - V_A - \frac{1}{\varepsilon}RI = 0$. The test that this is the case is that when the connections of the plates (supposed to be in a vacuum) with the wire MN are broken and the plates then separated to a great distance, there is no effect produced on the electrometer needle. Knowing R and I, we have $\varepsilon(V_B - V_A)$. The advantage of this method is that it does not require us to use the electrometer for purposes of measurement, but merely to show when a charge is zero, which it does with great accuracy. It also *avoids* the measurement of the small distance between the plates, and renders the exact parallelism of the plates unnecessary.

It is known that there is an abrupt change of

potential at the surface of contact of a conductor and a dielectric, such as air. Hence it follows that when the experiments of Prof. Clifton and M. Pellat are performed in air or a gas, the difference of potential of the plates is complicated with the properties of the dielectric by which the apparatus is surrounded. The results obtained are found to be different when different kinds of gas are used. We therefore speak of the differences of potential in air, hydrochloric acid, etc.

The part played in the experiments by air or gas may be illustrated as follows. Imagine the plates A, B, coated with mercury but otherwise situated in a *perfect vacuum*. Then we have two surfaces of mercury opposed to one another instead of two solid plates. If the whole system be at a uniform temperature, the electrification of the parallel opposite surfaces of mercury will be zero; and if we were unaware of the existence of the covering of mercury on the solid plates, we should naturally infer that there is no change of potential at the surface of contact of two conductors at the same temperature. If the plates be of the same metal but at different temperatures, there will be three junctions to take account of: (1) the colder junction of mercury and the solid metal, (2) the junction of the colder and warmer pieces of the solid metal, and (3) the warmer junction of the solid metal and mercury.

When the apparatus is exposed to the air, there will probably be a layer of air clinging to the plates, because, owing to the difference of potential, the air is attracted to the plates with a force which is perhaps 40,000 times as great[1] as the pressure of the atmosphere. If it were not for this difference of potential, there would be little error made in treating the air as a vacuum. Hence, as was first pointed out by Maxwell,[2] the experiments of Prof. Clifton and M. Pellat, when performed in the open air, do not give $V_B - V_A$, but $(V_B + b) - (V_A + a)$, or $V_B - V_A + (b - a)$, where a and b are the abrupt rises of

---

[1] In English measure, 262 tons per square inch.
[2] Muscart's "Electricity and Magnetism"

potential in passing from the plates A, B to the surrounding air.[1] If $b - a$ be comparable with $V_B - V_A$, the result of an experiment may be widely different from the required difference of potential $V_B - V_A$.

Exactly the same conclusion is arrived at in another way, which is described as follows in Mascart's "Electricity and Magnetism." The account will be understood if we remark that it is customary to write B | A for $V_B - V_A$—the abrupt rise of potential at the junction of two bodies A, B in contact when we pass from A to B.

"M. Garbe has deduced from the properties of surface tension, demonstrated by M. Lippmann, an ingenious method of measuring the electromotive forces of contact. This method, as Bichat and Blondlot observe, may furnish absolute values, independent of the external medium."

"In a capillary electrometer in the ordinary condition, the external electromotive force, which produces the maximum tension, is equal to the difference of potential of contact Hg | L of mercury with acidulated water. If the acidulated water is replaced by a liquid L′, we shall have, in like manner, the value of Hg | L′ by the electromotive force E′, which produces the maximum tension.

Let us now put the two liquids L and L′ in two vessels, above a layer of mercury, and let them be connected by a syphon filled with either of the two liquids, and provided with a diaphragm. By measuring the electromotive force E_1 of the couple thus formed, and of which the two layers of mercury are the electrodes, we have

$$E_1 = Hg \mid L + L \mid L' + L' \mid Hg = E + L \mid L' - E',$$

from which is deduced the value of L | L′ as a function of the three electromotive forces determined directly.

This method gives results which differ completely, not only in magnitude, but also in sign, from those which

---

[1] Air being a non-conductor, the layers of air which cling to the plates will not be at the same potential, as two metallic layers of the *same* metal would be. Thus $b - a$ is not zero.

the ordinary methods furnish. Such a divergence can only be explained by the fact that there is an electric difference between a liquid and air, as had been pointed out by Maxwell."

When the experiments of Prof. Clifton and M. Pellat are performed in the open air with plates of the same metal at different temperatures, it is found that the result obtained is not zero. It therefore follows from our theory that the presence of the air about the plates cannot be ignored. But it may perhaps be thought that the effects of the air may be made to disappear by a sufficient degree of exhaustion. On the contrary, M. Pellat only found very slight changes in the results on reducing the pressure of the air to 1 centimetre or 2 centimetres of mercury. It therefore appears that this degree of exhaustion produces only very slight changes in the differences of potential between the plates and air, and perhaps also in the layers of air clinging to the plates. To abolish the effects of the air, we must remove every particle of air which clings to the plates. Owing to the enormous force with which the air is attracted to the plates, the only way to remove the film of air may be to replace the air by some other substance. It would therefore be interesting to have the following experiments performed:—

Let the two plates A, B be placed in a glass vessel opening at the bottom by a very perfect stop-cock and completely filled with mercury. Remove every particle of air from the vessel by long-continued boiling of the mercury or otherwise. Then let every opening into the glass vessel except that of the stop-cock be closed with the blow-pipe. Lastly, let the stop-cock be opened and mercury allowed to run out without admitting air until the level of the mercury is below the plates. If the temperature be now reduced low enough, the pressure of the mercury vapour around the plates will be very small. At $0°$ C., the pressure is only ·02 millimetre of mercury.

If the apparatus be so arranged that the distance of the plates can be varied by means of magnets, the experiments of Prof. Clifton and M. Pellat can be repeated in mercury vapour. The results will depend on the nature of the layer of mercury which may adhere to the plates.

(1) There may be a perfect conducting layer of mercury covering each plate. In this case, if the system be at a uniform temperature, the difference of the potentials of the plates will appear to be zero. In other words, there will appear to be no difference of potential between two different metals in contact at the same temperature.

(2) There may be no layer of mercury on either plate. In this case, the difference of the potentials of two plates of the same metal at different temperatures will be correctly shown to be zero; and the experiment will also give correctly the difference of potential of two different metals in contact at the same temperature.

(3) Some other cases are also conceivable.

In concluding the article, it may be mentioned that the Peltier effect is very small. Now if the electromotive force of contact were independent of the temperature, the equation $P = \theta D_\theta$ shows that the Peltier effect would be zero. The fact, therefore, that the Peltier effect is found to be very small, does not prove that the electromotive force of contact is small, but only that it hardly varies with the temperature.

Art. 104. By means of the results established in the preceding parts of the work, we can compare the economies of the various methods of obtaining work from fuel. The system used in connection with the fuel for this purpose, must undergo a complete cycle of operations while the fuel is being consumed. There must also be no positive quantities of heat absorbed during the cycle nor a positive quantity of electric energy absorbed, otherwise than from the fuel. If these conditions are not fulfilled, the work obtained cannot be said to be derived from the fuel. For example, if the operations do not constitute a complete cycle, work may be obtained from the change of energy in the system without the consumption of fuel. Again, if positive quantities of heat or electric energy be absorbed otherwise than from the fuel, work may be obtained without the consumption of fuel by employing the system as a heat engine or an electromotor.

The operations undergone by the fuel and the system used in connection with it cannot form a complete cycle because the fuel is supposed to be consumed. Hence if W be the work obtained from the consumption of a quantity of fuel which would give out a quantity of heat Q if burnt at constant pressure, $\dfrac{W}{Q}$ cannot be called the "efficiency" of the process or method. We will therefore define it to be the "economy" of the method or process.

The most obvious way of obtaining work from any kind of fuel is to burn it in the furnace of a steam engine. Another way is to burn the fuel and to use the heat produced in maintaining a current in a thermoelectric circuit, as in Clamond's thermo-pile.[1]

It will be sufficient to consider a thermo-electric circuit of two pieces of metal only. On raising the junctions to different temperatures, a current will generally be produced. If there be an electro-motor in the circuit, the current will tend to turn it round and work may thus be obtained from the circuit. When the electro-motor, supposed perfect, runs round fast enough, the current will be stopped or rendered very feeble. If at the same time we can neglect the conduction of heat between unequally heated parts of the circuit, the whole of the process which the circuit experiences may be considered reversible. In what follows, we suppose the temperatures of all parts of the circuit kept constant, in which case the state of the circuit will be invariable.

In Clamond's thermo-pile, the state is kept invariable in an irreversible way. Heat is absorbed at a certain temperature and then conducted along the metal pieces of the circuit to places of lower temperature where it is wanted.

[1] Dunman's "Electricity and Magnetism."

When a thermo-electric circuit undergoes a cycle of any kind, Carnot's principle applies in the usual form $\frac{Q_1}{\theta_1} + \frac{Q_2}{\theta_2} + \frac{Q_3}{\theta_3} + \ldots \lessgtr 0$. In the simple reversible cycle described above, there are four places at which heat is absorbed or evolved. In the case of Clamond's thermopile, heat may be supposed to be absorbed at one temperature only and given out at another; and it is evident, from what has already been explained in this chapter, that the pile forms an imperfect heat engine working between these two temperatures. If we denote the positive quantity of heat absorbed at the higher temperature θ_a by Q_a, and the positive quantity of heat given out at the lower temperature θ_b by Q_b, we have, if W be the work obtained from the cycle, $Q_a - Q_b = W$, and since the cycle is irreversible, $\frac{W}{Q_a} < \frac{\theta_a - \theta_b}{\theta_a}$ or $W < Q_a \times \frac{\theta_a - \theta_b}{\theta_a}$. If therefore Q be the thermal value of the fuel burnt to supply the heat Q_a, the "economy" of Clamond's method, or $\frac{W}{Q}$, is $< \frac{Q_a}{Q} \times \frac{\theta_a - \theta_b}{\theta_a}$.

It will be seen that the "words efficiency" and "economy" have not quite the same meaning. The "efficiency" of the cycle undergone by the thermo-electric circuit is $\frac{W}{Q_a}$, and the "economy" of the method of obtaining work from fuel is $\frac{W}{Q}$. These are not identical unless Q_a happens to be the same as Q.

There are two conceivable methods of imparting to the circuit the heat obtained from the burning of the fuel. One of these, used by Clamond, is to allow heat to pass directly to the circuit from the furnace and the gaseous products of combustion. This method permits of θ_a being high and of the factor $\frac{\theta_a - \theta_b}{\theta_a}$ approaching its maximum value, unity. On the other hand, the value of $\frac{Q_a}{Q}$ is small, (1) because there is not sufficient time allowed for the passage of heat from the furnace to the circuit before the products of combustion are discharged into the chimney, and (2) because, according to the Theory of Exchanges, when the temperature of the junction is high, the junction rapidly radiates heat back into the furnace in addition to absorbing heat from it.

Another method would be to employ the fuel in heating a steam-boiler and to insulate the hot parts of the circuit and immerse them in the boiler. In this case the value of $\frac{Q_a}{Q}$ may be taken as $\frac{7}{10}$, but $\frac{\theta_a - \theta_b}{\theta_a}$ will be rather small. Supposing the hotter parts of the circuit to be at 195° C. and the colder parts at 45° C., $\frac{\theta_a - \theta_b}{\theta_a}$ will be equal to $\frac{150}{195 + 273}$, or $\frac{32}{100}$, nearly. Thus $\frac{Q_a}{Q} \times \frac{\theta_a - \theta_b}{\theta_a} = \frac{7}{10} \times \frac{32}{100} = \frac{9}{40}$. If we allow 20% for waste owing to the imperfections of the motor and the conduction of heat between unequally heat parts of the circuit, the "economy" of the method is found to be $\frac{4}{5} \times \frac{9}{40}$, or $\frac{18}{100}$.

It may be doubted whether Clamond's method of applying heat to the circuit could conveniently show so high an "economy" as 18%. We may therefore look upon this fraction as a superior limit to the "economy" of the thermo-electric method of obtaining work from fuel.

It will be noticed that in calculating the "efficiency" of the cycle of the thermo-electric circuit, we have made no allowance for the fact that an electric current is made use of. This is in accordance with the principles laid down in the preceding parts of the work. It has been seen that if two systems, one of which is electric and the other constantly free from electric properties, be made to undergo reversible cycles during which heat is absorbed and evolved at two given temperatures only, the "efficiency" is the same for both cycles, provided the electric properties of the electric system are all internal. Thus when the two temperatures at which heat is absorbed and evolved are given, the use of electric currents cannot make the efficiency of a cycle greater than that of a Carnot's perfectly reversible engine. In general, it will make it less. For if there be a sensible electric current in any homogeneous part of the system, the cycle will be irreversible, and the efficiency will be less than if the cycle had been reversible. What then is the advantage of employing an electric current during a cycle when the temperatures at which heat is absorbed and evolved are given? If an unelectrified heat engine can be made to undergo a reversible cycle during which heat is absorbed and evolved at the two given temperatures, there will be no advantage as regards efficiency to be derived from the use of a current. It may, however, happen not to be convenient for an unelectrified heat engine to perform a reversible cycle under the given conditions; and in this case it may not be practically possible to approach the theoretical maximum efficiency except when an electric current is made use of.

If a convenient method were found of using any kind of gas and in any quantity in a reversible gas battery, as in Grove's gas battery, we should have another method of obtaining work from fuel, far superior to the thermo-electric method. To this supposed method we now proceed.

If two platinum plates be immersed in a vessel of acidulated water, both plates will, of course, assume the same potential. Let the plates be so placed that an electric current cannot pass from one to the other except through the acidulated water. Then we know from the experiments of Helmholtz—one of the founders of the modern contact theory of electricity—that it is impossible for even a feeble current to pass from one plate to the other without decomposing the water into its constituent gases. The quantity of water decomposed being proportional to the quantity of electricity that passes, it follows that any amount of water may be decomposed by a feeble current by simply continuing the current long enough. Again, we know from Joule's law that a feeble current produces no thermal effects in a homogeneous conductor or electrolyte of uniform temperature. Let us make use of these two facts in applying the principle of energy to the system formed by the platinum plates, acidulated water, and products of decomposition. Consider an interval of time during which the potentials of the plates may be regarded as constant. Let E be ε times the excess of the potential of the plate where the current enters the system over the potential of the other plate, and let q be the quantity of electricity which a feeble current conveys through the system during the interval under consideration. Then the equation of energy is $\Delta U = W + Q + qE$. Here W is the negative work done on the system when the liberated gases force back the air. Again, if the temperature of the system be kept uniform and constant, Q refers only to the small thermal effects at the two junctions. Now the energy of water is much less than that of its constituent gases when these gases are in their ordinary form. Hence in this case, we have roughly, $\Delta U = qE$. It is found, however, that the gases are not liberated in their ordinary form when the apparatus is first set to work. But even in this case, the equation $\Delta U = qE$ is sufficient to explain the fundamental results of experiment.

When the apparatus first begins to work, the plates are at the same potential, or the value of E is zero. Thence it follows from the equation $\Delta U = qE$, that the gases first formed by electrolysis have the same energy as the water from which they are formed. This result is explained by experiment, which shows that these gases do not take the ordinary form, but form thin films on the platinum plates, there being a film of hydrogen firmly adhering to one plate, and a film of oxygen to the other. On continuing the current, the films and the value of E tend simultaneously to a maximum.[1] After this condition is attained, the gases formed by electrolysis appear in bubbles on the plates and rise to the surface of the water where they assume the ordinary gaseous form. The value of E in this case may be approximately found by means of the equation $\Delta U = q E$ by taking ΔU to be the excess of the energy of the gases in the ordinary state formed by the quantity of electricity q over that of the water from which they are formed. It appears from Thomson's calculations that the electromotive force required for this decomposition of water is 1·318 times that furnished by a single cell of Daniell's battery.

He says, "Hence at least two cells of Daniell's battery are required for the electrolysis of water; but fourteen cells of Daniell's battery connected in one circuit with ten electrolytic vessels of water with platinum electrodes would be sufficient to effect gaseous decomposition in each vessel."[2]

When the products of electrolysis are being freely evolved in their ordinary gaseous state, let them be caught in inverted test-tubes placed over the platinum plates. Then if we remove the machine or battery or thermo-electric couples which produce the current by which the water is decomposed, and connect the platinum plates by means of a wire, it will be found that there is a reverse current between the platinum plates through the acidulated water, and the gases will be observed to disappear from the test-tubes. For theoretical purposes, it will be best to suppose these experiments performed as follows: Let the platinum plates be connected by wires with the poles of a perfectly reversible dynamo or electro-motor. Then let the current be produced by forcibly turning the dynamo round. When sufficient water has been decomposed, let a reverse

[1] We are, of course, supposing that there is no "electrolytic convection." See Art. 102.
[2] Tait's "Sketch of Thermo-dynamics."

Applications to Electricity.

current be allowed, causing the gases to recombine to form water. If these operations be slow enough, they may be considered reversible; and if the system be kept at the same uniform and constant temperature during the two operations, the principle of Thomson and Clausius shows that the quantities of heat absorbed in the two operations are equal and opposite. Hence it is seen that the work obtained from the recombination of the gases is exactly equal to the work done in effecting the decomposition of the water.

Now oxygen and hydrogen, when not adhering to the platinum plates, form fuel. Let us therefore suppose that quantities of these gases whose thermal value is Q, are recombined in the reversible way just described; and let us calculate the amount of work obtained.

Denote by ΔU the increase of energy (a negative quantity) when the gases are changed into water at the same temperature, and let Δv be the increase of volume. Then if p be the constant pressure under which the gases are burned so as to become water at the same temperature, we have, since Q is the quantity of heat given out in the process,

$$\Delta U = - p\Delta v - Q.$$

Writing $- \Delta'U$ and $- \Delta'v$ for the negative quantities ΔU and Δv, we get

$$- \Delta'U = p\Delta'v - Q,$$

or

$$\Delta'U = Q - p\Delta'v.$$

Now from the table of Heats of Combination with Oxygen, and the table of volumes of gases, it is found that $p\Delta'v = \dfrac{Q}{83}$, nearly. Hence $\Delta'U = \tfrac{82}{83} Q$, nearly.

Next, if W be the negative work done on the system when the recombination of the gases is effected in a reversible way at the constant and uniform temperature θ, we have, if $\Delta \phi$ be the increase of entropy

$$\Delta U = W + \theta \Delta \phi.$$

But if W' be the positive work obtained from the motor, $W = -W' + p\Delta'v$. Thus the work obtained from the motor is equal to $\Delta U + \theta\Delta\phi + p\Delta'v$, or $Q + \theta\Delta\phi$. Now the change of entropy is insignificant in comparison with the change of energy. Taking $\Delta\phi$ to be negative and $\theta\Delta\phi$ numerically equal to 5% of Q, or $5\frac{5}{82}$% of $\Delta'U$, we find that the work obtained is equal to $\frac{19}{20}$ of Q. Hence the "economy" of the method is 95%.

In practice, the operation cannot be made reversible. Allowing for the imperfections of the motor, and for the heat developed according to Joule's law, the "economy" may be taken to be 80 to 85%.

If the circuit be allowed to stand long enough with its circuit interrupted, it will be found that the only effect of the reverse current on closing the circuit, is to cause the gaseous films on the platinum plates to disappear. The free gases in the test-tubes have therefore lost their power of recombination. If it were not for the disappearance of the films, the free gases, it is easy to see, would never lose their power of recombining and giving a reverse current, however long they were allowed to stand.

The results which have just been found with regard to the "economies" of the two methods of obtaining work from fuel illustrate the difference between "economy" and "efficiency." In the method of the gas battery there is nothing which can be called "efficiency"; for (1) the temperature of the system is uniform and constant during the operation, and (2) there is very little heat absorbed or evolved in any part of the system during the operation.

The superior "economy" of the gas battery is evidently a consequence of its reversibility, and in nowise due to the fact that the method is electrical. If we could consume the gases in a reversible non-electric method during which the system is at the same uniform and constant temperature as the gas battery, the "economy" of the method would be exactly equal to that of the gas battery.

In the animal body, what is usually called "efficiency" is clearly "economy." The fact that it is considerable does not prove the existence of electric processes within the body. To decide this point, recourse must be had to experiment. And in the majority of creatures there does not appear to be anything found particularly electrical in any part of the body. Even in the case of the electric eel, the comparison with an electro-magnetic engine is not appropriate. In the case of an electro-magnetic engine, electric energy is obtained from the consumption of *work*: in the case of the electric eel, electric energy is obtained from the consumption of *food*.

Art. 105. Let us consider the action between a current and a magnet. For this purpose, the equation of energy will be used in the form $\Delta U = W + Q + e$; and only some simple cases will be taken. It will be supposed:—

(1) That the strength and form of the magnet are invariable.

(2) That the wire which conveys the current is homogeneous and of an invariable form.

(3) That the current has the same constant value (I say) in all parts of the wire.

(4) That the potential at the place where the current enters the wire exceeds by a constant amount the potential at the place where the current leaves the wire. This excess multiplied by ϵ will be denoted by E.

(5) That the mechanical motions of the system are constant. Then the external forces which act on the system will just balance the action between the magnet and current; and the work which these forces do on the system will depend only on the change in the relative position of the magnet and current.

If t be the duration in seconds of the interval to which the equation of energy is applied, and q the quantity of electricity conveyed by the current, we have $q = It$ and $e = qE = ItE$. Again, it is supposed that Joule's law is universally true. Hence if the temperature of the system be kept uniform and constant, $Q = -RI^2 t$; and the equation of energy is

$$\Delta U = W - RI^2 t + ItE.$$

Let us first take the case in which the magnet and wire are both at rest. Then ΔU and W are both zero; and $-RI^2 t + ItE = 0$, or $E = RI$. Hence in this case, Ohm's law is deduced from Joule's law and the principle of energy.

Next, let us suppose the magnet at rest and let the wire be restricted to move in such a way that its position relative to the central line of the magnet remains the same. Then $\Delta U = 0$, and the equation of energy is $0 = W - RI^2 t + ItE$. Thus when W is not zero, Ohm's law $E = RI$ is no longer true.

The condition that the magnet should be at rest, and the relative positions of the wire and central line of the magnet invariable, may be realized as follows: Take a straight magnet; and let the wire be fastened at one end to a pivot at a point M in the axis of the magnet, and at the other end to a ring N which fits loosely around the middle part of the magnet. Let fixed wires be in connection with M and N, as in the figure, so that the potentials at these places can be kept constant.

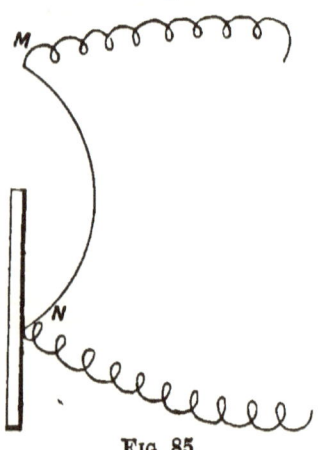

Fig. 85.

To find a convenient expression for W, let one direction in the wire be considered positive, the other negative. For simplicity, let the positive direction be so chosen that the electric energy absorbed in one second when the current is unity is $+ E$. Again, let that direction of rotation of the wire about the magnet be chosen as positive which requires a positive amount of work to be done on the wire when the current is positive.

Denote the positive amount of work done on the wire during a single rotation in the positive direction when the current is unity, by w. Then if n be the number of rotations per second and I the current, $W = ntIw$. This expression for W is true whether n and I are positive or negative. For $ntIw$ is positive or negative according as the signs of n and I are the same or different, and it is evident that the same is true of the sign of W.

Substituting $ntIw$ for W, the equation of energy becomes

$$0 = ntIw - RI^2t + ItE,$$

or

$$E - RI + nw = 0 \quad \ldots \quad (22).$$

If I_0 be the current when n is zero, or the wire at rest, we have $E - RI_0 = 0$, which is Ohm's law. Hence in any case

$$nw = RI - E = R(I - I_0).$$

When n is positive, I is positive and greater than I_0, and the work $ntIw$, done in turning the wire, is positive.

If n be the negative, I will be less than I_0 and may be negative. Put $-m$ for n, where m is positive. Then $R(I - I_0) = -mw$. Thus as m increases from zero, I continually diminishes. When $m = \dfrac{RI_0}{w}$, the current is absent. For a greater value m, I is negative, or the current flows in the opposite direction to I_0. Now the work done in turning the wire is $-mtIw$. This is negative when the current is positive, positive when the current is negative. Hence we see that if there be no friction and the system be abandoned to the constant electromotive force E, the wire will spin round faster and faster until a speed of $\dfrac{RI_0}{w}$ rotations per second is attained, at which speed the current just disappears.

The system just considered forms an ideal dynamo or motor, and the constant electromotive force E may be supposed to be produced by a battery whose internal resistance is so small that its poles M, N are always at the same potentials, whatever the current may be.

From the formula $mw = RI_0 - RI$, we see that if the rotation be in the negative direction and the speed below $\frac{RI_0}{w}$ revolutions per second, the current will be positive. A positive quantity of work will then be obtained from the system and a positive quantity of electric energy drawn from the battery, the "efficiency" of the motor being $\frac{mtIw}{ItE}$, or $\frac{mw}{E}$, or $\frac{mw}{RI_0}$. This is equal to $1 - \frac{I}{I_0}$. Thus it appears that the "efficiency" is practically unity when the speed is such that the current is practically absent. Again, if the rotation be in the negative direction and the speed above $\frac{RI_0}{w}$ revolutions per second, the current will be negative. A positive quantity of work will then be done on the system and a positive quantity of electric energy restored to the battery. The ratio of the electric energy restored to the battery to the work done in restoring it, is $\frac{-ItE}{-mtIw}$, or $\frac{E}{mw}$, or $\frac{RI_0}{mw}$, or $\frac{1}{1 - \frac{I}{I_0}}$. This is less than unity, since mw is greater than RI_0, or since I is negative; but it will be practically equal to unity when m is only just greater than $\frac{RI_0}{w}$, or the current only just existent.

We have thus the following important results. When the rotation is in the negative direction and the number of revolutions per second equal to $\frac{RI_0}{w}$ the current is absent. At a slightly lower speed, there is a feeble

NOTE.

A CURIOUS AND NOVEL ELECTRICAL PHENOMENON.

WHILE making experiments, the author obtained a curious and novel electrical phenomenon—a spasmodic current in a circuit apparently invariable.

The circuit contained two cells of Daniell's or Bunsen's battery, a galvanometer, and a voltameter with lead electrodes. The voltameter was a flat porous pot; and the electrodes, which were flat plates of lead, were placed outside the pot. One of these plates had a hole drilled in it, and a short piece of brass tube was fitted into the hole at right angles to the plate, and soldered. An india-rubber tube connected with a gas-holder, could be slipped on to the brass tube; and thus gas could be introduced between the electrode and the porous pot. To prevent waste of gas, the lead plate into which the piece of brass tube was soldered, was fitted air-tight at its edges to the porous pot.

In preparing the circuit for action, the porous pot was filled up with water (slightly acidulated), and the circuit allowed to stand until the moisture oozing through the porous earthenware made sufficient contact with the electrodes. When the circuit was then completed, the needle of the galvanometer was set in motion but soon came to rest, as usual. The current was then steady. Carbonic Acid gas (CO_2) was now introduced through the india-rubber and brass tubes between one of the electrodes and the porous pot. If the current happened to be passing through the voltameter in the right direction, a remarkable occurrence took place on the introduction of the gas. The current became spasmodic, and the needle of the galvanometer was set in motion and kept swinging. Sometimes, when the needle was in the midst of a swing, it was suddenly brought to rest; at other times, it was not merely brought to rest, but set swinging in the opposite direction.

APPENDIX.

Water at a Pressure of one Atmo.
(Everett's "Units and Physical Constants," also "Encyclopedia Britannica.")

Mass of one cubic centimetre in grammes, or of one cubic decimetre in kilogrammes.		Volume of one gramme in cubic centimetres, or of one kilogramme in cubic decimetres.	
0°	·999,884	0°	1·000,116
4°	1·000,013	4°	·999,987
5°	1·000,003	5°	·999,997
10°	·999,760	10°	1·000,240
15°	·999,173	15°	1·000,828
20°	·998,272	20°	1·001,731
25°	·997,108	25°	1·002,900
30°	·995,778	30°	1·004,240
35°	·994,69	35°	1·005,34
40°	·992,36	40°	1·007,70
45°	·990,38	45°	1·009,71
50°	·988,21	50°	1·011,93
55°	·985,83	55°	1·014,37
60°	·983,39	60°	1·016,89
65°	·980,75	65°	1·019,63
70°	·977,95	70°	1·022,55
75°	·974,99	75°	1·025,65
80°	·971,95	80°	1·028,86
85°	·968,80	85°	1·032,20
90°	·965,57	90°	1·035,66
95°	·962,09	95°	1·039,40
100°	·958,66	100°	1·043,12

Mercury at a Pressure of one Atmo (or less).

It has been found (Everett's "Units") that the density of mercury at 0° C. is 13·595,6 times that of water at 4° C. It is therefore 13·595,776,74 grammes per cubic centimetre. Hence the pressure produced (at Paris) by a column of mercury at 0° C. and one millimetre high, whose top is acted on by no force but the insignificant pressure of its own vapour, is 1,333·566,2 dynes per square centimetre. Also an atmo, or the pressure produced (at Paris) by a column of mercury at 0° C. and 760 millimetres high, is 1,013,510·335,6 dynes, or 1,033·279 grammes, per square centimetre.

According to Regnault, the specific heat of water (in calories) is as follows:—

at 0° C.	...	...	1·000,0
at 10° C.	...	...	1·000,5
at 20° C.	...	...	1·001,2
at 30° C.	...	...	1·002,0

Ice at the pressure of one atmo and at 0° C.
Mass of one cubic centimetre = ·920 gramme.
Volume of one gramme = 1·087 cubic centimetres.
Specific heat (in calories) at constant pressure = ·48 (Person).
Latent Heat of Fusion = 79·25 calories.

Mechanical Equivalent of Heat (Art. 12).

It has been found that a calorie is equivalent to about 42,350 gramme-centimetres, or 41,540,000 (roughly 42 million) ergs, or about 3 foot-pounds. In English measure, the heat required to raise the temperature of one pound of water under a pressure of one atmo from 0° C. to 1° C., is equivalent to about 1,390 foot-pounds.

Appendix. 307

HEAT AND ENERGY OF COMBINATION WITH OXYGEN (or the heat evolved on burning one gramme of the substance named in Oxygen under a constant pressure of one atmo, the final temperature being the same as the first).

(From Everett's "Units and Physical Constants.")

The numbers in the last column are the products of the numbers in the preceding column by 42 millions.

The authorities for these determinations are indicated by the initial letters A (Andrews), F (Favre and Silbermann), T (Thomsen). Where two initial letters are given, the number adopted is intermediate between those obtained by the two experimenters.

One gramme of	Compound formed.	Heat in calories.	Equivalent in ergs.
Hydrogen	H_2O	34,000 AF	1.43×10^{12}
Carbon	CO_2	8,000 AF	3.36×10^{11}
Sulphur	SO_2	2,300 AF	9.66×10^{10}
Phosphorus	P_2O_5	5,747 A	2.41×10^{11}
Zinc	ZnO	1,301 A	5.46×10^{10}
Iron	Fe_3O_4	1,576 A	6.62×10^{10}
Tin	SnO_2	1,233 A	5.18×10^{10}
Copper	CuO	602 A	2.53×10^{10}
Carbonic Oxide	CO_2	2,420 A	1.02×10^{11}
Marsh Gas	CO_2 and H_2O	13,100 AF	5.50×10^{11}
Olefiant Gas	,,	11,900 AF	5.00×10^{11}
Alcohol	,,	6,900 AF	2.90×10^{11}

COMBUSTION IN CHLORINE.

Hydrogen	HCl	23,000 FT	9.66×10^{11}
Potassium	KCl	2,655 A	1.12×10^{11}
Zinc	$ZnCl_2$	1,529 A	6.42×10^{10}
Iron	Fe_2Cl_6	1,745 A	7.33×10^{10}
Tin	$SnCl_4$	1,079 A	4.53×10^{10}
Copper	$CuCl_2$	961 A	4.04×10^{10}

DENSITIES OF GASES (Art. 19).

(In the last two columns, the pressure is supposed to be one atmo and the temperature 0° C.)

	Relative densities.	Relative specific volumes.	Mass of a litre in grammes.	Volume of a gramme in litres.	Volume of a pound in cubic feet.
Air	1	1	1·293,2	·773,3	12·39
Oxygen (O)	1·105,63	·904,46	1·429,8	·699,4	11·20
Hydrogen (H)	·069,26	14·438,3	·089,57	11·164,45	178·85
Nitrogen (N)	·971,35	1·029,45	1·256,15	·793,1	12·75
Carbonic Oxide (CO)	·954,5	1·047,6	1·234,4	·810,1	12·97
Carbonic Acid (CO_2)	1·529,07	·654,0	1·977,4	·505,7	8·10
Chlorine (Cl)	2·422,2	·412,8	3·132,8	·319,2	5·11
Cyanogen (NC_2)	1·801,9	·555,0	2·330,2	·429,1	6·87
Marsh Gas (CH_4)	·562	1·779	·727	1·375	22·04
Olefiant Gas (C_2H_4)	·982	1·018	1·270	·787	12·61
Ammonia (NH_3)	·595,2	1·680,1	·769,7	1·299,2	20·81

SPECIFIC HEATS OF GASES (Art. 22).

	At constant pressure.		At constant volume.		
	In calories. (By experiment.)	Compared with an equal volume of air.	In calories.	Compared with an equal volume of air.	Ratio of the specific heats, or k.
Air	·237,5	1	·168,4	1	1·410
Oxygen	·217,51	1·012	·155,01	1·018	1·403
Hydrogen	3·409,00	·994	2·411,4	·992	1·414
Nitrogen	·243,80	·997	·172,66	·996	1·412
Carbonic Oxide	·245,0	·985	·172,8	·978	1·418
Carbonic Acid	·216,9	1·396	·171,7	1·559	1·263
Chlorine	·120,99	1·234	·092,5	1·330	1·308
Marsh Gas	·592,9	1·403	·470,0	1·568	1·260
Olefiant Gas	·404,0	1·670	·333,7	1·946	1·211
Ammonia	·508,4	1·274	·392,3	1·386	1·296

Appendix.

MELTING POINTS AND LATENT HEATS OF FUSION (IN CALORIES) OF SOLIDS AT A PRESSURE OF ONE ATMO.

(From Watt's "Dictionary of Chemistry.")

	Melting points (C.).	Latent Heats.
Mercury	− 39°	2·82
Phosphorus	44°·2	5·0
Sulphur	115°	9·4
Iodine	107°	11·7
Lead	332°	5·4
Tin	235°	14·25
Silver	1,000°	21·1
Zinc	433°	28·1
Bismuth	270°	12·6
Nitrate of Potassium	339°	47·4
Nitrate of Sodium	310°·5	63·0

BOILING POINTS AND HEATS OF VAPORISATION AT A PRESSURE OF ONE ATMO.

(Everett's "Units.")

	Boiling points (C.).	Latent Heat of Vaporisation.	Observer.
Alcohol	77°·9	202·4	Andrews
Bisulphide of Carbon	46°·2	86·7	,,
Bromine	58°	45·6	,,
Ether	34°·9	90·4	,,
Mercury	350° (?)	62	Person
Sulphur	316° (?)	362	,,
Sulphurous Acid	− 10°·08		
Concentrated Sulphuric Acid	325°		

Temperatures and Pressures of Critical Points.
(Cagniard de la Tour.)

	Critical Temperature (C.).	Pressure in atmos.
Bisulphide of Carbon ...	262°·5	66·5
Ether	187°·5	37·5
Alcohol	258°·7	119
Water	411°·7	(?)

In the case of water, Maxwell estimates the Critical Temperature to be about 434° C., the Critical Pressure about 378 atmos, and the Critical Volume about 2·52 cubic centimetres per gramme.

Pressure of the Saturated Vapour of Water (Art. 60).
(Calculated from Regnault.)
Pressures.

Temperatures (C.).	In millimetres of mercury.	In grammes (at Paris) per square centimetre.	In dynes per square centimetre.	In atmos.	In pounds per square inch (at London).
− 30°	·4	·54	529·7		·008
− 25°	·6	·82	804·3		·012
− 20°	·9	1·22	1,196·7	·001	·017
− 15°	1·4	1·90	1,863·6	·002	·027
− 10°	2·1	2·85	2,795·5	·003	·041
− 5°	3·1	4·21	4,129·4	·004	·059
0°	4·600	6·254	6,134·40	·006	·089
5°	6·534	8·883	8,713·52	·008,6	·126
10°	9·165	12·460	12,222·1	·012,0	·177
15°	12·699	17·265	16,934·9	·016,8	·246
20°	17·391	23·644	23,192·0	·022,9	·336
25°	23·550	32·018	31,405·5	·030,9	·455
30°	31·548	42·892	42,071·3	·041,5	·610
35°	41·827	56·867	55,779·1	·055,0	·809
40°	54·906	74·649	73,220·8	·072,2	1·062
45°	71·390	97·060	95,203·3	·093,9	1·380
50°	91·980	125·054	122,661	·121,0	1·779

Appendix. 311

PRESSURES (*continued*).

Temperatures (C.).	In millimetres of mercury.	In grammes (at Paris) per square centimetre.	In dynes per square centimetre.	In atmos.	In pounds per square inch (at London).
55°	117·475	159·716	156,661	·154,5	2·27
60°	148·786	202·262	198,416	·195,7	2·88
65°	186·938	254·247	249,294	·246	3·61
70°	233·082	316·893	310,830	·306	4·51
75°	288·500	392·238	384,734	·379	5·58
80°	354·616	482·128	472,904	·467	6·86
85°	433·002	588·700	577,437	·570	8·37
90°	525·392	714·311	700,645	·704	10·16
95°	633·692	861·553	845,070	·833	12·25
100°	760	1,033·279	1,013,510	1	14·697
105°	906·41	1,232·33	1,208,760	1·19	17·53
110°	1,075·37	1,462·05	1,434,080	1·41	20·80
115°	1,269·41	1,725·86	1,692,840	1·67	24·55
120°	1,491·28	2,027·51	1,988,720	1·96	28·84
125°	1,743·88	2,370·94	2,325,580	2·29	33·72
130°	2,030·28	2,760·32	2,707,510	2·67	39·26
135°	2,353·73	3,200·08	3,138,850	3·09	45·5
140°	2,717·63	3,694·83	3,624,140	3·57	52·6
145°	3,125·55	4,249·43	4,168,130	4·11	60·4
150°	3,581·23	4,868·97	4,775,810	4·71	69·3
155°	4,088·56	5,558·71	5,452,370	5·38	79·1
160°	4,651·62	6,324·24	6,203,240	6·12	90·0
165°	5,275·54	7,171·15	7,033,950	6·94	102·0
170°	5,961·66	8,105·34	7,950,270	7·84	115·3
175°	6,717·43	9,132·87	8,958,140	8·84	129·9
180°	7,546·39	10,259·9	10,063,600	9·93	145·9
185°	8,453·23	11,492·8	11,272,900	11·12	163·5
190°	9,442·70	12,837·1	12,592,500	12·42	182·6·
195°	10,519·63	14,302·2	14,028,600	13·84	203·4
200°	11,688·96	15,892·0	15,588,000	15·38	226·0
205°	12,955·7	17,614·3	17,277,300	17·0	250·5
210°	14,324·8	19,475·7	19,103,100	18·8	277·0
215°	15,801·3	21,483·1	21,072,100	20·8	305·6
220°	17,390·4	23,643·6	23,191,200	22·9	336·3
225°	19,097·|	25,963·8	25,800,000	25·1	369·3
230°	20,926·4	28,451·1	27,906,700	27·5	404·7

OTHER RESULTS RELATING TO SATURATED STEAM (Art. 60).

The Latent Heat of a gramme of steam in calories is also the Latent Heat of a pound of steam in terms of a new unit of heat, viz. in terms of the heat required to raise 1 lb. of water from 0° C. to 1° C.

The volume of a gramme of saturated steam in cubic centimetres is also the ratio of the volume of a quantity of saturated steam to that of an equal quantity of water.

Temp. (Centi.).	Latent Heat of a gramme in calories.	Latent Heat of a gramme in ergs.	Volume of a gramme of saturated steam in cubic centimetres.	Volume of a pound of saturated steam in cubic feet.
0°	606·5	25193,860000	209,911	3,363
5°	603·0	25048,660000	150,271	2,407
10°	599·5	24905,070000	108,581	1,739
15°	596·0	24760,580000	79,348	1,271
20°	592·6	24616,030000	58,600	939
25°	589·1	24471,400000	43,901	703
30°	585·6	24326,680000	33,228	532
35°	582·1	24181,860000	25,398	407
40°	578·6	24032,780000	19,611	314
45°	575·1	23891,890000	15,301	245
50°	571·6	23746,720000	12,033	193
55°	568·1	23601,420000	9,552·4	153
60°	564·6	23455,970000	7,645·0	122
65°	561·1	23310,370000	6,164·5	98·7
70°	557·6	23164,610000	5,008·1	80·2
75°	554·1	23018,670000	4,098·2	65·7
80°	550·6	22872,560000	3,375·9	54·1
85°	547·1	22726,250000	2,799·3	44·8
90°	543·6	22579,740000	2,332·6	37·4
95°	540·1	22433,020000	1,955·2	31·3
100°	536·5	22286,080000	1,648·1	26·4
105°	533	22138,910000	1,396·5	22·4
110°	529	21991,510000	1,189·8	19·1
115°	526	21843,860000	1,018·3	16·3
120°	522	21695,950000	875·3	14·2
125°	519	21547,780000	756	12·1
130°	515	21399,330000	655	10·5
135°	512	21250,590000	570	9·1
140°	508	21101,570000	498	8·0
145°	504	20952,200000	437	7·0
150°	501	20802,590000	385	6·2
155°	498	20652,630000	340	5·4
160°	494	20502,330000	301	4·8
165°	490	20351,690000	267	4·3
170°	486	20200,410000	238	3·8
175°	483	20049,360000	213	3·4
180°	479	19897,650000	191	3·1
185°	475	19745,550000	172	2·7
190°	472	19593,070000	155	2·5
195°	468	19440,190000	140	2·2
200°	464	19286,910000	127	2·0

Pressure of Saturated Vapour of Mercury (in millimetres of mercury).

(From the "Encyclop. Brit.")

Temperatures (C.).	Pressures.	Temperatures (C.).	Pressures.	Temperatures (C.).	Pressures.
0°	·02	180°	11	360°	797·7
10°	·03	190°	14·8	370°	954·6
20°	·04	200°	19·9	380°	1,139·6
30°	·05	210°	26·3	390°	1,346·7
40°	·08	220°	34·7	400°	1,588
50°	·11	230°	45·3	410°	1,863·7
60°	·16	240°	58·8	420°	2,177·5
70°	·24	250°	75·7	430°	2,533
80°	·35	260°	96·7	440°	2,934
90°	·51	270°	123	450°	3,384·4
100°	·75	280°	155·2	460°	3,888·1
110°	1·07	290°	194·5	470°	4,449·4
120°	1·53	300°	242·2	480°	5,072·4
130°	2·18	310°	299·7	490°	5,761·3
140°	3·06	320°	368·7	500°	6,520·3
150°	4·27	330°	450·9	510°	7,253·4
160°	5·90	340°	548·3	520°	8,265
170°	8·09	350°	663·2		

Pressure of Saturated Vapour of Sulphur (in millimetres of mercury).

("Encyclop. Brit.")

Temps. (C.).	Pressures.	Temps. (C.).	Pressures.	Temps. (C.).	Pressures.
390°	272·3	460°	912·7	520°	2,133·3
400°	329	470°	1,063·2	530°	2,422
410°	395·2	480°	1,232·7	540°	2,739·2
420°	472·1	490°	1,422·9	550°	3,086·5
430°	561	500°	1,635·3	560°	3,465·3
440°	663·1	510°	1,871·6	570°	3,877·1
450°	779·9				

PRESSURES OF SATURATED VAPOURS (IN MILLIMETRES OF MERCURY).

("Encyclop. Brit.")

Temps. (C.)	Ammonia (NH_3)	Sulphuretted Hydrogen (H_2S)	Carbonic Acid (CO_2)	Nitrous Oxide (N_2O)
−30°	866·1			
−25°	1,104·3	3,749·3	13,007	15,694·9
−20°	1,392·1	4,438·5	15,142·4	17,586·6
−15°	1,736·5	5,196·5	17,582·5	19,684·3
−10°	2,144·6	6,084·6	20,340·2	22,008
−5°	2,624·2	7,066	23,441·3	24,579·2
0°	3,183·3	8,206·3	26,906·6	27,421
5°	3,830·3	9,490·8	30,753·8	30,558·6
10°	4,574	10,896·3	34,998·6	34,019·1
15°	5,423·4	12,447·9	39,646·9	37,831·7
20°	6,387·8	14,151·5	44,716·6	42,027·0
25°	7,477	16,012·4	50,207·3	46,641·4
30°	8,701	18,035·3	56,119	51,708·5
35°	10,070·2	20,224·3	62,447·3	57,268·1
40°	11,595·3	22,582·5	69,184·6	63,359·8
45°	13,287·3	24,954·3	76,314·6	
50°	15,158·3	27,814·8		
55°	17,219·8	30,690·7		
60°	19,482·1	33,740·2		
65°	21,965·1	36,961·5		
70°	24,675·5	40,353·2		
75°	27,630			
80°	30,843·1			
85°	34,330·9			
90°	38,109·2			
95°	42,195·7			
100°	46,608·2			

Appendix.

PRESSURES OF SATURATED VAPOURS (IN MILLIMETRES OF MERCURY).

("Encyclop. Brit.")

Temps. (C.).	Essence of Turpentine ($C_{10}H_6$).	Chloroform ($CHCl_3$).	Carbon Bisulphide (CS_2).	Sulphurous Acid (SO_2).
−30°	...	...	...	287·5
−25°	...	...	...	373·8
−20°	...	...	47·3	479·5
−15°	...	...	61·6	607·9
−10°	...	...	79·4	762·5
−5°	...	...	101·3	946·9
0°	2·1	...	127·9	1,165·1
5°	...	..	160	1,421·1
10°	2·9	...	198·5	1,719·5
15°	...	...	244·1	2,064·9
20°	4·4	106·5	298	2,462
25°	...	202·2	361·1	2,916
30°	6·9	247·5	434·6	3,431·8
35°	...	303·5	519·7	4,014·8
40°	10·8	369·3	617·5	4,670·2
45°	...	446	729·5	5,403·5
50°	17	535	857·1	6,220
55°	...	637·7	1,001·6	7,125
60°	26·5	755·4	1,164·5	8,123·8
65°	...	889·7	1,347·5	9,221·4
70°	40·6	1,042·1	1,552·1	
75°	...	1,214·2	1,779·9	
80°	61·3	1,407·6	2,032·5	
85°	...	1,624·1	2,311·7	
90°	90·6	1,865·2	2,619·1	
95°	...	2,132·8	2,966·3	
100°	131·1	2,428·5	3,325·1	
105°	...	2,754	3,727·2	
110°	186	3,111	4,164·1	
115°	...	3,501	4,637·4	
120°	257	3,925·7	5,148·8	
125°	...	4,386·6	5,699·7	
130°	349	4,885·1	6,291·6	
135°	...	5,422·5	6,925·9	
140°	464	6,000·2	7,604	
145°	...	6,619·2	8,326·9	
150°	605	7,280·6	9,095·9	
155°	686	7,985·3		
160°	775	8,734·2		
165°	...	9,527·8		

Chemical Action of a Current.[1]

Experiment shows that a coulomb or an ampere per second reduces 1·1173 mgr. of silver and decomposes ·09316 mgr. of water. Hence it follows that for the

Action of an Ampere.

			During a minute. Mgr.	During an hour Gr.
Silver reduced	...	Ag = 107·93	... 67·04	... 4·022
Copper „	...	Cu = 31·98	... 19·74	... 1·184
Water decomposed	...		5·59	... ·335,4
Hydrogen	...		·621,1	... ·037,26
			C.C.	C.C.
Volume of Hydrogen at 760 mm. and 0° C.	...	6·933	... 416·0
„ „ Detonating Gas			... 10·40	... 624

Unit of Resistance—Legal Ohm.[1]

The standard of resistance may be chosen arbitrarily. Jacobi proposed to use a copper wire of known dimensions, and, in order to avoid errors arising from the unequal purity of the metal, he proposed to distribute specimens of this wire among physicists.

For a long time telegraphists took as unit a kilometre or a mile of copper wire of known diameter, but at present more exact measurements are required for industrial purposes; for the least trace of foreign substances, and physical changes such as tempering or twisting, so greatly modify the conductivity of a metal that the nature and dimensions of a wire are insufficient to define the resistance; the temperature, moreover, has a considerable influence. Pouillet, who observed these different causes of variation, referred all measurements of conductivity from 1837 to that of distilled mercury. He took as a standard for comparison the column of mercury comprised within a cylindrical tube. The diameter was determined by weighings of mercury, and the ends terminated in two flasks of large aperture.

Werner Siemens has supplied for industrial purposes a great number of standards which represents very approximately a column of mercury at 0° C., a metre in length and a millimetre in cross section.

This is still an arbitrary unit. While retaining mercury as standard metal, it is more rational to choose a column the

[1] Mascart's "Electricity and Magnetism."

Appendix. 317

resistance of which is in a determinate ratio with the absolute unit. The International Commission on Electrical Units, assembled at Paris in 1884, adopted as *practical* unit, under the name of the *legal ohm*, the resistance of a column of mercury a square millimetre in cross section and 106 cm. in length at the temperature of melting ice. From numerous experiments made by different methods, this unit only differs by a few thousandths from the value, 10^9 absolute C.G.S. units, which represents the theoretical definition of the ohm.

RESISTANCE OF METALS AND ALLOYS AT $0°$ C.[1]

All the following resistances are arranged in reference to the legal ohm. If we assume that the resistance of 10^9 C.G.S. units were represented by a column of mercury of 106·25 cm. instead of 106 cm., we should multiply the conductivities by the ratio $\frac{106\cdot25}{106} = 1\cdot0024$ and divide the resistances by the same number.

	Value in C.G.S. units.		Resistance in Ohms.	
	Specific Resistance.	Specific Conductivity.	1 metre weighing 1 gramme.	100 metres 1 mm. in diameter.
Silver annealed	$1\cdot492 \times 10^3$	$67\cdot03 \times 10^{-5}$	0·151,7	1·899
„ hard drawn	1·620	61·73	0·165,0	2·062
Copper annealed	1·584	63·13	0·141,5	2·017
„ hard drawn	1·621	61·69	0·144,3	2·063
Gold annealed	2·041	49·00	0·400,7	2·598
„ hard drawn	2·077	48·14	0·407,6	2·644
Aluminium annealed	2·889	34·61	0·074,3	3·678
Zinc compressed	5·580	17·92	0·399,5	7·105
Platinum annealed	8·981	11·14	1·925	11·435
Iron „	9·636	10·38	0·751,8	12·27
Nickel „	12·356	8·093	1·052	15·73
Tin „	13·103	7·632	0·956,4	16·68
Lead compressed	19·465	5·137	2·217	24·78
Antimony „	35·21	2·84	2·370	44·83
Bismuth „	130·10	0·769	12·80	165·60
Mercury liquid	94·34	1·06	12·826	120·11
Alloy 2 Pt. + 1 Ag.	24·187	4·135	2·907	30·79
„ 2 Au. + 1 Ag.	10·776	9·280	1·638	13·72
„ 2 Pt. + 1 Ir.	21·633	4·627	4·651	27·54
German Silver	20·76	4·817	1·817	26·43

[1] Mascart's "Electricity and Magnetism."

The numbers of the first column, omitting the factor 10^3, represent the specific resistances in microhms. The same numbers represent in ohms the resistance of a wire 100 metres in length, having a section of a square millimetre.

N.B. A megohm = 10^6 ohms.
A microhm = 10^{-6} ohm.

INFLUENCE OF TEMPERATURE ON RESISTANCE OF METALS.[1]

The following formulæ are given by Benoit for the ratio of the specific resistance at $t°$ C. to that at $0°$ C. :—

Aluminium	$1 + ·003,876t + ·000,001,320t^2$
Copper	$1 + ·003,67t + ·000,000,587t^2$
Iron	$1 + ·004,516t + ·000,005,828t^2$
Magnesium	$1 + ·003,870t + ·000,000,863t^2$
Silver	$1 + ·003,972t + ·000,000,687t^2$
Tin	$1 + ·004,028t + ·000,005,826t^2$
Mercury in glass tube, apparent resistance, not corrected for expansion.	$1 + ·000,864,9t + ·000,001,12t^2$

RESISTANCE OF LIQUIDS.[2]

The conductivity of a solution increases at first with its strength, and passes generally through a maximum especially in the case of very soluble bodies. The conductivity, moreover, increases with the temperature.

In tables for liquids, the specific resistances are estimated in ohms.

The following results are obtained by M. Paalzow.

AQUEOUS SOLUTIONS.

Sulphuric Acid	SO_3HO	15°	9·146
	$SO_3HO + 14H_2O$	19°	1·336
	$SO_3HO + 13H_2O$	22°	1·256
	$SO_3HO + 499H_2O$	22°	17·431
Sulphate of Zinc	$ZnOSO_3 + 23H_2O$	23°	18·31
	$ZnOSO_3 + 24H_2O$	23°	18·02
	$ZnOSO_3 + 105H_2O$	23°	33·04
Sulphate of Copper	$CuOSO_3 + 45H_2O$	22°	19·10
	$CuOSO_3 + 105H_2O$	12°	31·42
Hydrochloric Acid	$HCl + 15H_2O$	23°	1·285
	$HCl + 500H_2O$	23°	8·177

[1] Everett's "Units and Physical Constants."
[2] Mascart's "Electricity and Magnetism." Also Kohlrausch's "Physical Measurements."

Specific Resistances of Solutions as a Function of the Density.[1]

Sulphuric Acid at 22° C. (Kohlrausch and Nippoldt.)

Density of the Solution.	Proportion of Acid.	Specific Resistance.	Relative Increase of conductivity for 1°C.
·998,5	0·0	70·41	0·47 × 10⁻²
1·000,0	0·2	41·05	0·47
1·050,4	8·3	3·252	0·653
1·098,9	14·2	1·787	0·646
1·143,1	20·2	1·414	0·979
1·204,5	28·0	1·239	1·317
1·263,1	35·2	1·239	1·259
1·316,3	41·5	1·347	1·410
1·354,7	46·0	1·487	1·674
1·399,4	50·4	1·672	1·582
1·448,2	55·2	1·962	1·417
1·502,6	60·3	2·412	1·794

Sulphate of Copper at 10° C. (Ewing and MacGregor.)[2]

Density.	Specific Resistance.	Density.	Specific Resistance.
1·016,7	164·4	1·138,6	35·0
1·021,6	134·8	1·143,2	34·1
1·031,8	98·7	1·167,9	31·7
1·062,2	59·0	1·182,9	30·6
1·085,8	47·3	1·205,1	29·3
1·117,4	38·1	(saturated).	

---

[1] Mascart's "Electricity and Magnetism." Also Kohlrausch's "Physical Measurements."
[2] Mascart's "Electricity and Magnetism."

Sulphate of Zinc at 10° C. (Ewing and MacGregor.)

Density.	Specific Resistance.	Density.	Specific Resistance.
1·014,0	182·9	1·270,9	28·5
1·018,7	140·5	1·289,1	28·3
1·027,8	111·1	1·289,5	28·5
1·054,0	63·8	1·298,7	28·7
1·076,0	50·8	1·328,8	29·2
1·101,9	42·1	1·353,0	31·0
1·158,2	33·7	1·405,3	32·1
1·184,5	32·1	1·417,4	33·4
1·218,6	30·3	1·422,0	33·7
1·256,2	29·2	(saturated).	

RESISTANCE OF SOLUTION OF NITRIC ACID.[1]

In reference to percentages of solution. (Kohlrausch and Nippoldt.)

Weight of Acid in 100 Parts	5	10	15	20	25	30	35	40	50	60	70	80
Resistance		3·92	2·19	1·65	1·42	1·31	1·29	1·31	1·38	1·59	1·96	2·55	3·78

For a density 1·36.[1]

Temp. (C.)	2°	4°	8°	12°	16°	20°	24°	28°
Specific Resistance	1·74	1·83	1·65	1·50	1·39	1·30	1·22	1·28

ELECTROMOTIVE FORCES OF CONTACT.[1]

According to Kohlrausch, the contact zinc-copper is about ·5 volt. M. Pellat has found, on the contrary, ·8 volt by direct electrostatic measures. Professors Ayrton and Perry estimate it at ·75 volt.—In practice, the volt is the electromotive force capable of maintaining a current of an ampere in a legal ohm.

If 100 represents the difference of the potential of the contact of zinc and copper, the following are the values for

[1] Mascart's "Electricity and Magnetism."

Appendix. 321

the differences of contact of zinc with different metals (in air):—

In air.	Volta.	Kohlrausch.	Hankel.	Ayrton and Perry.
Zinc | Platinum	...	123	123	131
„ Carbon	...	...	122	146
„ Palladium	...	...	115	...
„ Gold	...	115	110	...
„ Silver	109	109	118	...
„ Copper	100	100	100	...
„ Iron	82	75	84	80
„ Mercury	...	...	81	...
„ Bismuth	...	...	72	...
„ Antimony	...	...	69	...
„ Tin	55	...	23	37
„ Lead	45	...	44	28
„ Cadmium	...	...	24	...
„ Aluminium	...	...	− 25	...

PELTIER EFFECTS AT $25°$ C. IN WATTS PER COULOMB, OR THE ELECTROMOTIVE FORCES OF CONTACT IN VOLTS ACCORDING TO THE ASSUMPTION $P = D$.

(From Mascart's "Electricity and Magnetism.")

Copper \| Antimony (commercial) ... ... − ·005,6
„ Iron ... ... − ·002.9
„ Cadmium ... ... − ·000,53
„ Zinc ... ... − ·000,45
„ Argentan ... ... + ·002,87
„ Bismuth (pure) ... ... + ·022,2
Iron \| Zinc, at $13°·8$ C. ... ... + ·002,5
Copper \| Sulphate of Copper, at $12°$ C. ... ... + ·212
Zinc \| Sulphate of Zinc, at $12°$ C. ... ... + ·241

ELECTROMOTIVE FORCES OF CONSTANT BATTERIES (J. C. MAXWELL).

(From Kohlrausch's "Physical Measurements.")

					Concentrated Solution of		Volts.
Daniell I.	Amalgamated Zinc..	H_2SO_4	+ 4Aq.	..	$CuSO_4$	..Copper..	1·079
„ II.	„	„	+ 12Aq.	..	„	.. „ ..	·978
„ III.	„	„	+ „	..	$CuNO_3$..	„ ..	1·00
Bunsen I.	„	„	+ „	..	HNO_3	..Carbon..	1·964
II.	„	„	+ „	..	Nitric Acid of specific gravity 1·38	„ ..	1·888
Grove	„	„	+ 4Aq.	..	HNO_3	Platinum	1·956

P. D.   Y

ELECTROMOTIVE FORCE OF LECLANCHÉ'S BATTERY.

(From Mascart's "Electricity and Magnetism.")

Amalgamated Zinc—Solution of Sal-Ammoniac—Binoxide of Manganese and Carbon—1·45 volts.

THERMO-ELECTRIC POWERS OR HEIGHTS AT $t°$ C. IN C.G.S. UNITS.

(Everett's "Units and Physical Constants.")

The following table is based upon Prof. Tait's thermo-electric diagram joined with the assumption that a Grove's cell has electromotive force $1·97 \times 10^8$:—

Iron	$+ 1,734 - 4·87t$
Steel	$+ 1,139 - 3·28t$
Alloy, believed to be Platinum Iridium,	$+ 839$ at all temperatures.
Alloy, Pt. 95 + Ir. 5,	$+ 622 - ·55t$
„ „ 90 + „ 10,	$+ 596 - 1·34t$
„ „ 85 + „ 15,	$+ 709 - ·63t$
Soft Platinum,	$- 61 - 1·10t$
Alloy, platinum and nickel,	$+ 544 - 1·10t$
Hard Platinum,	$+ 260 - ·75t$
Magnesium,	$+ 244 - ·95t$
German Silver,	$- 1,207 - 5·12t$
Cadmium,	$+ 266 + 4·29t$
Zinc	$+ 234 + 2·40t$
Silver	$+ 214 + 1·50t$
Gold	$+ 283 + 1·02t$
Copper,	$+ 136 + ·95t$
Lead,	0
Tin,	$- 43 + ·55t$
Aluminium,	$- 77 + ·39t$
Palladium,	$- 625 - 3·59t$
Nickel to 175° C.,	$- 2,204 - 5·12t$
„ 250° to 310° C.,	$- 8,449 + 24·1t$
„ from 340° C.,	$- 307 - 5·12t$

The lower limit of temperature for the table is $-18°$ C. for all the metals in the list. The upper limit is $416°$ C., with the following exceptions:—Cadmium, $258°$ C.; zinc, $373°$ C.; German Silver, $175°$ C.

INDEX.

The numbers refer to the Articles.

Absolute temperature, 34.
Adams, tidal friction, 83.
Adiabatic curve, 23, 24.
Adiabatic compression and expansion of a saturated vapour, 61, 63, 64.
Air thermometer, 20.
Andrews, Dr., continuity of liquid and gaseous states, 58.
Animal life and Carnot's principle, 51.
Atoms, velocities and distances of, 35.
Avenarius and Tait, formula of, for the electromotive force of a thermo-electric couple, 99.
Axiom, fundamental, in Carnot's principle, 31; deductions from, 32.

Batteries, table of electromotive forces of, App.
Battery, fundamental formula of Gibbs and Helmholtz for the electromotive force of a, 101.
Bertrand, on the mathematical theory of induction, 105.
Bichat and Blondlot, on the electromotive force of contact, 103.
Boiling points, table of, App.
Boyle's law, 18.
Bravais and Martens, observation of the velocity of sound in air, 21.
Bunsen, calorimeter, 14; behaviour of spermaceti, paraffin, 66.

Cagniard de la Tour, continuity of liquid and gaseous states, 58; critical points, App.
Caloric theory of heat, 1.
Calorie, the usual unit of heat, 10.
  „   , mechanical value of a, 12.
Capacity, thermal, 10.
Capillarity, elementary principles of, 68; application by Sir W. Thomson to vegetation, 68.
Carbon bisulphide, 64; App.
Carnot, mechanical theory of heat, 1; efficiency of cycle, 37; principle, properly so-called, 37; engine, 37; extension of principle by Thomson and Clausius, 40, 41, 42; principle true for animal and vegetable life, 51, 52.
Cazin, experiments with vapours, 63, 64.
Charles, law of, 18.
Chemical action of a current, App.
Chloroform, 64; App.
Clamond's thermo-pile, 104.
Clapeyron, 28.
Clausius, on the energy of the universe, 7; specific heats of gases, 21, 22; on Carnot's principle, 28; fundamental axiom, 31; entropy, 43; on the concentration of radiant heat by lenses, 53; "specific heat" of saturated steam, 63; of other saturated vapours, 64.
Clausius and Thomson, principle

of, for a cyclical process, 40, 41; for a non-cyclical process, 42.
Clifton, experiments on the electromotive force of contact, 103.
Colding, 1.
Contact theories of electricity, new and old, 88.
Critical temperatures and pressures, App.
Current, thermal effect at a junction of an electric, 88, 90.
  „ , chemical action of a, App.
Currents, induced, 105.
Cycle, equation for a, in which heat is absorbed and given out at two temperatures only, 36; at any number of temperatures, 41.
Cyclical process, definition of, 9.

Davy, 1.
Degradation of energy, 47.
Delaunay, tidal friction, 83.
Densities of gases, 19, or App.
Density of mercury, App.
Diagram of work, or Watt's indicator diagram, 11.
  „ of heat, 45.
Dives and Lazarus, 79.
Duhem, the Thermo-dynamic Potential, 69; on the relation between the Peltier effect and the electromotive force of contact, 97.
Dynamos and motors, 105.

Earth, state of interior, 57.
"Economy" of the animal body, 104.
Efficiency of Carnot's cycle, 37; of electro-magnetic engine, 50.
Efficiency distinguished from economy, 104.
Electric eel, not an electro-magnetic engine, 104.
Electric energy, 8.
Electrical phenomenon, a curious and novel, NOTE at end of Chap. VI.
Electrical units, 87.
Electricity, specific heat of, 91.
Electro-magnetic engine, 50.
Electromotive force of contact, 88, 90; absent between two pieces of the same metal at different temperatures, 98; electrification and attraction due to, 100.
Electromotive force of contact and Peltier effect, relation between the, 97.
Electromotive forces of contact, table of, App.
Electromotive force of a thermoelectric circuit, definition of, 99.
Electromotive force of a thermoelectric couple and Peltier effect between the two metals, relation between the, by Thomson's method, 92; otherwise, 99.
Electromotive force of a battery, fundamental formula of Gibbs and Helmholtz for, 101.
Electromotive forces of batteries, table of, App.
Encke's comet and hypothesis, 74.
Energy, definition of kinetic, Intro.; of potential, 5; of mechanical and non-mechanical kinetic, 6.
Energy, definition and equation of, 3.
  „ , mutual, 5.
  „ , electric, 8.
  „ , conservation of, 7.
  „ of solar system, decreasing, 7.
  „ of universe, 7.
  „ of a perfect gas, 16, 23.
  „ , available, 47; of solar system, decreasing, 48; of universe, 48.
  „ , degradation of, 47.
  „ and entropy of water and its constituent gases, 44.
  „ , electric, may be transformed wholly into work, 50.
  „ of an electrified system, formula of Helmholtz for, 93.
Entropy, definition and fundamental properties of, 43.
  „ and energy of water and its constituent gases, 44.
  „ , independent of the motion of the centre of mass, 47.

Index. 325

Entropy of solar system, decreasing, 48.
„ of universe, 48.
„ test of stability, 69.
„ of an electrified system in electric equilibrium, formula for, 95.
Ether, frictionless, 54.
„ , affects motion, 54, 55.
Ether (common liquid), 64.
Evaporation, theory of, 59, 61.
Exchanges, Theory of, 49.

Final state of a system left to itself, 29, 72, 75.
Forbes, on glaciers, 57.
Freezing of water under pressure, 57, 65.
Friction, absent in the ether, 54.
Fuel, economies of the various methods of obtaining work from, 104.
Fusing points, table of, App.

Garbe, on the electromotive force of contact, 103.
Gas battery, 44, 104.
Gases, perfect, laws of, 16, 18.
„ „ , expansion or compression at constant temperature, 17.
„ „ , fundamental experimental results relating to, 19, or App.
„ „ , specific heats of, 17, 21, 22.
„ „ , tables of specific heats of, 22, App.
„ „ , general equations referring to, 23.
„ „ , adiabatic expansion of, 23.
Gay Lussac, experiment with compressed air, 16; law of perfect gases, 18.
Gibbs, thermo-dynamic potential, 69; geometrical method, 70; fundamental formula for the electromotive force of a galvanic battery, 101.
Glaciers, 57.

Governor for electric machinery, 105, footnote.
Gravitation, Intro.
Gravity at Paris, Intro.

Hall, satellite of Mars, 84.
Heat, Rumford and Davy, Mohr, Seguin, Mayer, Colding, Joule, on theory of, 1.
Heat and work, equivalence of, 3.
„ absorbed in any operation not generally defined entirely by the initial and final states, 4, 10.
„ , difference between scientific and popular meaning of, 4.
„ , unit of, 10.
„ , mechanical equivalent of, 12.
„ developed by adiabatic compression of a gas, 23.
„ , transformation into work, 25, 26, 37; into electric energy, 50.
„ engine, 37.
„ , diagram of, 45.
„ developed by the tides, 85.
Heats of combination, App.
Helmholtz, on glaciers, 57; expression for the energy of an electrified system, 93; fundamental formula for the electromotive force of a galvanic battery, 101; on the electrolysis of water, 101.
Hirn, experiments with vapours, 63, 64.
Hoar-frost line, 57.
Holtzmann, 21.
Hopkins, behaviour of spermaceti, wax, sulphur, stearine, 66.

Ice, effect of pressure on melting point, 57, 65.
„ line, 57.
„ , experimental results relating to, App.
Induced currents, 105.
Isothermal curve, 23.

Jolly, air thermometer, 20.
Joule, on the theory of heat, 1, 2; mechanical value of a calorie,

12; experiments with perfect gases, 16.
Joule's law for the heating of a wire by a current deduced from Ohm's law, 94.

Kinetic energy, definition of, Intro.
,,   ,, , mechanical and non-mechanical, 6.
Kirchhoff, on the concentration of radiant heat by lenses, 53.

Latent heat, old ideas of, 1.
,,   ,, and change of volume, equation between, 59.
,,   ,, of vaporisation, old theory, 62; new theory, 59, 61.
,,   ,, , experimental results, App.
Laws of thermo-dynamics, 3.
Liquefaction of ice, etc., by pressure, 57, 65, 66.
,,   of vapours worked expansively, 63, 64.

Magnus, law of, 99.
Mariotte, 18.
Massieu, the thermo-dynamic potential, 69.
Maxwell, example of available energy, 47.
Mayer, theory of heat, 1; vital force, 51.
Melloni's thermo-pile, 99.
Melting points, effects of pressure on, 57, 65, 66.
,,   ,, , table of, App.
Mercury, density of, App.
Metric system of units, Intro.
Mixtures, method of, 12.
Mohr, 1.
Moment of momentum, total amount of, unaltered by tidal friction, 79; comparison of the various parts of, 80.
Moon, effect of the tides on the, 84.
Mutual energy, 5.

Neutral point of two metals, 99.

Ohm's law, deduced from Joule's law, 105.

Pellat, experiments on the electromotive force of contact, 103.
Peltier effect, 88, 90.
,,   ,, and electromotive force of contact, relation between the, 97.
,,   ,, between two metals and electromotive force of the thermoelectric couple formed by them, relation between the, by Thomson's method, 92; otherwise, 99.
,,   ,, and specific heats of electricity in the two metals, relation between the, by Thomson's method, 92; otherwise, 99.
Peltier effects, table of, App.
Phenomenon, a curious and novel, NOTE at end of Chap. VI.
Potential, abrupt change of, at the surface of contact of two conductors, 88, 89.
Potential energy, 5.
Pressure of saturated steam, 60, or App.
,,   of other saturated vapours, App.

Rankine, specific heat of air, 21; adiabatic curve, 23; Carnot's principle, 28; on the concentration of radiant heat by lenses, 53; on the steam line, 58; on the "specific heat" of saturated steam, 63.
Regelation, 57.
Regnault, specific heats of gases, 21, 22; of water, 12, App.; properties of liquids and vapours, 63, 64, App.
Resistances, tables of, App.
Reversibility, 30.

Index. 327

Reversible cycle, property of, 33.
,, ,, , efficiency of, 37.
,, ,, , equation for, 36, 41.
Rumford, 1.

Saturated and superheated steam, 58.
Saturn's rings, 76.
Seguin, 1.
Specific heat, defined for a homogeneous body, 10.
,, ,, of water, 12, App.
,, ,, of any perfect gas, 17, 21, 22, App.
,, ,, of a saturated vapour, 61; of saturated steam, 63; of the vapour of ether, carbon bisulphide, and chloroform, 64.
Specific heat of electricity, 91, 99.
Specific heats of electricity in two metals and the Peltier effect, relation between the, by Thomson's method, 92; otherwise, 99.
Steam line, 57.
,, , properties of saturated, 58, 60, 63, App.
Stewart, Balfour, on Theory of Exchanges, 49.

Tait, on available energy of universe, 48.
,, , thermo-electric diagram, 99.
,, , thermo-electric power or height, 99, App.
,, and Avenarius, formula for the electromotive force of a thermo-electric couple, 99.
Temperature, test of equality of, 29.
,, , absolute scale of, 34.
Temperatures of fusion and boiling, App.
Thermal capacity, 10.
Thermo-dynamic potential, at constant pressure and constant volume, 69.
Thermo-dynamics, laws of, 3.
Thermo-electric couple, Sir W. Thomson on the general properties of, 92; proved in another way, 99.
Thermo-electric powers or heights, 99; table, App.
Thermometer, air, 20.
Thermo-pile, Clamond's, 104.
,, , Melloni's, 99.
Thomson, Prof. J. J., on the relation between the Peltier effect and the electromotive force of contact, 97.
Thomson, Prof. J., units, Intro.; re-arranges Carnot's cycle, 27; effect of pressure on melting point of ice, 57; the triple point, 57; on glaciers, 57.
Thomson, Sir W., Carnot's principle, 28; fundamental axiom, 31; absolute temperature, 34; available energy and degradation of energy, 47; on glaciers, 57; melting point of ice under pressure, 57, 65; discovers the Thomson effect or specific heat of electricity, 91; on the properties of a thermo-electric couple, 92.
Thomson and Clausius, principle of, for a cyclical process, 40, 41; for a non-cyclical process, 42.
Tidal friction, as a deduction from Carnot's principle, 49, 72; from the law of gravitation, 73.
,, ,, and Encke's comet, 74.
,, ,, , effect on the axis of rotation of the earth, 77.
,, ,, , comparative effects of, in the case of two bodies, one large, the other small, 79; effects on a general system of one large body and a number of small ones, 81.
,, ,, in the case of the sun, earth, and moon, 83, 84, 85.
Tides, heat developed by, 85.

Transformation of heat into work, 25, 26, 37; into electric energy, 50; of work into electric energy and conversely, 50.
Triple point, 57, 67.

Units, fundamental, Intro.
" , electrical, 87.

Vaporisation, theory of, 59, 61.
Vapours, tables of pressures of saturated, App.
Vegetable life and Carnot's principle, 51, 52.

Water, density, specific volume, and specific heat of, App.
Watt's diagram of work, 11; law of latent heat, 62.
Work, definition of, Intro.
" and heat, equivalence of, 3.
" depends generally on the path, 4, 10.
" , mechanical and non-mechanical, definitions of, 6.
" may be transformed wholly into electric energy, 50.

St. Dunstan's House, Fetter Lane,
London, E.C. 1894.

Select List of Books in all Departments of Literature

PUBLISHED BY

Sampson Low, Marston & Company, Ld.

ABBEY, C. J., *Religious Thought in Old English Verse*, 8s. 6d.
—— and PARSONS, *Quiet Life*, from drawings; motive by Austin Dobson, 31s. 6d.
ABERDEEN, EARL OF. See Prime Ministers.
ABNEY, CAPT., and CUNNINGHAM, *Pioneers of the Alps*, new ed. 21s.
About in the World. See Gentle Life Series.
—— *Some Fellows*, by "an Eton boy," 2s. 6d.; new edit. 1s.
ADAMS, CHARLES K., *Historical Literature*, 12s. 6d.
Ægean. See "Fitzpatrick."
AINSLIE, P., *Priceless Orchid*, new ed., 3s. 6d. and 2s. 6d.
ALBERT, PRINCE. See Bay. S.
ALCOTT, L. M., *Jo's Boys*, 5s.
—— *Comic Tragedies*, 5s.
—— *Life, Letters and Journals*, by Ednah D. Cheney, 6s.; 3s. 6d.
See also Low's Standard Series and Rose Library.
ALDAM, W. H., *Flies and Fly Making*, with actual flies on cardboard, 63s.
ALDEN, W. L. See Low's Standard Series.
ALFORD, LADY MARIAN, *Needlework as Art*, 21s.; l. p. 84s.

ALGER, J. G., *Englishmen in the French Revolution*, 6s.
—— *Glimpses of the French Revolution*, 6s.
Amateur Angler in Dove Dale, by E. M., 1s. 6d., 1s.
AMPHLETT, F. H., *Lower and Mid Thames, where and how to fish it*, 1s.
ANDERSEN, H.C., *Fairy Tales*, illust. by Scandinavian artists, 6s.
ANDERSON, W., *Pictorial Arts of Japan*, 4 parts, 168s.; artist's proofs, 252s.
Angler's strange Experiences, by Cotswold Isys, new edit., 3s. 6d.
ANNESLEY, C., *Standard Opera Glass*, 8th edit., 3s.
Annual American Catalogue of Books, 1886-93, each 15s., half morocco, 18s. each.
Antipodean Notes; a nine months' tour, by Wanderer, 7s. 6d.
APPLETON, *European Guide*, new edit., 2 parts, 10s. each.
Arcadia, Sidney's, new ed., 6s.
ARCHER, F., *How to write a Good Play*, buckram, 6s.
ARLOT'S *Coach Painting*, from the French by A. A. Fesquet, 6s.
ARMSTRONG, *South Pacific Fern Album*, actual fronds, 63s. net.
—— ISABEL J., *Two Roving Englishwomen in Greece*, 6s.

ARMYTAGE, Hon. Mrs., *Wars of Queen Victoria's Reign*, 5s.
ARNOLD, *On the Indian Hills, Coffee Planting, &c.*, new ed., 7s. 6d.
—— R., *Ammonia and Ammonium Compounds*, illust. 5s.
Artistic Japan, text, woodcuts, and coloured plates, vols. I.-VI., 15s. each.
ASHE, R. P., *Two Kings of Uganda*, 6s.; new ed. 3s. 6d.
—— *Uganda, England's latest Charge*, stiff cover, 1s.
ATCHISON, C. C., *Winter Cruise in Summer Seas;* "how I found" health, illust., new ed. 7s. 6d.
ATKINSON, J. B. *Overbeck.* See Great Artists.
ATTWELL, *Italian Masters*, especially in the National Gallery, 3s. 6d.
AUDSLEY, G. A., *Chromolithography*, 44 coloured plates and text, 63s.
—— *Ornamental Arts of Japan*, 2 vols. morocco, 23l. 2s.; four parts, 15l. 15s.
—— W. and G. A., *Outlines of Ornament in all Styles*, 31s. 6d.
AUERBACH, B., *Brigitta* (B. Tauchnitz), 2s.; sewed, 1s. 6d.
—— *On the Height* (B. Tauchnitz), 3 vols. 6s.; sewed, 4s. 6d.
—— *Spinoza* (B. Tauchnitz), a novel, 2 vols. 4s.
AUSTRALIA. See F. Countries.
AUSTRIA. See F. Countries.
BACH. See Great Musicians.
BACON. See Eng. Philosophers.
—— Delia, *Biography*, 10s. 6d.
BADDELEY, W. St. Clair, *Love's Vintage;* sonnets &c., 5s.
—— *Tchay and Chianti*, 5s.
—— *Travel-tide*, 7s. 6d.

BAKER, James, *John Westacott*, new edit. 3s. 6d.
—— *Foreign Competitors*, 1s. See also Low's Standard Novels.
—— R. Hindle, *Organist and Choirmaster's Diary*, 2s. 6d.
BALDWIN, James, *Story of Siegfried*, illust. 6s.
—— *Story of Roland*, illust. 6s.
—— *Story of the Golden Age*, illust. 6s.
Ballad Stories. See Bayard Series.
BALL, J. D., *Things Chinese*, new edit., 10s. 6d.
BALLANTYNE, T., *Essays.* See Bayard Series.
BAMFORD, A. J., *Turbans and Tails*, 7s. 6d.
BANCROFT, G., *History of America*, new edit. 6 vols. 73s. 6d.
—— *United S. Constitution, its Formation*, 2 vols., 24s.
Barbizon Painters. See Great Artists.
BARLOW, Alfred, *Weaving by Hand and Power*, new ed. 25s.
—— P. W., *Kaipara, New Z.*, 6s.
—— W., *Matter and Force*, 12s.
BARR, Amelia E., *Preacher's Daughter*, 5s.
BARROW, J., *Mountain Ascents (in England)*, new edit. 5s.
BARRY, J. W., *Corsican Studies*, 12s.; new edit. 6s.
BASSETT, *Legends of the Sea and Sailors*, 7s. 6d.
BATHGATE, A., *Waitaruna, a Story of New Zealand*, 5s.
Bayard Series, edited by the late J. Hain Friswell; flexible cloth extra, 2s. 6d. each.
Chevalier Bayard, by Berville.
St. Louis, by De Joinville.
Essays of Cowley.
Abdallah, by Laboullaye.

Bayard Series—continued.
Table-Talk of Napoleon.
Vathek, by Beckford.
Cavalier and Puritan Songs.
Words of Wellington.
Johnson's Rasselas.
Hazlitt's Round Table.
Browne's Religio Medici.
Ballad Stories of the Affections, by Robert Buchanan.
Coleridge's Christabel, &c.
Chesterfield's Letters.
Essays in Mosaic, by Ballantyne.
My Uncle Toby.
Rochefoucauld, Reflections.
Socrates, Memoirs from Xenophon.
Prince Albert's Golden Precepts.

BEACONSFIELD, *Public Life*, 3s. 6d.
—— See also Prime Ministers.

BEATTIE, T. R., *Pambaniso*, 6s.

BEAUGRAND, *Young Naturalists*, now edit. 5s.

BECKER, A.L., *First German Book*, 1s.; *Exercises*, 1s.; *Key to both*, 2s. 6d.; *Idioms*, 1s. 6d.

BECKFORD. See Bayard Series.

BEECHER, H. W., *Biography*, new edit. 10s. 6d.

BEETHOVEN. See Great Musicians.

BEHNKE, E., *Child's Voice*, 3s. 6d.

BELL, *Obeah, Witchcraft in the West Indies*, 2s. 6d., n. ed., 3s. 6d.

BERRY, C. A. See Preachers.

BERVILLE. See Bayard Series.

BIART, *Lucien*. See Low's Standard Books and Rose Library.

BICKERSTETH, ASHLEY, B.A., *Harmony of History*, 2s. 6d.
' *Outlines of Roman History*, 2s. 6d.
—— E. and F., *Doing and Suffering*, new ed., 2s. 6d.
—— E. H., Bishop of Exeter, *Clergyman in his Home*, 1s.

BICKERSTETH, E. H., Bishop of Exeter, *From Year to Year*, original poetical pieces, morocco or calf, 10s. 6d.; padded roan, 5s.; roan, 5s.; cloth, 3s. 6d.
—— *Hymnal Companion to the Common Prayer*, full lists post free.
—— *Master's Home Call*, new edit. 1s.
—— *Octave of Hymns*, sewn, 3d., with music, 1s.
—— *The Reef, Parables*, illust. 7s. 6d. and 2s. 6d.
—— *Shadowed Home*, n. ed. 5s.
—— MISS M., *Japan as we saw it*, illust. from photos., 21s.

BIGELOW, JOHN, *France and the Confederate Navy*, 7s. 6d.

BILLROTH, *Care of the Sick*, 6s.

BIRD, F. J., *Dyer's Companion*, 42s.
——H.E., *Chess Practice*, n.c., 1s.

BLACK, WILLIAM. See Low's Standard Novels.

BLACKBURN, C. F., *Catalogue Titles, Index Entries, &c.* 14s.
——*Rambles in Books*, cr. 8vo. 5s.; edit. de luxe, on handmade paper, only 50 copies, 15s.
—— H., *Art in the Mountains*, new edit. 5s.
—— *Artistic Travel*, 10s. 6d.
—— *Breton Folk*, n. e., 10s. 6d.

BLACKMORE, R. D, *Georgics of Virgil*, 4s. 6d.; cheap edit. 1s.
See also Low's Standard Novels.

BLAIKIE, *How to get Strong*, new edit. 5s.
—— *Sound Bodies for our Boys and Girls*, 2s. 6d.

BLOOMFIELD, ROBERT. See Choice Eds.

Bobby, a Story, by Vesper, 1s.

BOCK, *Temples & Elephants*, 21s.

BONWICK, JAMES, *Colonial Days*, 2s. 6d.

A Select List of Books

BONWICK, JAMES, Colonies, 1s. ea.; 1 vol. 5s.
—— Daily Life of the Tasmanians, 12s. 6d.
—— First Twenty Years of Australia, 5s.
—— Last of the Tasmanians, 16s.
—— Port Philip, 21s.
—— Lost Tasmanian Race, 4s.
BOSANQUET, C., Blossoms from the King's Garden, 6s.
—— Jehoshaphat, 1s.
——Lenten Meditations, Ser. I. 1s. 6d.; II. 2s.
—— Tender Grass for Lambs, 2s. 6d.
BOULTON, N. W. Rebellions, Canadian life, 9s.
BOURKE, On the Border with Crook, illust., roy. 8vo, 21s.
—— Snake Dance of Arizona, with coloured plates, 21s.
BOUSSENARD. See Low's Standard Books.
BOWEN, F., Modern Philosophy, new ed. 16s.
BOWER, G. S., and WEBB, Law of Electric Lighting, 12s. 6d.
BOWNE, R. P., Metaphysics, 12s. 6d.
BOYESEN, H. II., Against Heavy Odds, 5s.; also 3s. 6d.
—— History of Norway, 7s. 6d.
—— Modern Vikings, 6s.; also 3s. 6d.
Boy's Froissart, King Arthur, Percy, see "Lanier."
Boys, first yearly vol. 7s. 6d.
BRADSHAW, New Zealand as it is, 12s. 6d.
—— New Zealand of To-day, 14s.
BRANNT, Fats and Oils, 42s.
—— Scourer and Dyer, 10s. 6d.
—— Soap and Candles, 35s.
—— Vinegar, Acetates, 25s.

BRANNT, Distillation of Alcohol, 12s. 6d.
—— Metal Worker's Receipts, 12s. 6d.
—— Metallic Alloys, 12s. 6d.
—— and ANDRES, Varnishes, 12s. 6d.
—— and WAHL, Techno-Chemical Receipt Book, 10s. 6d.
BRETON, JULES, Life of an Artist, an autobiography, 7s. 6d.
BRETT, EDWIN J., Ancient Arms and Armour, 105s. nett.
BRISSE, Menus and Recipes, French and English, new edit. 5s.
Britons in Brittany, 2s. 6d.
BROOKS, NOAH, Boy Settlers, 6s.; new ed., 3s. 6d.
BROWN, A. J., Rejected of Men, and other poems, 3s. 6d.
—— A. S. Madeira and Canary Islands for Invalids, n. ed. 2s. 6d.
—— RICHARD, Northern Atlantic, for travellers, 4s. 6d.
—— ROBERT. See Low's Standard Novels.
Brown's South Africa, 2s. 6d.
BROWNE, LENNOX, and BEHNKE, Voice, Song, & Speech, 15s.; new edit. 5s.
—— Voice Use, 3s. 6d.
—— SIR T. See Bayard Series.
BRYCE, G., Manitoba, 7s. 6d.
—— Short History of the Canadian People, 7s. 6d.
BUCHANAN, R. See Bayard.
BULKELEY, OWEN T., Lesser Antilles, 2s. 6d.
BUNYAN. See Low's Standard Series.
BURDETT-COUTTS, Brookfield Stud, 5s.
—— BARONESS, Woman's Mission, Congress papers, 10s. 6d.

BURNABY, EVELYN, *Ride from Land's End to John o' Groats*, 3s. 6d.
—— MRS., *High Alps in Winter*, 14s.
BURNLEY, JAMES, *History of Wool and Wool-combing*, 21s.
BUTLER, COL. SIR W. F., *Campaign of the Cataracts*, 18s.
—— See also Low's Standard Books.
BUXTON, ETHEL M. WILMOT, *Wee Folk*, 5s.
BYNNER. See Low's Standard Novels.
CABLE, G. W., *Bonaventure*, 5s.
CADOGAN, LADY ADELAIDE, *Drawing-room Comedies*, illust. 10s. 6d., acting edit. 6d.
—— *Illustrated Games of Patience*, col. diagrams, 12s. 6d.
—— *New Games of Patience*, with coloured diagrams, 12s. 6d.
CAHUN. See Low's Standard Books.
CALDECOTT, RANDOLPH, *Memoir*, by Henry Blackburn, new edit. 7s. 6d. and 5s.
—— *Sketches*, pict. bds. 2s. 6d.
CALL, ANNIE PAYSON, *Power through Repose*, 3s. 6d.
CALLAN, H., M.A., *Wanderings on Wheel and Foot through Europe*, 1s. 6d.
CALVERT, EDWARD (*artist*), *Memoir and Writings*, imp. 4to, 63s. nett.
Cambridge Trifles, 2s. 6d.
Cambridge Staircase, 2s. 6d.
CAMPBELL, LADY COLIN, *Book of the Running Brook*, 5s.
—— T. See Choice Editions.
CANTERBURY, ARCHBISHOP. See Preachers.
Capitals of the World, plates and text, 2 vols., 4to, half morocco, gilt edges, 63s. nett.

CARBUTT, MRS., *Five Months Fine Weather; Canada, U.S. and Mexico*, 5s.
CARLETON, WILL, *City Ballads*, illust. 12s. 6d.
—— *City Legends*, ill. 12s. 6d.
—— *Farm Festivals*, ill. 12s. 6d.
—— *City Ballads*, 1s. ⎫
—— *City Legends*, 1s. ⎬ 1 vol., 3s. 6d.
—— *City Festivals*, 1s. ⎭
—— *Farm Ballads*, 1s. ⎫
—— *Farm Festivals*, 1s. ⎬ 1 vol., 3s. 6.
—— *Farm Legends*, 1s. ⎭
—— *Poems*, 6 vols. in case, 8s.
—— See also Rose Library.
CARNEGIE, ANDREW, *American Four-in-hand in Britain*, 10s. 6d.; also 1s.
—— *Triumphant Democracy*, 6s.; new edit. 1s. 6d.; paper, 1s.
CAROVE, *Story without an End*, illust. by E. V. B., 7s. 6d.
CARPENTER. See Preachers.
CAVE, *Picturesque Ceylon*, 21s. nett.
Celebrated Racehorses, fac-sim. portraits, 4 vols., 126s.
CELIÈRE. See Low's Standard Books.
Changed Cross, &c., poems, 2s. 6d.
Chant-book Companion to the Common Prayer, 2s.; organ ed. 4s.
CHAPIN, *Mountaineering in Colorado*, 10s. 6d.
CHAPLIN, J. G., *Bookkeeping*, 2s. 6d.
CHARLES, J. T. See Playtime Library.
CHATTOCK, *Notes on Etching*, new edit. 10s. 6d.
CHENEY, A. N., *Fishing with the Fly*, 12s. 6d.

CHERUBINI. See Great Musicians.

CHESTERFIELD. See Bayard Series.

Choice Editions of choice books, illustrated by Cope, Creswick, Birket Foster, Horsley, Harrison Weir, &c., cloth extra gilt, gilt edges, 2s. 6d. each; re-issue, 1s. each.
Bloomfield's Farmer's Boy.
Campbell's Pleasures of Hope.
Coleridge's Ancient Mariner.
Elizabethan Songs and Sonnets.
Goldsmith's Deserted Village.
Goldsmith's Vicar of Wakefield.
Gray's Elegy in a Churchyard.
Keats' Eve of St. Agnes.
Milton's Allegro.
Poetry of Nature, by H. Weir.
Rogers' Pleasures of Memory.
Shakespeare's Songs and Sonnets.
Tennyson's May Queen.
Wordsworth's Pastoral Poems.

CHURCH, W. C., *Life of Ericsson*, new ed., 16s.

CHURCHILL, LORD RANDOLPH, *Men, Mines and Animals in South Africa*, 21s.; new ed. 6s.

CLARK, Mrs. K. M., *Southern Cross Fairy Tale*, 5s.

—— *Persephone and other Poems*, 5s.

CLARKE, C. C., *Recollections of Writers, with Letters*, 10s. 6d.

—— PERCY, *Three Diggers*, 6s.

—— *Valley Council;* from T. Bateman's Journal, 6s.

Claude le Lorrain. See Great Artists.

COCHRAN, W., *Pen and Pencil in Asia Minor*, 21s.

COLERIDGE, S.T. See Choice Editions and Bayard Series.

COLLINGWOOD, H. See Low's Standard Books.

COLLYER, ROBERT, *Things Old and New*, Sermons, 5s.

CONDER, J., *Flowers of Japan and Decoration*, coloured Plates, 42s. nett.

—— *Landscape Gardening in Japan*, 52s.6d. nett.; supplement. 36s. nett.

CORDINGLEY, W. G., *Guide to the Stock Exchange*, 2s.

CORREGGIO. See Great Artists.

COWEN, JOSEPH, M.P., *Life and Speeches*, 14s.

COWLEY. See Bayard Series.

COX, DAVID. See Great Artists.

—— J. CHARLES, *Gardens of Scripture; Meditations*, 5s.

COZZENS, F., *American Yachts*, pfs. 21l.; art. pfs. 31l. 10s.

—— S. W. See Low's Standard Books.

CRADDOCK. See Low's Standard Novels.

CRAIK, D., *Millwright and Miller*, 21s.

CROCKER, *Education of the Horse*, 8s. 6d. nett.

CROKER, Mrs. B. M. See Low's Standard Novels.

CROSLAND, MRS. NEWTON, *Landmarks of a Literary Life*, 7s. 6d.

CROUCH, A. P., *Glimpses of Feverland* (West Africa), 6s.

—— *On a Surf-bound Coast*, 7s. 6d.; new edit. 5s.

CRUIKSHANK, G. See Great Artists.

CUDWORTH, W., *Abraham Sharp, Mathematician*, 26s.

CUMBERLAND, STUART. See Low's Standard Novels.

CUNDALL, J., *Shakespeare*, 3s. 6d., 5s. and 2s.

CURTIS, C. B., *Velazquez and Murillo*, with etchings, 31s. 6d.; large paper, 63s.

CURTIS, W. E., *Capitals of Spanish America*, 18s.
CUSHING, W., *Anonyms*, 2 vols. 52s. 6d.
—— W., *Initials and Pseudonyms*, 25s.; ser. II., 21s.
CUTCLIFFE, H. C., *Trout Fishing*, new edit. 3s. 6d.
DALY, Mrs. Dominic, *Digging, Squatting in N. S. Australia*, 12s.
DANTE, *Text-book in Four Languages*, illum. cover, 5s. nett.
D'ANVERS, N., *Architecture and Sculpture*, new edit. 5s.
—— *Elementary Art, Architecture, Sculpture, Painting*, new edit. 12s. and 10s. 6d.
—— *Painting*, new ed. by F. Cundall, 6s.
DAUDET, Alphonse, *Port Tarascon*, by H. James, 7s. 6d.; also 5s. and 3s. 6d.
DAVIES, C., *Modern Whist*, 4s.
DAVIS, C. T., *Manufacture of Leather*, 52s. 6d.
—— *Manufacture of Paper*, 28s.
—— *Manufacture of Bricks*, 25s.
—— *Steam Boiler Incrustation*, 8s. 6d.
—— G. B., *International Law*, 10s. 6d.
DAWIDOWSKY, *Glue, Gelatine, Veneers, Cements*, 12s. 6d.
Day of my Life, by an Eton boy, new edit. 2s. 6d.; also 1s.
Days in Clover, by the "Amateur Angler," 1s.; illust., 2s. 6d.
DELLA ROBBIA. See Great Artists.
Denmark and Iceland. See Foreign Countries.
DENNETT, R. E., *Seven Years among the Fjort*, 7s. 6d.
DERRY (B. of). See Preachers.
DE WINT. See Great Artists.

DIGGLE, J. W., *Bishop Fraser's Lancashire Life*, new edit. 12s. 6d.; popular ed. 3s. 6d.
—— *Sermons for Daily Life*, 5s.
DIRUF, O., *Kissingen*, 5s. and 3s. 6d.
DOBSON, Austin, *Hogarth*, with a bibliography, &c., of prints, illust. 24s.; l. paper 52s. 6d.; new ed. 12s. 6d.
—— See also Great Artists.
DOD, *Peerage, Baronetage, and Knightage, for 1894*, 10s. 6d.
DODGE, Mrs., *Hans Brinker*. See Low's Standard Books.
Doing and Suffering; memorials of E. and F. Bickersteth, new ed., 2s. 6d.
DONKIN, J. G., *Trooper and Redskin; Canada police*, 8s. 6d.
DONNELLY, Ignatius, *Atlantis, the Antediluvian World*, new edit. 12s. 6d.
—— *Cæsar's Column*, authorised edition, 3s. 6d.
—— *Doctor Huguet*, 3s. 6d.
—— *Great Cryptogram*, Bacon's Cipher in the so-called Shakspere Plays, 2 vols., 30s.
—— *Ragnarok: the Age of Fire and Gravel*, 12s. 6d.
DORÉ, Gustave, *Life and Reminiscences*, by Blanche Roosevelt, fully illust. 24s.
DOS PASSOS, J. R., *Law of Stockbrokers and Exchanges*, 35s.
DOUGALL, J. D., *Shooting Appliances, Practice*, n. ed. 7s. 6d.
DOUGLAS, James, *Bombay and Western India*, 2 vols., 42s. nett.
DU CHAILLU, Paul. See Low's Standard Books.
DUFFY, Sir C. G., *Conversations with Carlyle*, 6s.

DUNCKLEY ("Verax.") See Prime Ministers.
DUNDERDALE, GEORGE, *Prairie and Bush*, 6s.
Dürer. See Great Artists.
DYKES, J. Osw. See Preachers.
EBERS, G., *Per Aspera*, 2 vols., 21s.; new ed., 2 vols., 4s.
Echoes from the Heart, 3s. 6d.
EDEN, C. H. See For. Countries.
EDMONDS, C., *Poetry of the Anti-Jacobin*, new edit. 7s. 6d.; large paper, 21s.
EDWARDS, *American Steam Engineer*, 12s. 6d.
—— *Modern Locomotive Engines*, 12s. 6d.
—— *Steam Engineer's Guide*, 12s. 6d.
—— H. S. See Great Musicians.
—— M. B., *Dream of Millions*, &c., 1s.
—— See also Low's Standard Novels.
EDWORDS. *Camp Fires of a Naturalist*, N. Am. Mammals, 6s.
EGGLESTON, G. CARY, *Juggernaut*, 6s.
Egypt. See Foreign Countries.
Elizabethan Songs. See Choice Editions.
EMERSON, DR. P. H., *English Idylls*, new ed., 2s.
—— *Naturalistic Photography*, new edit. 5s.
—— *Pictures of East Anglian Life*; plates and vignettes, 105s.; large paper, 147s.
—— *Son of the Fens*, 6s.
—— See also Low's 1s. Novels.
—— and GOODALL, *Life on the Norfolk Broads*, plates, 126s.; large paper, 210s.
—— and GOODALL, *Wild Life on a Tidal Water*, copper plates, ord. edit. 25s.; édit. de luxe, 63s.
EMERSON, RALPH WALDO, *In Concord*, a memoir by E. W. Emerson, 7s. 6d.
English Catalogue, 1863-71, 42s.; 1872-80, 42s.; 1881-9, 52s. 6d.; 1890-93, 5s.
English Catalogue, Index vol. 1837-56, 26s.; 1856-76, 42s.; 1874-80, 18s.; 1881-89, 31s. 6d.
English Philosophers, edited by E. B. Ivan Müller, 3s. 6d. each.
Bacon, by Fowler.
Hamilton, by Monck.
Hartley and James Mill, by Bower.
Shaftesbury & Hutcheson; Fowler.
Adam Smith, by J. A. Farrer.
English Readers. See Low.
ERCKMANN-CHATRIAN. See Low's Standard Books.
ESLER, E. RENTOUL, *The Way they Loved at Grimpat*, 3s. 6d.
ESMARCH, F., *Handbook of Surgery*, with 647 new illust. 24s.
Essays on English Writers. See Gentle Life Series.
EVANS, G. E., *Repentance of Magdalene Despar*, &c., poems, 5s.
—— S. & F., *Upper Ten*, a story, 1s.
—— W. E., *Songs of the Birds, Analogies of Spiritual Life*, 6s.
EVELYN. See Low's Stand. Books.
—— JOHN, *Life of Mrs. Godolphin*, 7s. 6d.
EVES, C. W., *West Indies*, n. ed. 7s. 6d.
FAGAN, L., *History of Engraving in England*, illust. from rare prints, £25 nett.
FAIRBAIRN. See Preachers.
Faith and Criticism; Essays by Congregationalists, 6s.

Familiar Words. See Gentle Life Series.

FARINI, G. A., *Through the Kalahari Desert,* 21s.

FAWCETT, *Heir to Millions,* 6s.

—— See also Rose Library.

FAY, T., *Three Germanys,* 2 vols. 35s.

FEILDEN, H. St. J., *Some Public Schools,* 2s. 6d.

—— Mrs., *My African Home,* 7s. 6d.

FENN, G. Manville. *Black Bar,* illust. 5s.

—— *Fire Island,* 6s.

—— See also Low's Stand. Bks.

FFORDE, B., *Subaltern, Policeman, and the Little Girl,* 1s.

—— *Trotter, a Poona Mystery,* 1s.

FIELDS, James T., *Memoirs,* 12s. 6d.

—— *Yesterdays with Authors,* 16s.; also 10s. 6d.

Figure Painters of Holland. See Great Artists.

FINCK, Henry T., *Pacific Coast Scenic Tour,* fine pl. 10s. 6d.

FISHER, G. P., *Colonial Era in America,* 7s. 6d.

FITZGERALD. See Foreign Countries.

—— Percy, *Book Fancier,* 5s.; large paper, 12s. 6d.

FITZPATRICK, T., *Autumn Cruise in the Ægean,* 10s. 6d.

—— *Transatlantic Holiday,* 10s. 6d.

FLEMING, S., *England and Canada,* 6s.

Fly Fisher's Register of Date, Place, Time Occupied, Flies Observed, wind, weather, &c., 4s.

FOLKARD, R., *Plant Lore, Legends and Lyrics,* n. ed., 10s. 6d.

Foreign Countries and British Colonies, descriptive handbooks edited by F. S. Pulling, 3s. 6d.
Australia, by Fitzgerald.
Austria-Hungary, by Kay.
Denmark and Iceland, by E. C. Otté.
Egypt, by S. L. Poole.
France, by Miss Roberts.
Germany, by S. Baring Gould.
Greece, by L. Sergeant.
Japan, by Mossman.
Peru, by Clements R. Markham.
Russia, by Morfill.
Spain, by Webster.
Sweden and Norway, by Woods.
West Indies, by C. H. Eden.

FOREMAN, J., *Philippine Islands,* 21s.

FRA ANGELICO. See Great Artists.

FRA BARTOLOMMEO, ALBERTINELLI, and ANDREA DEL SARTO. See Great Artists.

FRANC, Maud Jeanne, *Beatrice Melton,* 4s.

—— *Emily's Choice,* n. ed. 5s.

—— *Golden Gifts,* 4s.

—— *Hall's Vineyard,* 4s.

—— *Into the Light,* 4s.

—— *John's Wife,* 4s.

—— *Little Mercy; for better, for worse,* 4s.

—— *Marian, a Tale,* n. ed. 5s.

—— *Master of Ralston,* 4s.

—— *Minnie's Mission, a Temperance Tale,* 4s.

—— *No longer a Child,* 4s.

—— *Silken Cords, a Tale,* 4s.

—— *Two Sides to Every Question,* 4s.

—— *Vermont Vale,* 5s.
A plainer edition is issued at 2s. 6d.

France. See Foreign Countries.

Frank's Ranche; or, My Holiday in the Rockies, n. ed. 5s.

FRASER, Sir W. A., *Hic et ubique,* 3s. 6d.; large paper, 21s.

FREEMAN, J., *Melbourne Life, lights and shadows*, 6s.
French and English Birthday Book, by Kate D. Clark, 7s. 6d.
French Readers. See Low.
Fresh Woods and Pastures New, by the Amateur Angler, 5s., 1s. 6d., 1s.
FRIEZE, *Duprè, Florentine Sculptor*, 7s. 6d.
FRISWELL. See Gentle Life.
Froissart for Boys. See Lanier.
FROUDE, J. A. See Prime Ministers.
Gainsborough and Constable. See Great Artists.
GARLAND, HAMLIN, *Prairie Folks*, 6s.
GASPARIN, *Sunny Fields and Shady Woods*, 6s.
GEFFCKEN, *British Empire*, translated, 7s. 6d.
Gentle Life Series, edited by J. Hain Friswell, sm. 8vo, 6s. per vol.; calf extra, 10s. 6d. ea.; 16mo, 2s. 6d., except when price is given.
Gentle Life.
About in the World.
Like unto Christ.
Familiar Words, 6s.; also 3s. 6d.
Montaigne's Essays.
Gentle Life, second series.
Silent hour; essays.
Half-length Portraits.
Essays on English Writers.
Other People's Windows, 6s. & 2s. 6d.
A Man's Thoughts.
Germany. See For. Countries.
GESSI, ROMOLO PASHA, *Seven Years in the Soudan*, 18s
GHIBERTI & DONATELLO. See Great Artists.
GIBBS, W. A., *Idylls of the Queen*, 1s., 5s., & 3s.; Prelude, 1s.
GIBSON, W. H., *Happy Hunting Grounds*, 31s. 6d.

GILES, E., *Australia Twice Traversed*, 1872-76, 2 vols. 30s.
GILL, J. See Low's Readers.
GILLIAT. See Low's Stand. Novels.
Giotto, by Harry Quilter, illust. 15s.
—— See also Great Artists.
GLADSTONE. See Prime Ministers.
GLAVE, E. J., *Congoland, Six Years' Adventure*, 7s. 6d.
Goethe's Faustus, in the original rhyme, by Alfred H. Huth, 5s.
—— *Prosa*, by C. A. Buchheim (Low's German Series), 3s. 6d.
GOLDSMITH, O., *She Stoops to Conquer*, by Austin Dobson, illust. by E. A. Abbey, 84s.
—— See also Choice Editions.
GOOCH, FANNY C., *Face to Face with the Mexicans*, 16s.
GOODMAN, E. J., *The Best Tour in Norway*, new edit., 7s. 6d.
GOODYEAR, W. H., *Grammar of the Lotus, Ornament and Sun Worship*, 63s. nett.
GORDON, E. A., *Clear Round, Story from other Countries*, 7s. 6d.
—— J. E. H., *Physical Treatise on Electricity and Magnetism*, 3rd ed. 2 vols. 42s.
—— *Electric Lighting*, 18s.
—— *School Electricity*, 5s.
—— Mrs. J. E. H., *Decorative Electricity*, illust. 12s.; n. ed. 6s.
—— *Eunice Anscombe*, 7s. 6d.
GOT (E.) *Comedie Française à Londres*, 3s.
GOULD, S. B. See Foreign Countries.
Gounod, Life and Works, 10s. 6d.
GOWER, LORD RONALD. See Great Artists.

GRAESSI, *Italian Dictionary*, 3s. 6d.; roan, 5s.
GRAY, T. See Choice Eds.
Great Artists, Illustrated Biographies, 3s. 6d. per vol. except where the price is given.
Barbizon School,2 vols.; 1vol.7s.6d.
Claude le Lorrain.
Correggio, 2s. 6d.
Cox and De Wint.
George Cruikshank.
Della Robbia and Cellini, 2s. 6d.
Albrecht Dürer.
Figure Painters of Holland. By Lord Ronald Gower.
Fra Angelico, Masaccio, &c.
Fra Bartolommeo; Leader Scott.
Gainsborough and Constable.
Ghiberti and Donatello, by Leader Scott, 2s. 6d.
Giotto, by H. Quilter; 4to, 15s.
Hogarth, by Austin Dobson.
Hans Holbein.
Landscape Painters of Holland.
Landseer, by F. G. Stephens.
Leonardo da Vinci, by J. P. Richter.
Little Masters of Germany, by W. B. Scott; éd. de luxe, 10s. 6d.
Mantegna and Francia.
Meissonier, 2s. 6d.
Michelangelo.
Mulready.
Murillo, by Ellen E. Minor, 2s. 6d.
Overbeck, by J. B. Atkinson.
Raphael, by N. D'Anvers.
Rembrandt, by J. W. Mollett.
Reynolds, by F. S. Pulling.
Romney and Lawrence, 2s. 6d.
Rubens, by Kett.
Tintoretto, by Osler.
Titian, by Heath.
Turner, by Monkhouse.
Vandyck and Hals, by P. R. Head.
Velasquez, by Edwin Stowe.
Vernet & Delaroche.
Watteau, by Mollett, 2s. 6d.
Wilkie, by Mollett.
Great Musicians, biographies, edited by F. Hueffer, 3s. each :—
Bach, by Poole.
Beethoven.

Great Musicians—continued.
Cherubini.
English Church Composers
Handel.
Haydn.
Mendelssohn.
Mozart.
Purcell.
Rossini, &c., by H. Sutherland Edwards.
Schubert.
Schumann.
Richard Wagner.
Weber.
Greece. See Foreign Countries.
GRIEB, *German Dictionary*, n. ed. 2 vols., fine paper, cloth, 21s.
GROHMANN, *Camps in the Rockies*, 12s. 6d.
GROVES. See Low's Std. Bks.
GROWOLL, A., *Profession of Bookselling*, pt. I., 9s. nett.
GULLIE. *Instruction and Amusements of the Blind*, ill., 5s.
GUIZOT, *History of England*, illust. 8 vols. re-issue, 10s. 6d. ea.
—— *History of France*, illust. re-issue, 8 vols. 10s. 6d. each.
—— Abridged by G. Masson, 5s.
GUYON, Madame, *Life*, 6s.
HADLEY, J., *Roman Law*, 7s. 6d.
HALE, *How to Tie Salmon-Flies*, 12s. 6d.
Half-length Portraits. See Gentle Life Series.
HALFORD, F. M., *Dry Fly-fishing*, n. ed. 25s.
—— *Floating Flies*, 15s. & 30s.
HALL, *How to Live Long*, 2s.
HALSEY, F. A., *Slide Valve Gears*, 8s. 6d.
HAMILTON. See English Philosophers.
—— E. *Fly-fishing for Salmon*, 6s.; large paper, 10s. 6d.
—— *Riverside Naturalist*, 14s.

HANDEL. See G. Musicians.
HANDS, T., *Numerical Exercises in Chemistry*, 2s. 6d.; without ans. 2s.; ans. sep. 6d.
Handy Guide to Dry-fly Fishing, by Cotswold Isys, new ed., 1s.
Handy Guide Book to Japanese Islands, 6s. 6d.
HARKUT. See Low's Stand. Novels.
HARLAND, MARION, *Home Kitchen, Receipts*, &c., 5s.
HARRIS, W. B., *Land of an African Sultan*, 10s. 6d.; large paper, 31s. 6d.
HARRISON, MARY, *Modern Cookery*, 6s.
—— *Skilful Cook*, n. ed. 5s.
—— W., *London Houses*, Illust. 6s. net; n. edit., 2s. 6d.
—— *Memor. Paris Houses*, 6s.
HARTLEY and MILL. See English Philosophers.
HATTON. See Low's Standard Novels.
HAWEIS, H. R., *Broad Church*, 6s.
—— *Poets in the Pulpit*, new edit. 6s.; also 3s. 6d.
—— Mrs., *Housekeeping*, 2s. 6d.
—— *Beautiful Houses*, n. ed. 1s.
HAYDN. See Great Musicians.
HAZLITT. See Bayard Ser.
HEAD, PERCY R. See Illus. Text Books and Great Artists.
HEARN, L., *Youma*, 5s.
HEATH, F. G., *Fern World*, col. plates, 12s. 6d., new edit. 6s.
—— GERTRUDE, *Tell us Why*, 2s. 6d.
HEGINBOTHAM, *Stockport*, I., II., III., IV., V., 10s. 6d. each.
HELDMANN, B. See Low's Standard Books for Boys.

HENTY, G. A. See Low's Standard Books for Boys.
—— RICHMOND, *Australiana*, 5s.
HERRICK, R., *Poetry Edited by Austin Dobson*, illust. by E. A. Abbey, 42s.
HERVEY, GEN., *Records of Crime, Thuggee, &c.*, 2 vols., 30s.
HICKS, C. S., *Our Boys, and what to do with Them; Merchant Service*, 5s.
—— *Yachts, Boats, and Canoes, Design and Construction*, 10s. 6d.
HILL, G. B., *Footsteps of Johnson*, 63s.; édition de luxe, 147s.
—— KATHARINE ST., *Grammar of Palmistry*, new ed., 1s.
HINMAN, R., *Eclectic Physical Geography*, 5s.
Hints on proving Wills without Professional Assistance, n. ed. 1s.
Historic Bindings in the Bodleian Library, many plates, 94s. 6d., 84s., 52s. 6d. and 42s.
HODDER, E., *History of South Australia*, 2 vols., 24s.
HOEY, Mrs. CASHEL. See Low's Standard Novels.
HOFFER, *Caoutchouc & Gutta Percha*, by W. T. Brannt, 12s. 6d.
HOGARTH. See Gr. Artists, and Dobson, Austin.
HOLBEIN. See Great Artists.
HOLDER, CHARLES F., *Ivory King*, 8s. 6d.; new ed. 3s. 6d.
—— *Living Lights*, n. ed. 3s. 6d.
HOLM, SAXE, *Draxy Miller*, See Low's Standard Series.
HOLMAN, T., *Life in the Navy*, 1s.
—— *Salt Yarns*, new ed., 1s.
HOLMES, O. WENDELL, *Before the Curfew*, 5s.
—— *Over the Tea Cups*, 6s.

HOLMES, O. WENDELL, *Iron Gate, &c., Poems*, 6s.
—— *Last Leaf*, holiday vol., 42s.
—— *Mechanism in Thought and Morals*, 1s. 6d.
—— *Mortal Antipathy*, 8s. 6d., 2s. and 1s.
—— *Our Hundred Days in Europe*, new edit. 6s. and 3s. 6d.; large paper, 15s.
—— *Poetical Works*, new edit., 2 vols. 10s. 6d.
—— *Works*, prose, 10 vols.; poetry, 3 vols.; 13 vols. 84s. Limited large paper edit., 14 vols. 294s. nett.
—— See also Low's Standard Novels and Rose Library.
Homer, *Iliad*, translated by A. Way, vol. I., 9s.; II., 9s.; Odyssey, in English verse, 7s. 6d.
Horace in Latin, with Smart's literal translation, 2s. 6d.; translation only, 1s. 6d.
HOSMER, J., *German Literature*, a short history, 7s. 6d.
How and where to Fish in Ireland, by Hi-Regan, 3s. 6d.
HOWARD, BLANCHE W., *Tony the Maid*, 3s. 6d.
—— See also Low's Standard Novels.
HOWELLS, W. D. *Undiscovered Country*, 3s. 6d. and 1s.
HOWORTH, SIR H. H., *Glacial Nightmare & the Flood*, 2 vols., 30s.
—— *Mammoth and the Flood*, 18s.
HUBERT. See Men of Achievem.
HUEFFER. E. See Great Musicians.
HUGHES, HUGH PRICE. See Preachers.
HUME, FERGUS, *Creature of the Night*, 1s. See also Low's Standard Novels and 1s. Novels.

HUMFREY, MARIAN, *Obstetric Nursing*, 3s. 6d.
Humorous Art at the Naval Exhibition, 1s.
HUMPHREYS, JENNET, *Some Little Britons in Brittany*, 2s. 6d.
HUNTINGDON, *The Squire's Nieces*, 2s. 6d. (Playtime Library.)
HYDE, *A Hundred Years by Post*, Jubilee Retrospect, 1s.
Hymnal Companion to the Book of Common Prayer, separate lists gratis.
Iceland. See Foreign Countries.
Illustrated Text-Books of Art-Education, edit. by E. J. Poynter, R.A., 5s. each.
  Architecture, Classic and Early Christian, by Smith and Slater.
  Architecture, Gothic and Renaissance, by T. Roger Smith.
  German, Flemish, and Dutch Painting.
  Painting, Classic and Italian, by Head, &c.
  Painting, English and American.
  Sculpture, modern; Leader Scott.
  Sculpture, by G. Redford.
  Spanish and French artists; Smith.
  Water Colour Painting, by Redgrave.
INDERWICK, F. A., *Interregnum*, 10s. 6d.
—— *Prisoner of War*, 5s.
—— *King Edward and New Winchelsea*.
—— *Sidelights on the Stuarts*, new edit. 7s. 6d.
INGELOW, JEAN. See Low's Standard Novels.
INGLIS, HON. JAMES, *Our New Zealand Cousins*, 6s.
—— *Sport and Work on the Nepaul Frontier*, 21s.
—— *Tent Life in Tiger Land*, with coloured plates, 18s.

IRVING, W., *Little Britain*, 10s. 6d. and 6s.
—— *Works*, "Geoffrey Crayon" edit. 27 vols. 16l. 16s.
JACKSON, John, *Handwriting in Relation to Hygiene*, 3d.
—— *New Style Vertical Writing Copy-Books*. Series I. 1—15, 2d. and 1d. each.
—— *New Code Copy-Books*, 25 Nos. 2d. each.
—— *Shorthand of Arithmetic*, Companion to Arithmetics, 1s. 6d.
—— *Theory and Practice of Handwriting*, with diagrams, 5s.
JALKSON, Lowis, *Ten Centuries of European Progress*, new ed., 5s.
JAMES, Croake, *Law and Lawyers*, new edit. 7s. 6d.
JAMES and MOLÉ'S *French Dictionary*, 3s. 6d. cloth; roan, 5s.
JAMES, *German Dictionary*, 3s. 6d. cloth; roan, 5s.
JANVIER, *Aztec Treasure House* 7s. 6d., See also Low's Standard Books.
Japan. See Foreign Countries.
Japanese Books, untearable.
1. Rat's Plaint, by Little, 5s.
2. Smith, Children's Japan, 3s. 6d.
3. Bramhall, Niponese Rhymes, 5s.
4. Princess Splendor, fairy tale. 2s.
JEFFERIES, Richard, *Amaryllis at the Fair*, 7s. 6d.
—— See also Low's Stan. Books.
JEPHSON, A. J. M., *Emin Pasha relief expedition*, 21s.
—— *Stories told in an African Forest*, 8s. 6d.
JOHNSTON, H. H., *The Congo, from its Mouth to Bolobo*, 21s.
JOHNSTON-LAVIS, H. J., *South Italian Volcanoes*, 15s.

JOHNSTONE, D. L., *Land of the Mountain Kingdom*, new edit. 3s. 6d. and 2s. 6d.
JOINVILLE. See Bayard Ser.
JULIEN, F., *Conversational French Reader*, 2s. 6d.
—— *English Student's French Examiner*, 2s.
—— *First Lessons in Conversational French Grammar*, n. ed. 1s.
—— *French at Home and at School*, Book I. accidence, 2s.; key, 3s.
—— *Petites Leçons de Conversation et de Grammaire*, n. ed. 3s.
—— *Petites Leçons*, with phrases, 3s. 6d.
—— *Phrases of Daily Use*, separately, 6d.
KARR, H. W. Seton, *Shores and Alps of Alaska*, 16s.
KAY. See Foreign Countries.
Keene (C.), *Life*, by Layard, 24s.; l.p., 63s. nett; n. ed., 12s. 6d.
KENNEDY, E. B., *Blacks and Bushrangers*, new edit. 5s., 3s. 6d. and 2s. 6d.
KERSHAW, S. W., *Protestants from France in their English Home*, 6s.
Khedives and Pashas, 7s. 6d.
KILNER, E. A., *Four Welsh Counties*, 5s.
King and Commons. See Cavalier in Bayard Series.
KINGSLEY, R. G., *Children of Westminster Abbey*, 5s.
KINGSTON. See Low's Standard Books.
KIPLING, Rudyard, *Soldiers Three, &c.*, stories, 1s.
—— *Story of the Gadsbys*, new edit. 1s.

KIPLING, RUDYARD, *In Black and White, &c.*, stories, 1s.
—— *Wee Willie Winkie, &c.*, stories, 1s.
—— *Under the Deodars, &c.*, stories, 1s.
—— *Phantom Rickshaw, &c.*, stories, 1s.
*** The six collections of stories may also be had in 2 vols. 3s. 6d. each, in cloth.
—— *Stories*, Library Edition, 2 vols. 6s. each.
KIRKALDY, W. G., *David Kirkaldy's Mechanical Testing*, 84s.
KNIGHT, E. F., *Cruise of the Falcon*, 7s. 6d.; new edit. 3s. 6d.
KNOX, T. W., *Boy Travellers with H. M. Stanley*, new edit. 5s.
—— *John Boyd's Adventures*, 6s.
KUNHARDT, C. P., *Small Yachts*, new edit. 50s.
—— *Steam Yachts*, 16s.
KWONG, *English Phrases*, 21s.
LABILLIERE, *Federal Britain*, 6s.
LALANNE, *Etching*, 12s. 6d.
LAMB, CHAS., *Essays of Elia*, with designs by C. O. Murray, 6s.
Landscape Painters of Holland. See Great Artists.
LANDSEER. See Great Artists.
LANIER, S., *Boy's Froissart*, 7s. 6d.; *King Arthur*, 7s. 6d.; *Percy*, 7s. 6d.
LANSDELL, HENRY, *Through Siberia*, 2 vols., 30s. See also Low's Standard Library.
—— *Russian Central Asia*, 2 vols. 42s.
—— *Through Central Asia*, 12s.
—— *Chinese Central Asia*, 2 vols., fully illustrated, 36s.
LARDEN, W., *School Course on Heat*, 5th ed., entirely revised, 5s.

LAURIE, A. See Low's Stand. Books.
LAWRENCE, SERGEANT, *Autobiography*, 6s.
LAWRENCE. See Romney in Great Artists.
LAYARD, MRS., *West Indies*, 2s. 6d.
—— G.S., *His Golf Madness*, 1s.
—— See also Keene.
LEA, H. C., *Inquisition in the Middle Ages*, 3 vols., 42s.
LEARED, A., *Morocco*, n. ed. 16s.
LEFFINGWELL, W. B., *Shooting*, 18s.
—— *Wild Fowl Shooting*, 10s. 6d.
LEFROY, W., DEAN OF NORWICH. See Preachers of the Age.
Leo XIII. Life, 18s.
Leonardo da Vinci. See Great Artists.
—— *Literary Works*, by J. P. Richter, 2 vols. 252s.
LIEBER, *Telegraphic Cipher*, 42s. nett.
Like unto Christ. See Gentle Life Series.
Lincoln, Abraham, true story of a great life, 2 vols., 12s.
LITTLE, ARCH. J., *Yang-tse Gorges*, n. ed., 10s. 6d.
—— See also Japanese Books.
LITTLE, W. J. KNOX. See Preachers of the Age.
Little Masters of Germany. See Great Artists.
LONG, JAMES, *Farmer's Handbook*, 4s. 6d.
LONGFELLOW, *Maidenhood*, with coloured plates, 2s. 6d.; gilt edges, 3s. 6d.
—— *Nuremberg*, photogravure illustrations, 31s. 6d.

LONGFELLOW, *Song of Hiawatha*, illust., 21s.

LOOMIS, E., *Astronomy*, n. ed. 8s. 6d.

LORD, Mrs. FREWEN, *Tales from Westminster Abbey*, 2s. 6d.; new edition, 1s.

LORNE, MARQUIS OF, *Canada and Scotland*, 7s. 6d.

—— See also Prime Ministers.

Louis, St. See Bayard Series.

Low's French Readers, edit. by C. F. Clifton, I. 3d., II. 3d., III. 6d.

—— *German Series.* See Goethe, Meissner, Sandars, and Schiller.

—— *London Charities*, annually, 1s. 6d.; sewed, 1s.

—— *Illustrated Germ. Primer*, 1s.

—— *Infant Primers*, I. illus. 3d.; II. illus. 6d.

—— *Pocket Encyclopædia*, with plates, 3s. 6d.; roan, 4s. 6d.

—— *Readers*, Edited by John Gill, I., 9d.; II., 10d.; III., 1s.; IV., 1s. 3d.; V., 1s. 4d; VI., 1s. 6d.

Low's Stand. Library of Travel (unless price is stated), vol. 7s. 6d.

Ashe (R. P.) Two Kings of Uganda, 3s. 6d.; also 6s.

Butler, Great Lone Land; also 3s. 6d.

—— Wild North Land.

Knight, Cruise of the *Falcon*, also 3s. 6d.

Lansdell (H). Through Siberia, unabridged, 10s. 6d.

Low's Stand. Libr.—continued.

Marshall (W.) Through America.

Schweinfurth's Heart of Africa, 2 vols. 3s. 6d. each.

Spry (W. J. J., R.N.), *Challenger* cruise.

Stanley (H. M.) Coomassie, 3s. 6d.

—— How I Found Livingstone; also 3s. 6d.

—— Through the Dark Continent, 1 vol. illust., 12s. 6d.; also 3s. 6d.

Thomson, Through Masai Land.

Low's Standard Novels, Library Edition (except where price is stated), cr. 8vo., 6s.; also popular edition, small post 8vo, 2s. 6d.; paper bds. 2s.

Baker, John Westacott.

—— Mark Tillotson.

Black (William) Adventures in Thule.

—— The Beautiful Wretch.

—— Daughter of Heth.

—— Donald Ross.

—— Green Pastures & Piccadilly.

—— In Far Lochaber.

—— In Silk Attire.

—— Judith Shakespeare.

—— Kilmeny.

—— Lady Silverdale's Sweetheart.

—— Macleod of Dare.

—— Madcap Violet.

—— Maid of Killeena.

—— New Prince Fortunatus.

—— The Penance of John Logan.

—— Princess of Thule.

—— Sabina Zembra.

—— Shandon Bells.

—— Stand Fast, Craig Royston!

—— Strange Adventures of a House Boat.

—— Strange Adventures of a Phaeton.

—— Sunrise.

—— Three Feathers.

—— White Heather.

—— White Wings.

—— Wise Women of Inverness.

—— Wolfenberg.

—— Yolande.

In all Departments of Literature.

Low's Stand. Novels—continued.

Blackmore (R. D.) Alice Lorraine.
—— Christowell.
—— Clara Vaughan.
—— Cradock Nowell.
—— Cripps the Carrier.
—— Erema, or My Father's Sin.
—— Kit and Kitty.
—— Lorna Doone.
—— Mary Anerley.
—— Springhaven.
—— Tommy Upmore.
Bremont, Gentleman Digger.
Brown (Robert) Jack Abbott's Log.
Bynner, Agnes Surriage.
—— Begum's Daughter.
Cable (G. W.) Bonaventure, 5s.
Coleridge (C. R.) English Squire.
Craddock, Despot of Broomsedge.
Croker (Mrs. B. M.) Some One Else.
Cumberland (Stuart) Vasty Deep.
DeLeon, Under the Stars & Crescent.
Edwards (Miss Betham) Half-way.
Eggleston, Juggernaut.
Emerson (P. H.), Son of the Fens.
French Heiress in her own Chateau.
Gilliat, Story of the Dragonnades.
Harkut, The Conspirator.
Hatton, Old House at Sandwich.
—— Three Recruits.
Hoey (Mrs. Cashel) Golden Sorrow.
—— Out of Court.
—— Stern Chase.
Holmes (O. W.), Guardian Angel.
—— Over the Teacups.
Howard (Blanche W.) Open Door.
Hume (Fergus), Fever of Life.
Ingelow (Jean) Don John.
—— John Jerome, 5s.
—— Sarah de Berenger.
Lathrop, Newport, 5s.
Macalpine, A Man's Conscience.
Mac Donald (Geo.) Adela Cathcart.
—— Guild Court.
—— Mary Marston.
—— Orts.
—— Stephen Archer, &c.
—— The Vicar's Daughter.
—— Weighed and Wanting.
Macmaster, Our Pleasant Vices.

Low's Stand. Novels—continued.

Martin, Even Mine Own Familiar Friend.
Musgrave (Mrs.) Miriam.
Oliphant, Innocent.
Osborn, Spell of Ashtaroth, 5s.
Prince Maskiloff.
Riddell (Mrs.) Alaric Spenceley.
—— Daisies and Buttercups.
—— Senior Partner.
—— Struggle for Fame.
Russell (W. Clark) Betwixt the Forelands.
—— The Emigrant Ship.
—— Frozen Pirate.
—— Jack's Courtship.
—— John Holdsworth.
—— The Lady Maud.
—— The Little Loo.
—— Mrs. Dines' Jewels, 2s. 6d. and 2s. only.
—— My Watch Below.
—— The Ocean Free Lance.
—— A Sailor's Sweetheart.
—— The Sea Queen.
—— A Strange Voyage.
—— Wreck of the *Grosvenor*.
Ryce, Rector of Amesty.
Steuart, Kilgroom.
Stockton (F. R.) Ardis Claverden.
—— Bee-man of Orn, 5s.
—— Dusantes and Mrs. Lecks and Mrs. Aleshine, 1 vol.
—— Hundredth Man.
—— The late Mrs. Null.
Stoker (Bram) Snake's Pass.
Stowe (Mrs.) Dred.
—— Old Town Folk.
—— Poganuc People.
Thomas, House on the Scar.
Thomson (Joseph) Uln.
Tourgee, Murvale Eastman.
Tytler (S.) Duchess Frances.
Vane, From the Dead.
—— Polish Conspiracy.
Walford (Mrs.), Her Great Idea.
Warner, Little Journey in the World.
Wilcox, Senora Villena.
Woolson (Constance F.) Anne.
—— East Angels.
—— For the Major, 5s
—— Jupiter Lights.

Low's Shilling Novels.

Edwards, Dream of Millions.
Emerson, East Coast Yarns.
—— Signor Lippo.
Evans, Upper Ten.
Forde, Subaltern, &c.
—— Trotter: a Poona Mystery.
Hewitt, Oriel Penhaligon.
Holman, Life in the Navy.
—— Salt Yarns.
Hume (F.), Creature of the Night.
—— Chinese Jar.
Ignotus; Visitors' Book.
Layard, His Golf Madness.
Married by Proxy.
Rux, Roughing it after Gold.
—— Through the Mill.
Vane, Lynn's Court Mystery.
Vesper, Bobby, a Story.

Low's Standard Books for Boys, with numerous illustrations, 2s. 6d. each; gilt edges, 3s. 6d.

Ainslie, Priceless Orchid.
Biart (Lucien) Young Naturalist.
—— My Rambles in the New World.
Boussenard, Crusoes of Guiana.
—— Gold Seekers, a sequel.
Butler (Col. Sir Wm.) Red Cloud.
Cahun (Leon) Captain Mago.
—— Blue Banner.
Célière, Exploits of the Doctor.
Chaillu (Paul) Equator Wild Life.
Collingwood, Under the Meteor Flag
—— Voyage of the *Aurora*.
Cozzens (S.W.) Marvellous Country.
Dodge (Mrs.) Hans Brinker.
Du Chaillu (Paul) Gorilla Country.
Erckmann-Chatrian, Bros. Rantzau.
Evelyn, Inca Queen.
Fenn (G. Manville) Off to the Wilds.
—— Silver Cañon.
Groves (Percy) Charmouth Grange.
Heldmann (B.) *Leander* Mutiny.
Henty (G. A.) Cornet of Horse.
—— Jack Archer.
—— Winning his Spurs.
Hyne, Sandy Carmichael.
Janvier, Aztec Treasure House.
Jefferies (Richard) Bevis, Story of a Boy.

Low's Stand. Books for Boys—continued.

Johnstone, Mountain Kingdom.
Kennedy, Blacks and Bushrangers.
Kingston (W. H. G.) Ben Burton.
—— Captain Mugford.
—— Dick Cheveley.
—— Heir of Kilfinnan.
—— Snowshoes.
—— Two Supercargoes.
—— With Axe and Rifle.
Laurie (A.) Axel Ebersen.
—— Conquest of the Moon.
—— New York to Brest.
—— Secret of the Magian.
MacGregor (John) *Rob Roy* Canoe.
—— *Rob Roy* in the Baltic.
—— Yawl *Rob Roy*.
Maclean, Maid of the *Golden Age*.
Malan (A. N.) Cobbler of Cornikeranium.
Meunier, Great Hunting Grounds.
Muller, Noble Words and Deeds.
Norway (G.) How Martin Drake found his Father.
Perelaer, The Three Deserters.
Reed (Talbot Baines) Roger Ingleton, Minor.
—— Sir Ludar.
Reid (Mayne) Strange Adventures.
Rousselet (Louis) Drummer-boy.
—— King of the Tigers.
—— Serpent Charmer.
—— Son of the Constable.
Russell (W. Clark) Frozen Pirate.
Stanley, My Kalulu.
Tregance, Louis, in New Guinea.
Verne, Adrift in the Pacific.
—— Purchase of the North Pole.
Winder (F. H.) Lost in Africa.

Low's Standard Series of Books by popular writers, cloth gilt, 2s.; gilt edges, 2s. 6d. each.

Alcott (L. M.) A Rose in Bloom.
—— An Old-Fashioned Girl.
—— Aunt Jo's Scrap Bag.
—— Eight Cousins, illust.
—— Jack and Jill.
—— Jimmy's Cruise.

Low's Stand. Series of Books—continued.

Alcott (L. M.) Little Men.
—— Little Women & L.Wo.Wedded
—— Lulu's Library, illust.
—— Recollections of Childhood.
—— Shawl Straps.
—— Silver Pitchers.
—— Spinning-Wheel Stories.
—— Under the Lilacs, illust.
—— Work and Beginning Again, ill.
Alden (W. L.) Jimmy Brown, illust.
—— Trying to Find Europe.
Bunyan, Pilgrim's Progress, 2s.
De Witt (Madame) An Only Sister.
Franc (Maud J.), Stories, 2s. 6d. edition, see page 9.
Holm (Saxe) Draxy Miller's Dowry.
Robinson (Phil) Indian Garden.
—— Under the Punkah.
Roe (E. P.) Nature's Serial Story.
Saintine, Picciola.
Samuels, Forecastle to Cabin, illust.
Sandeau (Jules) Seagull Rock.
Stowe (Mrs.) Dred.
—— Ghost in the Mill, &c.
—— My Wife and I.
—— We and our Neighbours.
Tooley (Mrs.) Harriet B. Stowe.
Warner, In the Wilderness.
—— My Summer in a Garden.
Whitney (Mrs.) Leslie Goldthwaite.
—— Faith Gartney's Girlhood.
—— The Gayworthys.
—— Hitherto.
—— Real Folks.
—— We Girls.
—— The Other Girls: a Sequel.

** A new illustrated list of books for boys and girls, with portraits of celebrated authors, sent post free on application.

LOWELL, J. R., *Among my Books*, I. and II., 7s. 6d. each.
—— *Vision of Sir Launfal*, illus. 63s.

LUMMIS, C. F., *Tramp, Ohio to California*, 6s.
—— *Land of Poco Tiempo* (New Mexico), 10s. 6d., illust.

MACDONALD, D., *Oceania*, 6s.
—— GEORGE. See Low's Stand. Novels.
—— SIR JOHN A., *Life*, 16s.

MACGOUN, *Commercial Correspondence*, 5s.

MACGREGOR, J., *Rob Roy in the Baltic*, n. ed. 3s. 6d. and 2s. 6d.
—— *Rob Roy Canoe*, new edit., 3s. 6d. and 2s. 6d.
—— *Yawl Rob Roy*, new edit., 3s. 6d. and 2s. 6d.

MACKENNA, *Brave Men in Action*, 10s. 6d.

MACKENZIE, SIR MORELL, *Fatal Illness of Frederick the Noble*, 2s. 6d.
—— *Essays*, 7s. 6d.

MACKINNON and SHADBOLT, *S. African Campaign*, 50s.

MACLAREN, A. See Preachers.

MACLEAN, H. E. See Low's Standard Books.

MACMASTER. See Low's Standard Novels.

MACMULLEN, JOHN MERCER, *History of Canada*, 3rd ed., 2 vols., 25s.

MACMURDO, E., *History of Portugal*, 21s.; II. 21s.; III. 21s.

MAEL, PIERRE, *Under the Sea to the North Pole*, 5s.

MAHAN, CAPT. A. T., *Admiral Farragut*, 6s.
—— *Influence of Sea Power on the French Revolution*, 2 vols. (British naval history), 30s.
—— *Sea Power in History*, 18s.

MAIN, MRS., *My Home in the Alps*, 3s. 6d.
—— See also Burnaby, Mrs.

MALAN. See Low's Stand. Books
—— C. F. DE M., *Eric and Connie's Cruise*, 5s.

Manchester Library, Reprints of Classics at nett prices, per vol., 6d.; sewed, 3d.
  *** List on application.
Man's Thoughts. See Gentle Life Series.
MANLEY, *Notes on Fish and Fishing*, 6s.
MANTEGNA and FRANCIA. See Great Artists.
MARBURY, *Favourite Flies*, with coloured plates, &c., 24s. nett.
MARCH, F. A., *Comparative Anglo-Saxon Grammar*, 12s.
—— *Anglo-Saxon Reader*, 7s. 6d.
MARKHAM, ADM., *Naval Career during the old war*, 14s.
—— CLEMENTS R., *Peru*. See For. Countries.
—— *War Between Peru and Chili*, 10s. 6d.
MARSH, A. E. W., *Holiday in Madeira*, 5s.
MARSH, G. P., *Lectures on the English Language*, 18s.
—— *Origin and History of the English Language*, 18s.
MARSHALL, W. G., *Through America*, new edit. 7s. 6d.
MARSTON, E., *How Stanley wrote "In Darkest Africa,"* 1s.
—— See also Amateur Angler, Frank's Ranche, and Fresh Woods.
MARSTON, WESTLAND, *Eminent Recent Actors*, n. ed., 6s.
MARTIN, J. W., *Float Fishing and Spinning*, new edit. 2s.
MATHESON, ANNIE, *Love's Music, and other lyrics*, 3s. 6d.
MATTHEWS, J. W., *Incwadi Yami, Twenty Years in S. Africa*, 14s.
MAUCHLINE, ROBERT, *Mine Foreman's Handbook*, 21s.
MAURY, M. F., *Life*, 12s. 6d.

MAURY, M. F., *Physical Geography and Meteorology of the Sea*, new ed. 6s.
MEISSNER, A. L., *Children's Own German Book* (Low's Series), 1s. 6d.
—— *First German Reader* (Low's Series), 1s. 6d.
—— *Second German Reader* (Low's Series), 1s. 6d.
MEISSONIER. See Great Artists.
MELBOURNE, LORD. See Prime Ministers.
MELIO, G. L., *Swedish Drill*, 1s. 6d.
Member for Wrottenborough, 3s. 6d.
Men of Achievement, 8s. 6d. each.
 Noah Brooks, *Statesmen*.
 Gen. A. W. Greeley, *Explorers*.
 Philip G. Hubert, *Inventors*.
 W. O. Stoddard, *Men of Business*.
MENDELSSOHN. *Family, 1729-1847, Letters and Journals*, new edit., 2 vols., 30s.
—— See also Great Musicians.
MERIWETHER, LEE, *Mediterranean*, new ed., 6s.
MERRIFIELD, J., *Nautical Astronomy*, 7s. 6d.
MESNEY, W., *Tungking*, 3s. 6d.
Metal Workers' Recipes and Processes, by W. T. Brannt, 12s. 6d.
MEUNIER, V. See Low's Standard Books.
Michelangelo. See Great Artists.
MIJATOVICH, C., *Constantine*, 7s. 6d.
MILL, JAMES. See English Philosophers.
MILLS, J., *Alternative Chemistry*, answers to the ordinary course, 1s.
—— *Alternative Elementary Chemistry*, 1s. 6d.; answers, 1s.

MILLS, J., *Chemistry for students*, 3s. 6d.

MILNE, J., AND BURTON, *Volcanoes of Japan*, collotypes by Ogawa, part i., 21s. nett.

MILTON'S *Allegro.* See Choice Editions.

MITCHELL, D.G.(Ik. Marvel) *English Lands, Letters and Kings*, 2 vols. 6s. each.

—— *Writings*, new edit. per vol. 5s.

MITFORD, J., *Letters*, 3s. 6d.

—— Miss, *Our Village*, illus. 5s.

MODY, Mrs., *German Literature*, outlines, 1s.

MOFFATT, W., *Land and Work*, 5s.

MOINET. See Preachers.

MOLLETT. See Great Artists.

MOLONEY, J. A., *With Captain Stairs to Katanga*, 8s. 6d.

MONKHOUSE. See G. Artists.

Montaigne's Essays, revised by J. Hain Friswell, 2s. 6d.

MONTBARD (G.), *Among the Moors*, illust., 16s. ; ed. de Luxe, 63s.

MOORE, J.M., *New Zealand for Emigrant, Invalid, and Tourist*, 5s.

MORLEY, HENRY, *English Literature in the Reign of Victoria*, 2s. 6d.

—— *Five Centuries of English Literature*, 2s.

MORSE, E. S., *Japanese Homes*, new edit. 10s. 6d.

MORTEN, H., *Hospital Life*, 1s.

—— & GETHEN, *Tales of the Children's Ward*, 3s. 6d.

MORTIMER, J., *Chess Player's Pocket-Book*, new edit. 1s.

MOSS, F. J., *Great South Sea, Atolls and Islands*, 8s. 6d.

MOTTI, PIETRO, *Elementary Russian Grammar*, 2s. 6d.

—— *Russian Conversation Grammar*, 5s. ; Key, 2s.

MOULE, H.C.G. See Preachers.

MOXLY, *West India Sanatorium ; Barbados*, 3s. 6d.

MOXON, W., *Pilocereus Senilis*, 3s. 6d.

MOZART. See Gr. Musicians.

MULLER, E. See Low's Standard Books.

MULLIN, J.P., *Moulding and Pattern Making*, 12s. 6d.

MULREADY. See Gt. Artists.

MURDOCH, *Ayame San*, a Japanese Romance, with photos. reproduced by Ogawa, 30s. nett.

MURILLO. See Great Artists.

MURPHY, *Beyond the Ice*, from Farleigh's Diary, 3s. 6d.

MUSGRAVE, Mrs. See Low's Standard Novels.

My Comforter, &c., Religious Poems, 2s. 6d.

Napoleon I. See Bayard Series.

NELSON, WOLFRED, *Panama, the Canal, &c.*, 6s.

Nelson's Words and Deeds, 3s. 6d.

NETHERCOTE, *Pytchley Hunt*, 8s. 6d.

New Zealand, chromos, by Barraud, text by Travers, 168s.

NICHOLS, W. L., *Quantocks*, 5s.; large paper, 10s. 6d.

Nineteenth Century, a Monthly Review, 2s. 6d. per No.

NISBET, HUME, *Life and Nature Studies*, illustrated, 6s.

NIVEN, R., *Angler's Lexicon*, 6s.

NORMAN, C. B., *Corsairs of France*, 18s.

NORMAN, J. H., *Monetary Systems of the World*, 10s. 6d.
—— *Ready Reckoner of Foreign and Colonial Exchanges*, 2s. 6d.
NORWAY, 50 photogravures by Paul Lange, text by E. J. Goodman, 52s. 6d.
NOTTAGE, C. G., *In Search of a Climate*, illust. 25s.
Nuggets of the Gouph, 3s.
O'BRIEN, *Fifty Years of Concession to Ireland*, vol. i. 16s.; vol. ii. 16s.
OGAWA, *Open-Air Life in Japan*, 15s. nett; *Out of doors Life in Japan*, 12s. nett.
OGDEN, J., *Fly-tying*, 2s. 6d.
Ohrwalder's Ten Years' Captivity; Mahdi's Camp, new ed., 6s.
Orient Line Guide, new edit. by W. J. Loftie, 2s. 6d.
ORTOLI, *Evening Tales*, done into English by J. C. Harris, 6s.
ORVIS, C. F., *Fly Fishing*, with coloured plates, 12s. 6d.
OSBORN, H. S., *Prospector's Guide*, 8s. 6d.
Other People's Windows. See Gentle Life Series.
OTTÉ, *Denmark and Iceland.* See Foreign Countries.
Our Little Ones in Heaven, 5s.
Out of Doors Life in Japan, Burton's photos., from paintings by Ogawa, 12s. net.
Out of School at Eton, 2s. 6d.
OVERBECK. See Great Artists.
OWEN, *Marine Insurance*, 15s.
Oxford Days, by an M.A., 2s. 6d.
PAGE, T. N., *Marse Chan*, illust. 6s.
—— *Meh Lady*, a Story of Old Virginian Life, 6s.

PALGRAVE, R. F. D. *Chairman's Handbook*, 10th edit. 2s.
—— *Oliver Cromwell*, 10s. 6d.
PALLISER, Mrs. BURY, *China Collector's Companion*, 5s.
—— *History of Lace*, n. ed. 21s.
PANTON, *Homes of Taste*, 2s. 6d.
PARKE, *Emin Pasha Relief Expedition*, 21s., new ed.
—— T. H., *Health in Africa*, 5s.
PARKER, E. H., *Chinese Account of the Opium War*, 1s. 6d.
PARKS, LEIGHTON, *Winning of the Soul, &c.*, sermons, 3s. 6d.
Parliamentary Pictures and Personalities (from the Graphic), illust., 5s.; ed. de luxe, 21s. nett.
PARSONS, J., *Principles of Partnership*, 31s. 6d.
—— T. P., *Marine Insurance and General Average*, 2 vols. 63s.
PEACH, *Annals of Swainswick, near Bath*, 10s. 6d.
Peel. See Prime Ministers.
PELLESCHI, G., *Gran Chaco of the Argentine Republic*, 8s. 6d.
PEMBERTON, C., *Tyrol*, 1s. 4d.
PENNELL, *Fishing Tackle*, 2s.
—— *Sporting Fish*, 15s. & 30s.
Penny Postage Jubilee, 1s.
Peru. See Foreign Countries.
PHELPS, E. S., *Struggle for Immortality*, 5s.
—— SAMUEL, *Life*, by W. M. Phelps & Forbes-Robertson, 12s.
PHILBRICK, F. A., AND WESTOBY, *Post and Telegraph Stamps*, 10s. 6d.
PHILLIMORE, C. M., *Italian Literature*, new. edit. 3s. 6d.
—— See also Gt. Artists, *Fra An.*
PHILLIPS, L. P., *Dictionary of Biographical Reference*, n.e. 25s.
—— W., *Law of Insurance*, 2 vols. 73s. 6d.

PHILPOT, H. J., *Diabetes*, 5s.
—— *Diet Tables*, 1s. each.
Playtime Library, 2s. 6d. each.
Charles, Where is Fairy Land?
Humphreys, Little Britons.
Huntingdon, Squire's Nieces.
Pleasant History of Reynard the Fox, trans. by T. Roscoe, illus. 7s. 6d.
PLUNKETT (solid geometry) *Orthographic Projection*, 2s. 6d.
POE, by E. C. Stedman, 3s. 6d.
—— *Raven*, ill. by G. Doré, 63s.
Poems of the Inner Life, 5s.
Poetry of Nature. See Choice Editions.
Poetry of the Anti-Jacobin, 7s. 6d. large paper, 21s.
PORCHER, A., *Juvenile French Plays*, with Notes, 1s.
PORTER, NOAH, *Memoir*, 8s. 6d.
Portraits of Racehorses, 4 vols. 126s.
POSSELT, *Structure of Fibres, Yarns and Fabrics*, 63s.
—— *Textile Design*, illust. 28s.
POYNTER. See Illustrated Text Books.

Preachers of the Age, 3s. 6d. ea.
Living Theology, by His Grace the Archbishop of Canterbury.
The Conquering Christ, by Rev. A. Maclaren.
Verbum Crucis, by the Bishop of Derry.
Ethical Christianity, by Hugh P. Hughes.
Knowledge of God, by the Bishop of Wakefield.
Light and Peace, by H. R. Reynolds.
Journey of Life, by W. J. Knox-Little.
Messages to the Multitude, by C. H. Spurgeon.
Christ is All, by H. C. G. Moule, M.A.
Plain Words on Great Themes, by J. O. Dykes.

Preachers of the Age—cont.
Children of God, by E. A. Stuart.
Christ in the Centuries, by A. M. Fairbairn.
Agoniæ Christi, by Dr. Lefroy.
The Transfigured Sackcloth, by W. L. Watkinson.
The Gospel of Work, by the Bishop of Winchester.
Vision and Duty, by C. A. Berry.
The Burning Bush; Sermons, by the Bishop of Ripon.
Good Cheer of Jesus Christ, by C. Moinet, M.A.
A Cup of Cold Water, by J. Morlais Jones.
The Religion of the Son of Man, by E. J. Gough, M.A.
PRICE, *Arctic Ocean to Yellow Sea*, illust., new ed., 7s. 6d.
Prime Ministers, a series of political biographies, edited by Stuart J. Reid, 3s. 6d. each.
Earl of Beaconsfield, by J. Anthony Froude.
Viscount Melbourne, by Henry Dunckley ("*Verax*").
Sir Robert Peel, by Justin McCarthy.
Viscount Palmerston, by the Marquis of Lorne.
Earl Russell, by Stuart J. Reid.
Right Hon. W. E. Gladstone, by G. W. E. Russell.
Earl of Aberdeen, by Sir Arthur Gordon.
Marquis of Salisbury, by H. D. Traill.
Earl of Derby, by G. Saintsbury.
*** *An edition, limited to 250 copies, is issued on hand-made paper, medium 8vo, half vellum, cloth sides, gilt top, 9 vols. 4l. 4s. nett.*
Prince Maskiloff. See Low's Standard Novels.
Prince of Nursery Playmates, new edit. 2s. 6d.
PRITT, T. N., *North Country Flies*, coloured plates, 10s. 6d.
Reynolds. See Great Artists.

Purcell. See Great Musicians.
PYLE, HOWARD, *Robin Hood*, 10s. 6d.
QUILTER, HARRY, *Giotto, Life, &c.* 15s.
—— See also Great Artists.
RAMBAUD, *History of Russia*, new edit., 3 vols. 21s.
RAPHAEL. See Great Artists.
READ, OPIE, *Emmett Bonlore*, 6s.
REDFORD, *Sculpture.* See Illustrated Text-books.
REDGRAVE, *Century of English Painters*, new ed., 7s. 6d.
REED, SIR E. J., *Modern Ships of War*, 10s. 6d.
—— T. B. See Low's St. Bks.
REID, MAYNE, CAPTAIN. See Low's Standard Books.
—— STUART J. See Prime Min.
Remarkable Bindings in British Museum, 168s.; 94s. 6d.; 73s. 6d. and 63s.
REMBRANDT. See Gr. Artists.
REYNOLDS. See Gr. Artists.
—— HENRY R. See Preachers.
RICHARDS, J. W., *Aluminium*, new edit. 21s.
RICHTER, *Italian Art in the National Gallery*, 42s.
—— See also Great Artists.
RIDDELL, MRS. J. H. See Low's Standard Novels.
RIFFAULT, *Colours for Painting*, 31s. 6d.
RIPON, BP. OF. See Preachers.
RIVIÈRE, J., *Recollections*, 3s. 6d.
ROBERTS. See For. Countries.
—— W., *English Bookselling*, earlier history, 7s. 6d.; n. ed. 3s. 6d.
ROBERTSON, AL., *Fra Paolo Sarpi, the Greatest of the Venetians*, 6s.
—— *Count Campello*, 5s.

ROBIDA, A., *Toilette*, coloured plates, 7s. 6d.; new ed. 3s. 6d.
ROBINSON, PHIL., *Noah's Ark*, n. ed. 3s. 6d.
—— *Sinners & Saints*, 10s. 6d.; new ed. 3s. 6d.
—— See also Low's Stan. Ser.
—— SERJ., *Wealth and its Sources*, 5s.
—— J. R., *Princely Chandos*, illust., 12s. 6d.
—— *Last Earls of Barrymore*, 12s. 6d.
—— "*Romeo*" *Coates*, 7s. 6d.
ROCHEFOUCAULD. See Bayard Series.
ROCKSTRO, *History of Music*, new ed. 14s.
RODRIGUES, *Panama Can.*, 5s.
ROE, E. P. See Low's St. Ser.
ROGERS, S. See Choice Eds.
ROLFE, *Pompeii*, n. ed., 7s. 6d.
—— H. L., *Fish Pictures*, 15s.
ROMNEY. See Great Artists.
ROOPER, G., *Thames and Tweed.*
ROSE, J., *Mechanical Drawing Self-Taught*, 16s.
—— *Key to Engines*, 8s. 6d.
—— *Practical Machinist*, new ed. 12s. 6d.
—— *Steam Engines*, 31s. 6d.
—— *Steam Boilers*, 12s. 6d.
Rose Library. Per vol. 1s., unless the price is given.
Alcott (L. M.) Eight Cousins, 2s.
—— Jack and Jill, 2s.
—— Jimmy's cruise in the *Pinafore*, 2s.; cloth, 3s. 6d.
—— Little Women.
—— Little Women Wedded; Nos. 4 and 5 in 1 vol. cloth, 3s. 6d.
—— Little Men, 2s.; cl. gt., 3s. 6d.
—— Old-fashioned Girls, 2s.; cloth, 3s. 6d.
—— Rose in Bloom, 2s.; cl. 3s. 6d.

Rose Library— Continued.
Alcott (L. M.) Silver Pitchers.
—— Under the Lilacs, 2s.; cl.3s.6d.
—— Work, 2 vols. in 1, cloth, 3s.6d.
Stowe (Mrs.) Pearl of Orr's Island.
—— Minister's Wooing.
—— We and Our Neighbours, 2s.
—— My Wife and I, 2s.
Dodge (Mrs.) Hans Brinker, 1s.; cloth, 5s.; 3s. 6d.; 2s. 6d.
Holmes, Guardian Angel, cloth, 2s.
Stowe (Mrs.) Dred,2s.; cl. gt.,3s.6d.
Carleton (W.) City Ballads, 2 vols. in 1, cloth gilt, 2s. 6d.
—— Legends, 2 vols. in 1, cloth gilt, 2s. 6d.
—— Farm Ballads, 6d. and 9d.; 3 vols. in 1, cloth gilt, 3s. 6d.
—— Farm Festivals, 3 vols. in 1, cloth gilt, 3s. 6d.
—— Farm Legends, 3 vols. in 1, cloth gilt, 3s. 6d.
Biart, Bernagius' Clients, 2 vols.
Howells, Undiscovered Country.
Clay (C. M.) Baby Rue.
—— Story of Helen Troy.
Whitney, Hitherto, 2 vols. 3s. 6d.
Fawcett (E.) Gentleman of Leisure.
Butler, Nothing to Wear.
ROSSETTI. See Wood.
ROSSINI, &c. See Great Mus.
Rothschilds, by J. Reeves,7s. 6d.
Roughing it after Gold, by Rux, new edit. 1s.
ROUSSELET. See Low's Standard Books.
Royal Naval Exhibition,illus.1s.
RUBENS. See Great Artists.
RUSSELL, G.W. E.,*Gladstone.* See Prime Ministers.
—— H., *Ruin of Soudan*, 21s.
—— W. CLARK, *Mrs. Dines' Jewels*, boards, 2s.
—— *Nelson's Words and Deeds*, 3s. 6d.
—— *Sailor's Language*, 3s. 6d.
—— See also Low's Standard Novels.

RUSSELL, W. HOWARD,*Prince of Wales' Tour*, ill. 52s. 6d. and 84s.
Russia. See Foreign Countries.
Russia's March towards India, by an Indian Officer, 2 vols., 16s.
Saints and their Symbols, 3s. 6d.
SAINTSBURY, G., *Earl of Derby.* See Prime Ministers.
SAINTINE. See Low's Stan. Series.
SALISBURY, LORD. See Prime Ministers.
SAMUELS. See Low's Standard Series.
SAMUELSON, JAMES, *Greece, her Condition and Progress*, 3s.6d.
SANBORN, KATE, *A Truthful Woman in Southern California*, 3s. 6d.
SANDARS,*German Primer*, 1s.
SANDEAU. See Low's Stand. Series.
SANDLANDS, *How to Develop Vocal Power*, 1s.
SAUER, *European Commerce*, 5s.
—— *Italian Grammar* (Key, 2s.), 5s.
—— *Spanish Dialogues*, 2s. 6d.
—— *Spanish Grammar* (Key, 2s.), 5s.
—— *Spanish Reader*, new edit. 3s. 6d.
Scenes from Open-Air Life in Japan, photo plates by Ogawa, text by Murdoch, 15s. net.
SCHAACK, *Anarchy*, 16s.
SCHERER, *Essays in English Literature*, by G. Saintsbury, 6s.
SCHILLER'S *Prosa*, 2s. 6d.
SCHUBERT. See Great Mus.
SCHUMANN. See Great Mus.
SCHWAB, *Age of the Horse* ascertained by the teeth, 2s. 6d.

SCHWEINFURTH. See Low's Standard Library.
Scientific Education of Dogs, 6s.
SCOTT, LEADER, Renaissance of Art in Italy, 31s. 6d.
—— See also Great Artists and Illust. Text Books.
—— SIR GILBERT, Autobiography, 18s.
—— W. B. See Great Artists.
Scribner's Magazine, monthly, 1s.; half-yearly volumes, 8s. 6d.
Sea Stories. See Russell in Low's Standard Novels.
SEVERN, JOSEPH, Life, Letters, and Friendships, by Sharp, 21s.
Shadow of the Rock, 2s. 6d.
SHAFTESBURY. See English Philosophers.
SHAKESPEARE, ed. by R. G. White, 3 vols. 36s.; l. paper, 63s.
—— Annals; Life & Work, 2s.
—— Hamlet, 1603, also 1604, 7s. 6d.
—— Heroines, by living painters, 105s.; artists' proofs, 630s.
—— Macbeth, with etchings, 105s. and 52s. 6d.
—— Songs and Sonnets. See Choice Editions.
SHEPHERD, British School of Painting, 2nd edit. 5s.; also sewed, 1s.
SHOCK, W. H. Steam Boilers, 73s. 6d.
Silent Hour. See Gentle Life Series.
SIDNEY, SIR PHILIP, Arcadia, new ed., 6s.
SIMSON, Ecuador and the Putumayor River, 8s. 6d.
SKOTTOWE, Hanoverian Kings, new edit. 3s. 6d.
SLOANE, T. O., Home Experiments in Science, 6s.

SLOANE, W. M., French War and the Revolution, 7s. 6d.
SMITH, CHARLES W., Theories and Remedies for Depression in Trade, &c., 2s.
—— Commercial Gambling the Cause of Depression, 3s. 6d.
—— G., Assyria, 18s.
—— Chaldean Account of Genesis, new edit. by Sayce, 18s.
—— GERARD. See Illustrated Text Books.
—— MRS. See Japanese.
——T. ASSHETON, Reminiscences by Sir J. E. Wilmot, n. ed.,2s.6d. and 2s.
—— T. ROGER. See Illustrated Text Books.
—— W. A., Shepherd Smith, the Universalist, 8s. 6d.
—— HAMILTON, and LEGROS' French Dictionary, 2 vols. 16s., 21s., and 22s.
Socrates. See Bayard Series.
SNOWDEN (J. KEIGHLEY) Tales of the Yorkshire Wolds, 3s. 6d.
SOMERSET, Our Village Life, with coloured plates, 6s.
Spain. See Foreign Countries.
SPIERS, French Dictionary, new ed., 2 vols. 18s., half bound, 21s.
SPRY. See Low's Standard Library.
SPURGEON, C. H. See Preachers.
STANLEY, H. M., Congo, 2 vols. 42s. new ed., 2 vols., 21s.
—— Emin's Rescue, 1s.
—— In Darkest Africa, 2 vols., 42s.; new edit. 1 vol. 10s. 6d.
—— My Dark Companions and their Strange Stories, illus. 7s. 6d.
—— See also Low's Standard Library and Low's Stand. Books.

START, *Exercises in Mensuration*, 8d.

STEPHENS. See Great Artists.

STERNE. See Bayard Series.

STERRY, J. Ashby, *Cucumber Chronicles*, 5s.

STEUART, J. A., *Letters to Living Authors*, new edit. 2s. 6d.; édit. de luxe, 10s. 6d.

—— See also Low's Standard Novels.

STEVENI (N. B.). *Through Famine-Stricken Russia*, 3s. 6d.

STEVENS, J. W., *Leather Manufacture*, illust. 18s.

—— T., *Around the World on a Bicycle*, 16s.; part II. 16s.

STEWART, Dugald, *Outlines of Moral Philosophy*, 3s. 6d.

STOCKTON, F. R., *Ardis Claverden*, 6s.

—— *Clocks of Rondaine, and other Stories*, 7s. 6d.

—— *Mrs. Lecks*, 1s.

—— *The Dusantes*, a sequel to *Mrs. Lecks*, 1s.

—— *Personally Conducted*, (tour in Europe), illust. 7s. 6d.

—— *Rudder Grangers Abroad*, 2s. 6d.

—— *Schooner Merry Chanter*, 2s. 6d. and 1s.

—— *Squirrel Inn*, illust. 6s.

—— *Story of Viteau*, 5s., 3s. 6d.

—— *Three Burglars*, 2s. & 1s.

—— See also Low's Standard Novels.

STOKER, Bram, *Under the Sunset*, Christmas Stories, 6s.

STORER, F. H., *Agriculture and Chemistry*, 2 vols., 25s.

Stories from Scribner, illust., 6 vols., transparent wrapper. 1s. 6d. each; cloth, top gilt, 2s. each.
I. Of New York.
II. Of the Railway.
III. Of the South.
IV. Of the Sea.
VI. Of Italy.
V. Of the Army.

STOWE, Mrs., *Flowers and Fruit from Her Writings*, 3s. 6d.

—— *Life . . . her own Words . . . Letters, &c.*, 15s.

—— *Life*, for boys and girls, by S. A. Tooley, 5s., 2s. 6d. and 2s.

—— *Little Foxes*, cheap edit. 1s.; also 4s. 6d.

—— *Minister's Wooing*, 1s.

—— *Pearl of Orr's Island*, 3s. 6d. and 1s.

—— *Uncle Tom's Cabin*, with 126 new illust. 2 vols. 16s.

—— See also Low's Standard Novels and Low's Standard Series.

STRACHAN, J., *New Guinea, Explorations*, 12s.

STRANAHAN, *French Painting*, 21s.

STRICKLAND, F., *Engadine*, new edit. 5s.

STRONGE, S. E., & EAGAR, *English Grammar*, 3s.

STUART, E. A. See Preachers.

—— Esmé, *Claude's Island*, 6s.

STUTFIELD, *El Maghreb*, 8s. 6d.

SUMNER, C., *Memoir*, vols. iii., iv., 36s.

Sweden and Norway. See Foreign Countries.

Sylvanus Redivivus, 10s. 6d.; new ed., 3s. 6d.

SZCZEPANSKI, *Technical Literature*, a directory, 2s.

TAINE, H. A., *Origines*, I. Ancient Regime and French Revolution, 3 vols., 16s. ea.; Modern, I. and II., 16s. ea.

TAYLER, J., *Beyond the Bustle*, 6s.

TAYLOR, Hannis, *English Constitution*, 18s.

—— Mrs. Bayard, *Letters to a Young Housekeeper*, 5s.

—— R. L., *Analysis Tables*, 1s.

—— *Chemistry*, n. ed., 2s.

—— *Students' Chemistry*, 5s.

—— and S. PARRISH, *Chemical Problems, with Solutions*, 2s.6d.

Techno-Chemical Receipt Book, by Brannt and Wahl, 10s. 6d.

TENNYSON. See Choice Eds.

THANET, *Stories of a Western Town (United States)*, 6s.

THAUSING, *Malt & Beer*, 45s.

THEAKSTON, *British Angling Flies*, illust., 5s.

Thomas à Kempis Birthday-Book, 3s. 6d.

—— *Daily Text-Book*, 2s. 6d.

—— See also Gentle Life Series.

THOMAS, Bertha, *House on the Scar, Tale of South Devon*, 6s.

THOMSON, Joseph. See Low's Stan. Lib. and Low's Stan. Novs.

—— W., *Algebra*, 5s.; without Answers, 4s. 6d.; Key, 1s. 6d.

THORNTON, W. Pugin, *Heads, and what they tell us*, 1s.

THORODSEN, J P., *Lad and Lass*, 6s.

TILESTON, Mary W., *Daily Strength*, 5s. and 3s. 6d.

TINTORETTO. See Gr. Art.

TITIAN. See Great Artists.

TODD, Alpnaeus, *Parliamentary Government in England*, 2 vols., 15s.

TOLSTOI, A. K., *The Terrible Czar, a Romance of the time of Ivan the Terrible*, new ed. 6s.

TOMPKINS, *Through David's Realm*, illust. by author., n. ed., 5s.

TOURGEE. See Low's Standard Novels.

TRACY, A., *Rambles Through Japan without a Guide*, 6s.

TRAILL. See Prime Ministers.

TURNER, J. M. W. See Gr. Artists.

TYACKE, Mrs., *How I shot my Bears*, illust., 7s. 6d.

TYTLER, Sarah. See Low's Standard Novels.

UPTON, H., *Dairy Farming*, 2s.

Valley Council, by P. Clarke, 6s.

VANDYCK and HALS. See Great Artists.

VAN DYKE, J. C., *Art for Art's Sake*, 7s. 6d.

VANE, Denzil, *Lynn's Court Mystery*, 1s.

—— See also Low's St. Nov.

Vane, Young Sir Harry, 18s.

VAN HARE, *Showman's Life, Fifty Years*, new ed., 2s. 6d.

VELAZQUEZ. See Gr. Artists.
—— and MURILLO, by C. B. Curtis, with etchings, 31s. 6d.; large paper, 63s.
VERNE, J., *Works by.* See last page but one.
Vernet and Delaroche. See Great Artists.
VERSCHUUR, G., *At the Antipodes*, 7s. 6d.
VINCENT, Mrs. Howard, 40,000 *Miles over Land and Water*, 2 vols. 21s.; also 3s. 6d.
—— *Newfoundland to Cochin China*, new ed. 3s. 6d.
Visitors' Book in a Swiss Hotel, 1s.
WAGNER. See Gr. Musicians.
WAHNSCHAFFE, *Scientific Examination of Soil*, by Branut, 8s. 6d.
WALERY, *Our Celebrities*, vol. II. part i., 30s.
WALFORD, Mrs. L. B. See Low's Standard Novels.
WALL, *Tombs of the Kings of England*, 21s.
WALLACE, L., *Ben Hur*, 2s.
WALLER, *Silver Sockets*, 6s.
WALTON, Iz., *Angler*, Lea and Dove edit. by R. B. Marston, with photos., 210s. and 105s.
—— T. H., *Coal-mining*, 25s.
WARBURTON, Col., *Race-horse, How to Buy, &c.*, 6s.
WARDROP, Ol., *Kingdom of Georgia*, 14s.
WARNER, C. D. See Low's Stand. Novels and Low's Stand. Series.
WARREN, W. F., *Paradise Found*, illust. 12s. 6d.

WATKINSON. See Preachers.
WATTEAU. See Great Artists.
WEBER. See Great Musicians.
WELLINGTON. See Bayard Series.
WELLS, H. P., *Salmon Fisherman*, 6s.
WELLS, H. P., *Fly-rods and Tackle*, 10s. 6d.
WENZEL, *Chemical Products of the German Empire*, 25s.
West Indies. See Foreign Countries.
WESTGARTH, *Australasian Progress*, 12s.
WESTOBY, *Postage Stamps*, 5s.
WHITE, R. Grant, *England Without and Within*, new edit. 10s. 6d.
—— *Every-day English*, 10s. 6d.
—— *Studies in Shakespeare*, 10s. 6d.
—— *Words and their Uses*, new edit. 5s.
—— W., *Our English Homer, Shakespeare and his Plays*, 6s.
WHITNEY, Mrs. See Low's Standard Series.
WHITTIER, *St. Gregory's Guest*, 5s.
—— *Text and Verse for Every Day in the Year*, selections, 1s. 6d.
WILCOX, Marrion. See Low's Standard Novels.
WILKIE. See Great Artists.
WILLS, *Persia as it is*, 8s. 6d.
WILSON, *Health for the People*, 7s. 6d.

WINCHESTER, Bishop of. See Preachers of the Age.

WINDER, *Lost in Africa.* See Low's Standard Books.

WINGATE. See Ohrwalder.

WINSOR, J., *Columbus,* 21s.

—— *Cartier to Frontenac,* A Study of Geographical Discovery in the Interior of North America, 1534—1700, with reproductions of old maps, 15s.

—— *History of America,* 8 vols. per vol. 30s. and 63s.

WITTHAUS, *Chemistry,* 16s.

Woman's Mission, Congress Papers, edited by the Baroness Burdett-Coutts, 10s. 6d.

WOOD, Esther, *Dante Gabriel Rossetti and the Preraphaelite Movement,* with illustrations from Rossetti's paintings, 12s. 6d.

—— Sir Evelyn, *Life,* by Williams, 14s.

WOODS, *Sweden and Norway.* See Foreign Countries.

WOOLSEY, *Communism and Socialism,* 7s. 6d.

—— *International Law,* 6th ed. 18s.

—— *Political Science,* 2 v. 30s.

WOOLSON, C. Fenimore. See Low's Standard Novels.

WORDSWORTH. See Choice.

Wreck of the " Grosvenor," 4to, paper cover, 6d.

WRIGHT, H., *Friendship of God,* 6s.

—— T., *Town of Cowper,* 3s. 6d.

WRIGLEY, *Algiers Illustrated,* 100 views in photogravure, 45s.

Written to Order, 6s.

YOUNGHUSBAND, Capt. G. J., *On Short Leave to Japan,* illust. 6s.

BOOKS BY JULES VERNE.

WORKS. (LARGE CROWN 8VO.)	Containing 350 to 600 pp. and from 50 to 100 full-page illustrations.		Containing the whole of text with some illustrations	
	Handsome cloth binding, gilt edges.	Plainer binding, plain edges.	Cloth binding, gilt edges, smaller type.	Limp cloth
	s. d.	s. d.	s. d.	s. d.
20,000 Leagues under the Sea. Parts I. and II.	10 6	5 0	3 6	2 0
Hector Servadac	10 6	5 0	3 6	2 0
The Fur Country	10 6	5 0	3 6	2 0
The Earth to the Moon and a Trip round it	10 6	5 0	2 vols., 2s. ea.	2 vols., 1s. ea.
Michael Strogoff	10 6	5 0	3 6	2 0
Dick Sands, the Boy Captain	10 6	5 0	3 6	2 0
Five Weeks in a Balloon	7 6	3 6	2 0	1 0
Adventures of Three Englishmen and Three Russians	7 6	3 6	2 0	1 0
Round the World in Eighty Days	7 6	3 6	2 0	1 0
A Floating City	7 6	3 6	2 0	1 0
The Blockade Runners			2 0	1 0
Dr. Ox's Experiment	—	—	2 0	1 0
A Winter amid the Ice	—	—	2 0	1 0
Survivors of the "Chancellor"	7 6	3 6	3 6	2 0
Martin Paz			2 0	1 0
The Mysterious Island, 3 vols.:—	22 6	10 6	6 0	3 0
I. Dropped from the Clouds	7 6	3 6	2 0	1 0
II. Abandoned	7 6	3 6	2 0	1 0
III. Secret of the Island	7 6	3 6	2 0	1 0
The Child of the Cavern	7 6	3 6	2 0	1 0
The Begum's Fortune	7 6	3 6	2 0	1 0
The Tribulations of a Chinaman	7 6	3 6	2 0	1 0
The Steam House, 2 vols.:—				
I. Demon of Cawnpore	7 6	3 6	2 0	1 0
II. Tigers and Traitors	7 6	3 6	2 0	1 0
The Giant Raft, 2 vols.:—				
I. 800 Leagues on the Amazon	7 6	3 6	2 0	1 0
II. The Cryptogram	7 6	3 6	2 0	1 0
The Green Ray	5 0	3 6	2 0	1 0
Godfrey Morgan	7 6	3 6	2 0	1 0
Kéraban the Inflexible:—				
I. Captain of the "Guidara"	7 6	3 6	2 0	1 0
II. Scarpante the Spy	7 6	3 6	2 0	1 0
The Archipelago on Fire	7 6	3 6	2 0	1 0
The Vanished Diamond	7 6	3 6	2 0	1 0
Mathias Sandorf	10 6	5 0	3 3	2 vols 1 0 e
Lottery Ticket	7 6	3 6	2 0	1 0
The Clipper of the Clouds	7 6	3 6	2 0	1 0
North against South	7 6	3 6	—	2 vols 1 0 e
Adrift in the Pacific	6 0	2 6		
The Flight to France	7 6	3 6	1 0	
The Purchase of the North Pole	6 0	2 6		
A Family without a Name	6 0			
César Cascabel	6 0			
Mistress Branican	6 0			
Castle of the Carpathians	6 0			

*** *Special issue in eight cases of five books each, in a box, 4s. per box.*

CELEBRATED TRAVELS AND TRAVELLERS. 3 vols. 8vo, 600 pp., 100 full-page illustrations, 7s gilt edges, 9s. each:—(1) THE EXPLORATION OF THE WORLD. (2) THE GREAT NAVIGATORS OF

SAMPSON LOW, MARSTON & CO.'S
PERIODICAL PUBLICATIONS.

THE NINETEENTH CENTURY.

A Monthly Review. Edited by JAMES KNOWLES. Price Half-a-Crown.

Amongst the Contributors the following representative names may be mentioned: The Right Hon. W. E. Gladstone, Mr. J. A. Froude, Mr. Ruskin, Mr. G. F. Watts, R.A., Earl Grey, the Earl of Derby, Lord Acton, Mr. Herbert Spencer, Mr. Frederick Harrison, Mr. Algernon C. Swinburne, Mr. Leslie Stephen, Professor Huxley, Sir Theodore Martin, Sir Edward Hamley, Professor Goldwin Smith, and Sir Samuel Baker.

SCRIBNER'S MAGAZINE.

A Superb Illustrated Monthly. Price One Shilling.

Containing Contributions from the pens of many well-known Authors, among whom may be mentioned, Thomas Hardy, J. M. Barrie, Walter Besant, Bret Harte, Henry James, Thomas Bailey Aldrich, Sir Edwin Arnold, Andrew Lang, Sarah Orne Jewett, H. M. Stanley, Robert Louis Stevenson, R. H. Stoddard, Frank R. Stockton.

THE PUBLISHERS' CIRCULAR

and Booksellers' Record of British and Foreign Literature. Weekly. Every Saturday. Price Three-Halfpence. Subscription: Inland, Twelve Months (post free), 8s. 6d.; Countries in the Postal Union, 11s.

THE FISHING GAZETTE.

A Journal for Anglers. Edited by R. B. MARSTON, Hon. Treas. of the Fly Fishers' Club. Published Weekly, price 2d. Subscription, 10s. 6d. per annum.

The *Gazette* contains every week Twenty folio pages of Original Articles on Angling of every kind. The paper has recently been much enlarged and improved.

"Under the editorship of Mr. R. B. Marston the *Gazette* has attained a high standing."—*Daily News.* "An excellent paper."—*The World.*

BOYS.

A new High-class Illustrated Journal for our Lads and Young Men. Weekly, One Penny; Monthly, Sixpence; Annual Vol., 7s. 6d.

LONDON: SAMPSON LOW, MARSTON & COMPANY, LIMITED,
ST. DUNSTAN'S HOUSE, FETTER LANE, FLEET STREET, E.C.

www.ingramcontent.com/pod-product-compliance
Lightning Source LLC
Chambersburg PA
CBHW031418230426
43668CB00007B/355